D1137287

Electronic Engineering Semiconductors
and Devices
*A Second Edition of Electronic Engineering
Materials and Devices*

Electronic Engineering Semiconductors and Devices

A second edition of Electronic Engineering Materials and Devices

JOHN ALLISON

Reader
Department of Electronic and Electrical Engineering
University of Sheffield

McGRAW-HILL BOOK COMPANY

London · New York · St Louis · San Francisco · Auckland
Bogotá · Guatemala · Hamburg · Lisbon · Madrid · Mexico
Montreal · New Delhi · Panama · Paris · San Juan · São Paulo
Singapore · Sydney · Tokyo · Toronto

Published by
McGRAW-HILL Book Company (UK) Limited
Shoppenhangers Road
Maidenhead, Berkshire, England SL6 2QL
Telephone Maidenhead (0628) 23432
Fax 0628 35895

British Library Cataloguing in Publication Data
Allison, John, (date)
  Electronic engineering semiconductors and devices.
  2nd ed.
  1. Electronic equipment
  I. Title II. Allison, John, (date). Electronic
  engineering materials and devices
  621.381

  ISBN 0-07-084194-2

Library of Congress Cataloging-in-Publication Data
Allison, John.
  Electronic engineering semiconductors and devices/John Allison.
  –2nd ed.
  p.   cm.
  Rev. ed. of: Electronic engineering materials and devices. [1971].
  ISBN 0-07-084194-2
  1. Electronics–Materials. 2. Semiconductors. I. Allison, John.
  Electronic engineering materials and devices. II. Title.
  TK7871.A44 1989
  621.381--dc20   89-12813

12345 IL 92310

Typeset, printed and bound in Malta by
Interprint Limited.

TO
VICTORIA

We hope she is amused

'Perhaps a frail memorial, but sincere,
Not scorn'd in heav'n, though little notic'd here.'

William Cowper (1731–1800)

Contents

Preface

A tremendous flurry of exciting progress in electronic engineering has occurred in the two decades since the first edition of this book appeared. We have seen, for example, the development of rudimentary integrated circuits containing a few components into today's version containing a complete computer, incorporating millions of active devices, on a single chip. The apparently daunting task for the undergraduate student and the electronic device engineer (or the prospective author!) of keeping abreast of such an exponentially expanding technological development is somewhat ameliorated only because the principles of operation of constituent semiconductor components have remained constant. In spite of dramatic changes in the manner in which devices are constructed and interconnected, a flavour of which can be obtained from a quick glance at the later chapters, the essential device science is consequently as relevant today as it was 20 years ago. So, fortunately, is the basic philosophy on which the earlier book was founded, namely to concentrate essentially on timeless electronic fundamentals, such as the description of charge transport in semiconductor junctions, using the operating principles of recent devices as engineering examples of the application of such quintessential electronic concepts.

In the early days of electronic engineering, the essentials of vacuum electronic devices could be understood without too much difficulty and many of their characteristics could be derived theoretically by classical methods. Today, not only is the physical nature of the transport of charge in modern electronic devices more complicated, but many of the more recent devices have properties that cannot be explained satisfactorily without recourse to quantum electronics.

Many textbooks are available that describe, for example, the theoretical physics of the solid state, often in great detail and not without some degree of mathematical complexity. Other, usually engineering, texts often merely provide a cursory description of device behaviour as a preliminary to a detailed discussion of their circuit application. This book is an attempt to close the gap between the two extremes. It provides a physical description of the properties of electronic materials that is sufficiently detailed to allow complete characterization of the electrical performance of modern electronic devices in a manner that can be fully understood by the engineer.

The author is acutely aware of the rapid advances being made in electronic engineering. There is nothing to suggest that the rate of technological development observed over the last few decades will not continue; indeed the pace will probably increase. New devices are continually being developed and absorbed into the technology, causing some textbooks on the subject to become obsolete almost as soon as they are published. It is hoped that, by concentrating on fundamental processes occurring in electronic materials and by discussing contemporary devices as specific examples of these processes, this book will avoid such a fate. Although lack of space prevents complete in-depth coverage of all devices, sufficient insight into the basic properties of elementary devices is given to enable the reader to progress to the study of whatever new or perhaps as yet undeveloped device may be his or her own particular interest.

The text is based in the main on a series of lecture courses given to university electronic and electrical engineering undergraduates in their first, second and (part of) final years. For this reason a choice had to be made between subdividing the subject matter so as to make the level of treatment progressively more difficult or arranging the material in a more logically acceptable way. The former method is usual primarily in teaching, where the presentation has to be geared to the mathematical ability of the student, but it suffers the disadvantage of lack of continuity, each device being described on several occasions, each time with an increasing depth of treatment. In this book the latter course has been adopted in an attempt to provide a text that unifies much of the electronic materials and device teaching over the complete subject range, so emphasizing the relevance of each topic. This arrangement need not be a disadvantage, as few textbooks are read, in the first instance, straight through from beginning to end. The main advantage is that, while the students are able to cover the contents by any one of many routes, dictated by their own ability or as directed by their tutors, they will at the same time possess a book that amalgamates all the material into a coherent whole.

It might be supposed that this textbook is aimed solely at electronic and electrical engineering students, but practicing engineers who feel the need for retraining should also find it helpful, as well as students of associated disciplines in applied technology, physics and chemistry.

I am most grateful to my colleagues at Sheffield for many enjoyable and useful discussions, in particular Dr Clive Woods, whose meticulous reasoning, based on sound engineering principles, has been a constant source of inspiration. I am also most appreciative of friends in industry and in-house for providing additional information and illustrations and to the help of generations of students for their patient forbearance. I would also like to express my sincere thanks to Miss Margaret Eddell and her staff for their unstinting help in preparing the manuscript, especially Miss Elaine Jessop for her ability to read the author's mind, as well as decipher his hieroglyphics so efficiently.

Finally, my warmest thanks are extended to my wife, Judith, for her support and for providing endless coffee and a few cherished moments of silence, and also to Victoria, who was only a twinkle in her father's eye at the beginning of this project but is now able to offer her assistance, without which this manuscript might have been produced much earlier!

May I venture to hope that readers of this book will find it as interesting and informative as I have found it enjoyable and exciting to write.

John Allison

Physical constants

	Symbol		Units
Permittivity of free space	ϵ_0	8.854×10^{-12}	$F\,m^{-1}$
Permeability of free space	μ_0	$4\pi \times 10^{-7}$	$H\,m^{-1}$
Electronic charge	e	1.602×10^{-19}	C
Electronic rest mass	m	9.108×10^{-31}	kg
Electronic charge/mass ratio	e/m	1.759×10^{11}	$C\,kg^{-1}$
Proton rest mass		$1836m$	kg
Planck's constant	h	6.625×10^{-34}	J s
Boltzmann's constant	k	1.380×10^{-23}	$J\,K^{-1}$
kT at room temperature		0.0259	eV

Properties of some common semiconductors at room temperature

	Si	Ge	GaAs	InSb
Atomic weight	28.09	72.59	–	–
Atomic density (m^{-3})	5.02×10^{28}	4.42×10^{28}	–	–
Lattice constant, a(nm)	0.543	0.565	0.563	0.645
Relative permittivity, ϵ_r	11.8	16.0	13.5	11.5
Energy gap, E_g(eV)	1.08	0.66	1.58	0.23
Electron mobility, $\mu_e(m^2\,V^{-1}\,s^{-1})$	0.13	0.38	0.85	7.0
Hole mobility, $\mu_h(m^2\,V^{-1}\,s^{-1})$	0.05	0.18	0.04	0.10
Intrinsic concentration, $n_i(m^{-3})$	1.38×10^{16}	2.5×10^{19}	9×10^{12}	1.6×10^{22}
Electron diffusion constant, $D_e(m^2\,s^{-1})$	0.0031	0.0093	0.020	0.0093
Hole diffusion constant, $D_h(m^2\,s^{-1})$	0.0007	0.0044	–	–
Density of states at conduction band edge, $N_c(m^{-3})$	2.8×10^{25}	1.0×10^{25}	4.7×10^{23}	–
Density of states at valence band edge, $N_v(m^{-3})$	1.0×10^{25}	6.0×10^{24}	7.0×10^{24}	–
Melting point (°C)	1420	936	1250	523

1 The quantum behaviour of waves and particles

1.1 Introduction

Throughout our discussion of the electronic properties of materials and the application of these properties to a physical understanding of the operation of electronic devices, we shall constantly be referring to the interaction of particles of atomic size, for example electrons, with other particles or with waves. We shall discover that in some instances elementary mechanics is no longer adequate to describe the dynamics of microscopic particles and that this theory has to be supplemented by one that is more generally applicable, the so-called *quantum* or *wave* mechanics.

Classical mechanics is based on laws developed by Newton, for example

$$F = \mathrm{d}p/\mathrm{d}t \tag{1.1}$$

Newton's laws, together with a classical electromagnetic theory as summarized by Maxwell's equations, proved adequate for the quantitative explanation of most experiments done before the beginning of the twentieth century. Equation (1.1) was found to be quite satisfactory for predicting the dynamics of large-scale systems. This is, of course, also true today. The term 'large-scale systems' in this context applies equally well to normal engineering laboratory experiments and to the more obviously large systems such as collections of planets.

A series of experiments conducted at the beginning of this century exposed a basic limitation of classical Newtonian mechanics, which is its inability to predict correctly events that take place on a microscopic or atomic scale. In the sections that follow we shall discuss the experiments that led to this failure in classical mechanics, while at the same time we will attempt to lay a general foundation for later discussions of a more general quantum-mechanical theory. It should be stressed that quantum mechanics does not entirely supplant Newtonian mechanics but rather augments it in that it is more widely applicable. We shall show that, within the limits of laboratory-sized objects, however, the newer mechanics reduces to the classical theory, which, because it is simpler to apply, is still to be preferred.

1.2 Black-body radiation

One of the earliest experiments to defy analysis by classical methods was the determination of the frequency spectrum of emitted radiation of an incandescent radiator or 'black body'. In this experiment the intensity of emitted radiation is measured as a function of frequency or wavelength for a fixed temperature, with typical results as indicated diagrammatically in Fig. 1.1.

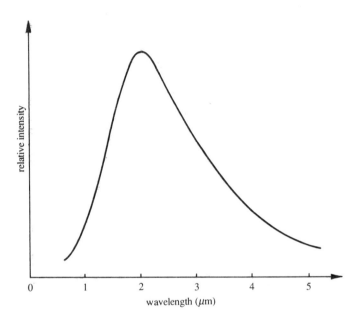

Fig. 1.1 **Relative intensity of radiation emitted from a black body as a function of wavelength.**

The various earlier theories attempting to explain this experimental evidence were based on classical mechanical ideas incorporated in thermodynamic theory. These theories were never successful in agreeing with the experiment, particularly in the short-wavelength limit. We know now that they broke down because of a fundamental misconception that atomic oscillators are capable of emitting or absorbing energy in continuously variable amounts. It was not until 1901 that Max Planck discredited this false notion by correctly predicting the intensity of radiation at all frequencies. His theory involved the hypothesis that energy could only be absorbed or emitted by the black body in discrete amounts. He assumed that the energy of light waves, for example, is transported in packets or bundles, called *photons* or *quanta*. He further assumed the energy of a photon to be given by

$$E = hf \text{ joules} \tag{1.2}$$

where f is the frequency of the radiation. Planck's constant, h, is a universal constant, which is found for the black-body radiator and other experiments to have a value $h = 6.626 \times 10^{-34}$ J s. Equation (1.2) is sometimes more conveniently written as

$$E = \hbar\omega \qquad (1.3)$$

where ω is the angular frequency of the radiation, $2\pi f$, and $\hbar$ equals $h/2\pi$. Thus, the total energy of the black-body radiator was envisaged to exist only in discrete allowed energy states:

$$0, \quad \hbar\omega, \quad 2\hbar\omega, \quad 3\hbar\omega, \quad \dots, \quad n\hbar\omega$$

transition between these states being brought about by absorption or emission of one or more photons of radiation, each of energy $\hbar\omega$.

It is not surprising that such a theory, being completely opposed to the existing continuously variable energy theories, was not readily accepted, even though it explained the experimental findings most satisfactorily. However, Planck's quantum hypothesis, which forms the basis of modern quantum mechanics, was further vindicated by later experimental evidence, as will be discussed in subsequent sections.

1.3 The photoelectric effect

If light of sufficiently short wavelength impinges on the surface of certain solids, then it is possible for electrons to be emitted from the solid. This is called the photoelectric effect. In the early twentieth century, Einstein reinforced Planck's photon concept of light by providing a satisfactory quantitative explanation of the effect.

The experimental evidence for the effect may be obtained using apparatus of the type shown diagrammatically in Fig. 1.2. Light of frequency f illuminates

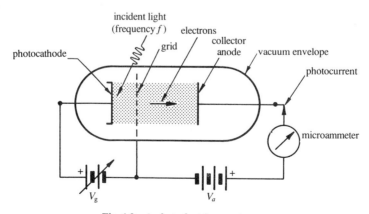

Fig. 1.2 A photoelectric experiment.

a cold cathode situated inside a vacuum envelope. If electrons are emitted, then, provided they have sufficient energy to overcome the retarding force field set up by the voltage V_g between grid and cathode, they will be swept to the positive collecting anode and a current will be registered on the microammeter in series with it.

The first thing that would be noticed when carrying out such an experiment is that, unless the frequency of the incident light is greater than some critical value, f_0, which is dependent on the material of the cathode, no emission is observed, *no matter how intense the light*. For constant light frequency, and provided f is greater than f_0, the photocurrent can be measured as a function of grid voltage V_g and light intensity, keeping the anode voltage constant, to give typical collector current data of the form shown in Fig. 1.3. The surprising

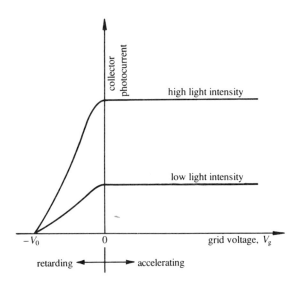

Fig. 1.3 Variation of photocurrent with grid voltage in the photoelectric experiment.

result is that, no matter what the intensity of the light, there is some constant retarding voltage, in this case $-V_0$, that entirely inhibits emission. This implies that the maximum kinetic energy of emitted electrons is constant and independent of the intensity of the incident light. However, as the light intensity is increased, the photocurrent increases in sympathy. Thus the number of emitted photoelectrons is a function of the intensity of the light but their maximum energy is constant.

Such experimental results cannot be explained by a classical wave theory of light and a satisfactory explanation can only be obtained by considering the light energy to be quantized. That is, the light energy is transported in discrete packets or photons.

Before discussing either theory we must digress a little to discuss briefly the reasons for the emission of electrons from a metal surface. We shall see later that a metal contains many highly mobile electrons, which can participate in the electrical conduction process, but these are confined to the interior of the metal by a binding energy. Thus, no conduction electrons can leave the surface of a metal unless they are in some way provided with additional energy to enable them to overcome this binding energy. The minimum energy required for an electron to be just emitted from a metal surface is called the *workfunction* of the particular metal and is usually designated $e\phi$, where ϕ is in volts.

In a classical theory of photoemission, conduction electrons in the cathode are accelerated by the electric field of the light wave and, if the light is bright enough, can gain sufficient energy to be emitted. Any surplus energy over and above the workfunction appears as kinetic energy of the emitted electron. Thus the brighter the light, the more energy is left over after overcoming the binding energy and the greater the kinetic energy of the emitted electron. This result is clearly at variance with the experimental evidence that the emitted electrons have a constant maximum energy. Further, a classical wave theory would not predict a threshold frequency f_0, which again is contrary to the experimental results.

If now we turn to a quantum theory based on Planck's photon hypothesis, we can give a simple explanation for the observed effects. We assume that the incident light is composed of discrete quanta or photons, each of energy hf. When the light impinges on the metal of the photocathode, each photon can transfer energy hf to a conduction electron. Some of the energy is used to overcome the binding forces and the remainder is converted to kinetic energy of the emitted electron. Thus

kinetic energy of emitted electron = photon energy – workfunction

or

$$\tfrac{1}{2}mv^2 = hf - e\phi \qquad (1.4)$$

The limiting case occurs when an electron is just emitted with no kinetic energy. Then

$$f = f_0 = e\phi/h \qquad (1.5)$$

At frequencies less than this critical value, the photon's energy, hf, is not even sufficient to overcome the workfunction and no emission occurs. Further, if the intensity is increased, the number of incident photons is increased but their energy remains constant, provided the frequency remains constant. Thus the kinetic energy of the emitted electrons, as given by Eq. (1.4), stays constant, which is again borne out by experiment. Incidentally, Eq. (1.4) gives the maximum kinetic energy at some particular frequency and in our experiment this is entirely converted to potential energy when the photocurrent is reduced to zero by a retarding voltage $-V_0$. Hence

$$eV_0 = hf - e\phi \qquad (1.6)$$

We see that V_0 does not vary with light intensity, a fact that we have already noted experimentally.

Thus, when considering the interaction of light with electrons, as in discussing the photoelectric effect, the quantum theory of light must be used in preference to the classical wave theory.

It will be useful to contrast this situation with the earlier experiments with light displaying such phenomena as diffraction, refraction, and so on. These effects could all be explained by a classical theory, which relied on a wave-like description of the light radiation. As an example, let us remind ourselves of the situation when light is diffracted by a mirror diffraction grating. A schematic diagram of the essential elements of the experiment is shown in Fig. 1.4. A light

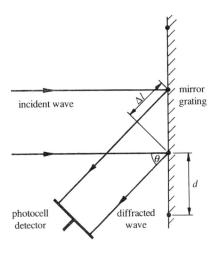

Fig. 1.4   **Diffraction of light by a grating.**

wave, wavelength λ, is incident normally on a reflecting diffraction grating of the type used in optical spectrographs, which has a grating spacing d. The light wave is diffracted by the grating and the diffracted wave is detected at some angle θ to the normal. The detector might, for example, be a photocell similar to the one just described. Experimentally, what is observed is that, as θ is varied, the intensity at the detector varies cyclically from maximum to minimum values. These are the well known diffraction fringes. A typical result might be as shown in Fig. 1.5. The appearance of the fringes can be explained quantitatively by invoking the wave description of light. The path length between light beams diffracted from adjacent rulings is indicated by Δl on Fig. 1.4. Whenever the angle θ is such that this length equals an integral

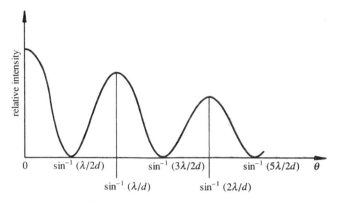

Fig. 1.5   Detected light intensity as a function of the angle of diffraction.

number of wavelengths, the diffracted waves interfere constructively and a bright fringe is observed. Thus, when

$$\Delta l = d \sin \theta = n\lambda$$

where n is an integer, there is constructive interference. This occurs at angles such that

$$\theta = \sin^{-1} (n\lambda/d) \tag{1.7}$$

Conversely, when the path difference Δl is equal to an odd number of half-wavelengths, which is equivalent to a 180° phase difference, there is destructive interference and a dark band is observed. This occurs at angles given by

$$\theta = \sin^{-1}\{[(2m+1)/2]\ \lambda/d\} \tag{1.8}$$

Both these results agree with the experimental evidence and a wave theory is entirely adequate to explain this phenomenon.

We see, then, that light can in some circumstances be considered to possess wave-like properties but on other occasions it must be treated in a quantized manner, its energy being transported by discrete photons, which are particle-like units each having energy hf. This wave–particle property of light radiation is sometimes referred to as the 'dual nature of light'.

If a light wave can behave as a particle, can a particle (say, for example, an electron) behave as a wave? The answer to this question will become apparent as we discuss further the historical development of quantum theory.

1.4 The Bohr atom

An excited hydrogen atom emits radiation at a discrete set of frequencies only. In 1913 Bohr produced a theoretical model that very accurately accounted for the observed sharp-line radiation spectrum from atomic hydrogen.

Let us consider the hydrogen atom to consist of a central nucleus with an electron travelling in a circular orbit round it, at some radius, r, as shown in Fig. 1.6(a). We shall see later that this description of an orbit is not very precise, neither is the orbit necessarily circular, but this simple model will suffice to

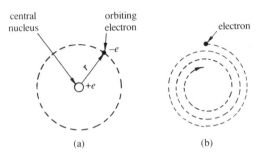

(a) (b)

Fig. 1.6 **(a) A possible model for a hydrogen atom and (b) spiral path due to radiated energy.**

demonstrate the inadequacy of a classical theory. The Coulomb force on the electron due to the electric field of the positive nucleus is just sufficient to provide inward acceleration for circular motion at a constant radius, r.

It follows that

$$F = -\frac{e^2}{4\pi\epsilon_0 r^2} = -\frac{mv^2}{r} \tag{1.9}$$

Now, the total energy of the electron, E, is the sum of its potential energy, V, and its kinetic energy, T. Further

$$V = -e^2/(4\pi\epsilon_0 r)$$

and from Eq. (1.9)

$$T = \tfrac{1}{2}mv^2 = e^2/(8\pi\epsilon_0 r)$$

Therefore, the total electron energy is

$$E = -e^2/(8\pi\epsilon_0 r) \tag{1.10}$$

Now, the electron in the circular orbit is constantly being accelerated and it can be shown by electromagnetic theory that such an accelerated charge radiates electromagnetic energy, with a corresponding loss of energy. Classical theory thus indicates that radiation can occur, its frequency corresponding to the periodic frequency of the circular motion. This frequency can be shown to be plausible by observing the electron's motion in the plane of the orbit. The electron will be seen to oscillate sinusoidally about a central position where the nucleus is located. The resultant sinusoidally varying current can then be likened to that occurring in an ordinary radio transmitting aerial, which radiates electromagnetic waves.

The frequency of the radiated wave from the classical atom is thus

$$f = v/(2\pi r)$$

which, using Eq. (1.9), gives

$$f = \frac{e}{4(\pi^3 \epsilon_0 m r^2)^{1/2}} \tag{1.11}$$

Now, the conservation of energy indicates that, as the electron radiates energy, its total energy, E, must decrease. Thus the radius of the orbit must also decrease, as shown by Eq. (1.10). This would lead to a continual loss of energy and spiralling of the electron towards the nucleus, as shown in Fig. 1.6(b). This, in turn, would indicate that the frequency of the emitted radiation is continuously varying according to the dependence of f on the radius r as set out in Eq. (1.11). This is clearly in complete disagreement with the experimentally observed discrete frequency spectrum.

To overcome this difficulty, Bohr postulated that the electron could only exist in discrete energy levels, corresponding to certain allowed stable orbits, without radiating any energy. He further argued that radiation from the atom occurs only when the electron makes a transition from one allowed energy level to another, when the energy lost by the atom is converted into the energy of a single photon. Thus, if the electron is transferred from one stable orbit, corresponding to a total energy E_1, to another allowed orbit with lower energy E_2, a photon of radiation is emitted whose frequency, f_{12}, is given by

$$E_1 - E_2 = hf_{12} \tag{1.12}$$

Thus, since only a discrete set of energy levels is postulated, only a discrete set of characteristic frequencies is present in the output spectrum.

In order to calculate the value of the discrete allowed energy levels, Bohr was obliged to postulate, in a rather intuitive way, that the angular momentum, L, associated with a gyrating electron is quantized such that

$$L = mvr = n\hbar \qquad \text{where } n = 1, 2, 3, \ldots \tag{1.13}$$

We can now eliminate v from Eqs (1.9) and (1.13) to obtain an expression for the radii of allowed orbits. From (1.9) and (1.13)

$$v^2 = \frac{2e^2}{m8\pi\epsilon_0 r} = \frac{n^2\hbar^2}{m^2 r^2}$$

or

$$r_n = \frac{4\pi n^2 \hbar^2 \epsilon_0}{e^2 m} = \frac{n^2 h^2 \epsilon_0}{\pi e^2 m} \simeq 0.05 n^2 \text{ nm} \tag{1.14}$$

Thus, the first possible orbit, when $n = 1$, has a radius of about 0.05 nm ($1 \text{ nm} \equiv 10^{-9} \text{ m}$). Other possible orbits, corresponding to the various integer values of n, are 0.2 nm ($n = 2$), 0.45 nm ($n = 3$), and so on.

Each discrete orbit has a corresponding allowed energy level associated with it, which is evaluated by substituting the values of r_n in Eq. (1.10) to give

$$E_n = -\frac{e^2}{8\pi\epsilon_0}\frac{\pi e^2 m}{n^2 h^2 \epsilon_0} = -\frac{me^4}{8\epsilon_0^2 h^2 n^2} \simeq -\frac{13.6}{n^2}\,\text{eV} \qquad (1.15)$$

Here the energy is expressed in electronvolts, a common practice in electronic engineering. One electronvolt corresponds to the energy acquired by an electron that has been accelerated through a potential difference of 1 V.

We see from Eq. (1.15) that the system energy is restricted to discrete levels corresponding to the various values of n. The lowest energy level or *ground state* for hydrogen is $E_1 = -13.6\,\text{eV}$. Further allowed energy values are $E_2 = -3.4\,\text{eV}$, $E_3 = -1.51\,\text{eV}$, and so on. The allowed energy levels are usually represented in an energy level diagram, which for the hydrogen atom is shown in Fig. 1.7 (Note that in this representation the horizontal scale has no physical

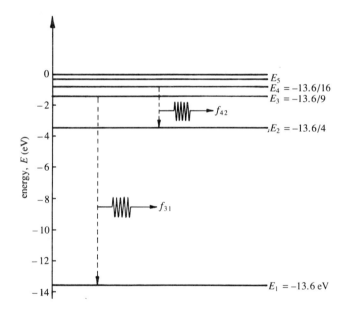

Fig. 1.7   Energy level diagram for the hydrogen atom.

significance.) As explained previously, it is possible for an electron to undergo a transition between any of the allowed energy levels as indicated by arrows on the diagram. The corresponding frequency of the emitted or absorbed radiation for such an occurrence is given by Eq. (1.12).

Although the characteristic emission frequencies predicted by the Bohr theory are very close to those observed for the hydrogen atom, the theory is only extendable to deal with hydrogen-like atoms and it fails for atoms with more than one orbiting electron. A further limitation is the *ad hoc* manner in which the assumptions of quantized angular momentum and the relationship between energy change and frequency are introduced. We shall see later that there are further fundamental inadequacies in the theory. Meanwhile, it is well to remember. that the Bohr theory, in a historical context, was a great step forward, in that it not only accounted for the predicted hydrogen frequency spectrum but, more important, it also clearly demonstrated a certain discreteness in some of the physical properties of matter, which is quite at variance with previous classical theory. A further advantage of the theory is that it indicates the importance of Planck's constant, h, for determining details of atomic structure.

1.5 Particle–wave duality

Let us return to our discussion of the possible dual nature of matter, in which waves can sometimes behave as particles and conversely particles can in some circumstances be considered to have wave-like properties.

We will first consider a photon of light, frequency f and energy hf. The photon travels at the velocity of light, c, which is related to frequency and wavelength, λ, by

$$c = f\lambda \tag{1.16}$$

Now, if energy E is transported with velocity c, the momentum of the photon is given by

$$p = E/c \tag{1.17}$$

This expression could be derived by finding the radiation pressure on a plate caused by an incident electromagnetic light wave and equating to the photon flux, but this is clearly too difficult, at this stage. Instead, we offer the following somewhat crude argument. Suppose that photons are subjected to an external force F, which acts over some distance dx. The change in photon energy is then

$$dE = F\,dx$$

Also, if the photon momentum is p, then, by Newton's law,

$$F = dp/dt$$

and

$$dE = \frac{dp\,dx}{dt} = \frac{dx}{dt}dp = c\,dp$$

Integrating, we get

$$p = E/c$$

as before. Equations (1.16) and (1.17) can now be combined to give

$$p = \frac{E}{c} = \frac{E}{f\lambda} = \frac{hf}{f\lambda}$$

or

$$p = h/\lambda \qquad\qquad (1.18)$$

Now, in 1924 de Broglie argued that if photons of light with wavelength λ have momentum $p = h/\lambda$, it might be possible for *particles* with momentum p to have some associated wavelength λ also and behave in a wave-like manner, under some circumstances. He further suggested that Eq. (1.18) might also be the correct relationship between p and λ for a particle.

Let us assume for the moment that this hypothesis, which as we shall see later can be substantiated by experimental evidence, is correct, and calculate the wavelength associated with various bodies. First, consider a classical Newtonian particle—an apple! If we let its mass be $m = 0.2$ kg, and its velocity $v = 10$ m s^{-1} then its momentum

$$p = mv = 0.2 \times 10 = 2 \text{ kg m s}^{-1}$$

and its associated wavelength is

$$\lambda = \frac{h}{p} = \frac{6.6 \times 10^{-34}}{2} \approx 10^{-34} \text{ m}$$

Effects due to such a wavelength are much too small to be detected in ordinary laboratory experiments!

As a further example, consider an electron, mass m, charge $-e$, accelerated through a potential difference, V. Equating the gain in kinetic energy of the electron to its loss in potential energy,

$$\tfrac{1}{2}mv^2 = eV$$

where v is its final velocity, gives

$$v = (2eV/m)^{1/2}$$

The momentum of the electron is then

$$p = mv = (2eVm)^{1/2}$$

and its associated wavelength is

$$\lambda = \frac{h}{p} = \frac{h}{(2eVm)^{1/2}} = \frac{1.225}{V^{1/2}} \text{ nm} \qquad\qquad (1.19)$$

For example, if $V = 50$ volts, $\lambda = 1.7 \times 10^{-7}$ mm, which again is small but, as we shall see, can produce measurable effects.

Soon after de Broglie suggested that particles could exhibit wave-like characteristics, Davisson and Germer provided experimental confirmation of his ideas by diffracting electrons and producing interference between electron waves. Their experiment is similar to the grating experiment with light, which we have already discussed, a beam of electrons rather than light being diffracted by a 'grating'. We have seen that the wavelengths, λ, associated with electrons are small and, for a reasonable angular spread of the diffracted wave, the periodicity of the grating, d, should be of the same order as λ. The regular array of atoms within a single crystal of a metal satisfies this condition and can behave as a grating for electron diffraction. The apparatus for such an experiment is shown diagrammatically in Fig. 1.8(a). An electron gun accelerates a beam of electrons through a potential V and the beam then impinges

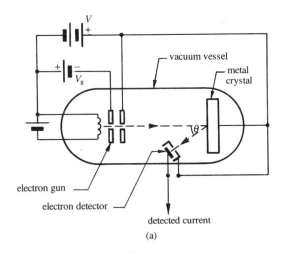

(a)

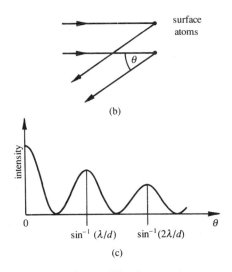

(b)

(c)

Fig. 1.8 **An electron diffraction experiment.**

normally on a plane single-crystal metal target. The detector is biased to provide a retarding field and only detects electrons that have been scattered with negligible loss of energy. It can be moved in angular direction, so as to measure the electron current diffracted by the target as a function of θ.

Let us now consider that electrons behave as waves, which are diffracted by the regular array of atoms on the surface of the target, spaced distance d apart, as illustrated in Fig. 1.8(b). By analogy with the mirror grating experiment, we would expect a maximum detected electron current at an angle θ such that the path difference between adjacent waves is an integral number of wavelengths and the first maximum detected signal to be given by

$$\sin \theta_{max} = \lambda/d$$

Now if de Broglie's hypothesis is correct, the wavelength of the electron is given by Eq. (1.19). Thus

$$\theta_{max} = \sin^{-1} [h/(2emVd^2)^{1/2}] \tag{1.20}$$

Clearly, our simple model is not complete since waves are also diffracted from atomic planes within the body of the crystal. Further, when these waves leave the crystal they are refracted at the crystal–vacuum interface. When Eq. (1.20) is modified to account for these additional factors, as was done by Davisson and Germer, excellent agreement between theory and experiment is obtainable. In all their experiments, electrons were found to behave as waves with wavelength given by Eq. (1.19). Incidentally, electron diffraction apparatus has since become an important analytical tool to study such things as interatomic spacing and the structure of molecules.

Suppose the experiment is now extended by replacing the detector with a cathode-ray screen. In this way an average measurement of electron current over a long time can be take by observing the intensity of illumination at the screen versus θ. A typical result would be as shown in Fig. 1.8(c). We see that there is a strong tendency for electrons to come off at angles near $\theta = 0$, $\sin^{-1}(\lambda/d)$, $\sin^{-1}(2\lambda/d)$, etc., and the probability of finding electrons at $\theta \simeq \sin^{-1}(\lambda/2d)$, $\sin^{-1}(3\lambda/2d)$, etc., is very small. In this respect, the electrons behave like light waves.

A further experiment might be to replace the detector by an electron multiplier, connected via a high-gain amplifier to a loudspeaker. Each time a diffracted electron is detected a sharp 'click' is heard on the loudspeaker, each click being equally loud. Further, for a fixed value of θ, the number of electrons arriving when averaged over some time is constant, although the individual arrival times are erratic. Moving the detector alters the rate of clicks but their size, as measured by their loudness, remains constant. Lowering the cathode temperature or making the grid voltage in the electron gun more negative reduces the number of emitted electrons and hence alters the click rate but the loudness of each still stays the same. Thus, whatever is being detected arrives in discrete amounts and in this respect the electrons are quantized and behave as particles.

We see, then, that the electrons are quantized but that they are in some way 'guided' by 'matter' waves (sometimes called probability waves, de Broglie waves, or ψ waves). Loosely speaking, wherever the matter waves have a large amplitude the probability of finding electrons is high, the converse being also true. Notice that we have had to abandon the absolute determinacy implicit in Newtonian mechanics since there is uncertainty as to where a particular electron will go after diffraction by the crystal lattice. We are forced to revert to discussing the probability of an electron being in a certain position at a given time.

The Davisson and Germer experiment, and a similar diffraction experiment carried out independently and almost concurrently by G. P. Thomson, clearly demonstrated the dual nature of matter and provide conclusive proof of the de Broglie wavelength relationship, $\lambda = h/p$. Whether a wave behaves as a particle or, conversely, whether a particle behaves as a wave is not only dependent on the type of experiment performed but also on the magnitude of the energies and momenta involved. Tables 1.1 and 1.2 are included to indicate under what conditions wave or particle properties become dominant.

Table 1.1   The electromagnetic spectrum of waves.

Frequency (Hz)	Wavelength (m)	Typical wave type	Remarks
	10^{-12}		
10^{20}		gamma rays	photon energy ($\hbar\omega$)
		X-rays	and momentum ($\hbar/\lambda$) increases—particle characteristics become important
	10^{-8}		
10^{16}		ultraviolet	
		visible	
	10^{-4}		
10^{12}		infrared	
		millimetre waves	wave characteristics
		microwaves	dominant
	1		
10^{8}			
	10^{2}		
10^{6}		radio waves	

Table 1.2 Particle spectrum.

Mass (kg)	Typical particle	Remarks
10^{20}	star planet car football dust grain	total energy increases— particle behaviour predominates
10^{-20}	molecule atom electron	wavelength (h/p) increases as p decreases—wave
10^{-40}	 neutrino	characteristics become important

1.6 Wavepackets: group and phase velocities of particles

It may be helpful to consider a geometric representation of how an object may simultaneously possess both wave and particle properties. We do this by studying the addition of waves of differing wavelength to produce a constructive interference pattern that has particle properties.

First, consider two waves of slightly different wavelength travelling in the same direction. These add together and produce regions of constructive interference that are periodically positioned in space, as in Fig. 1.9(a). This phenomenon is analogous to the beating of two sound waves to produce interference in the time domain. If, now, three waves, again slightly differing in frequency, are added, the interference maxima are not only larger but are spread at wider intervals, Fig. 1.9(b). The repetition in space of the regions of constructive interference is characteristic of the interference between finite numbers of waves. For an infinite number of waves, only one region of constructive interference exists; this is called a *wavepacket*, Fig. 1.9(c). The wavepacket geometrically represents an object with wave and particle properties. It obviously has wave properties since it is constructed from waves and has a wave-like form but it also behaves as a particle because of its localization in space. This is similar to the bow-wave of a ship, which has wave properties but travels with the ship and is always located relative to it. We have, so far, only considered waves travelling in the same direction but it is straightforward to extend the idea by describing a particle as consisting of an infinite number of waves travelling in all directions and constructively interfering at some point in space where the particle is positioned.

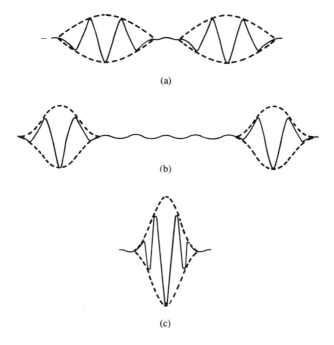

(a)

(b)

(c)

Fig. 1.9   Formation of a two-dimensional wavepacket.

What is the velocity of a wavepacket? In order to find this out, it will be necessary to digress slightly to summarize briefly the definitions and methods for calculating the various velocities associated with waves. A wave travelling in the positive x direction may be represented by an expression

$$A_0 \cos(\omega t - \beta x) \tag{1.21}$$

where A_0 is the amplitude of the wave, ω is its angular frequency and β is the phase constant, which is related to the wavelength by

$$\beta = 2\pi/\lambda \tag{1.22}$$

It is usually more convenient, mathematically, to represent the wave by an equivalent exponential function:

$$A_0 \, \mathrm{Re} \exp[\mathrm{j}(\omega t - \beta x)] \tag{1.23}$$

In most texts the Re is omitted but it is understood that only the real part of the result of any subsequent operation is valid.

The propagation of a wave is characterized by two velocities, the *phase velocity*, v_{ph} and the *group velocity*, v_{g}. Phase velocity is defined as the velocity of planes of constant phase along the propagation direction of the wave. To obtain an expression for this velocity we must examine the motion of a point of constant phase, which is given from Eq. (1.21) or (1.23) by the condition

$$\omega t - \beta x = \text{constant}$$

We obtain the phase velocity by differentiating this equation with respect to time:

$$\omega - \beta \, dx/dt = 0$$

or

$$v_{\text{ph}} = \omega/\beta \qquad (1.24)$$

It is important to remember that this is the velocity at which some arbitrary phase propagates. Nothing material propagates at this velocity; indeed, it is possible for v_{ph} to be greater than the speed of light without violating any physical laws.

Now, let us investigate what happens when two waves of equal amplitude but with slightly different wavelengths propagate simultaneously in the x direction, as discussed qualitatively earlier. Let the small differences of frequency and phase constant be $\delta\omega$ and $\delta\beta$. The resultant wave can then be represented by the sum

$$\begin{aligned}
A_0 \cos(\omega t - \beta x) &+ A_0 \cos[(\omega + \delta\omega)t - (\beta + \delta\beta)t] \\
&= 2A_0 \cos\{\tfrac{1}{2}[(2\omega + \delta\omega)t - (2\beta + \delta\beta)t]\} \cos[\tfrac{1}{2}(\delta\omega t - \delta\beta x)] \\
&\simeq 2A_0 \cos[\tfrac{1}{2}(\delta\omega t - \delta\beta x)] \cos(\omega t - \beta x) \qquad (1.25)
\end{aligned}$$

since $\delta\omega \ll 2\omega$ and $\delta\omega \ll 2\beta$ by our original assumption.

Thus, the resultant total wave consists of a high-frequency wave, varying as $\cos(\omega t - \beta x)$, whose amplitude varies at a much slower frequency rate represented by the other cosine term in (1.25); it is *modulated* by destructive and constructive interference effects, as shown in Fig. 1.10. The high-frequency wave has phase velocity ω/β as before. The variation of wave amplitude is called the envelope of the wavegroup; it varies sinusoidally with time and distance and is a travelling wave with the relatively long wavelength, $2\pi/\delta\beta$. Group velocity is defined as the velocity of propagation of a plane of constant

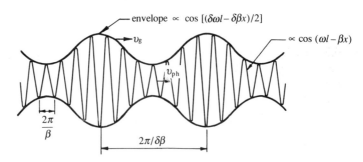

Fig. 1.10   Interaction of two waves of slightly different frequencies.

phase on the envelope. It corresponds to the velocity of the group or packet of waves along the direction of propagation. A plane of constant phase on the envelope is given by

$$\delta\omega t - \delta\beta x = \text{constant}$$

Thus, using the same procedure used to evaluate the phase velocity, we see that the group velocity is given by

$$v_g = \partial\omega/\partial\beta \tag{1.26}$$

Furthermore, it can be shown that v_g is the velocity at which energy is transmitted along the direction of propagation.

We can now return to the discussion of the velocities of particle waves. The hypotheses of de Broglie and Planck, corroborated by experiment, suggest that the momentum and kinetic energy of a particle are given by

$$p = mv = h/\lambda$$
$$T = \tfrac{1}{2}mv^2 = hf \tag{1.27}$$

Now, we know that an infinite plane wave travelling in the x direction has the form $A_0 \exp[-j(\omega t - \beta x)]$. Equation (1.27) indicates that for a particle wave we might write the equivalent phase constant and frequency:

$$\beta = \frac{2\pi}{\lambda} = \frac{2\pi p}{h} = \frac{p}{\hbar}$$
$$\omega = \frac{2\pi T}{h} = \frac{T}{\hbar} \tag{1.28}$$

This suggests that it might be possible to represent a particle by a function ψ, called a *wavefunction*, where

$$\psi = A_0 \exp[-j(Tt - px)/\hbar] \tag{1.29}$$

From (1.24) and (1.28) the phase velocity of such a wave is

$$v_{ph} = \frac{\omega}{\beta} = \frac{T}{p} = \frac{\tfrac{1}{2}mv^2}{mv} = \frac{v}{2}$$

Note that this result is not valid for a single particle, however, since we have seen that this must be represented by a discrete wavepacket and the concept of phase velocity is applicable only to infinite wavetrains.

A valid group velocity for the particle can, however, be found. From Eqs (1.27) and (1.28)

$$\partial\omega = (mv/\hbar)\partial v \quad \text{and} \quad \partial\beta = (m/\hbar)\partial v$$

which, using Eq. (1.26), gives

$$v_g = \partial\omega/\partial\beta = v \tag{1.30}$$

Thus, a single electron or bunch of electrons can be represented by a wavepacket travelling with the same velocity as the electron. This seems physically reasonable when we remember the alternative definition of group velocity as being the rate at which energy is transported by waves.

Our discussion so far on wave–particle duality is summarized in Fig. 1.11, where the behaviour of light waves and electrons is compared diagrammatically.

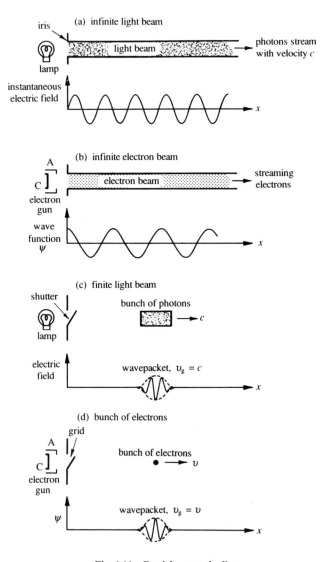

Fig. 1.11   Particle–wave duality.

1.7 The Schrödinger wave equation

So far, we have discussed experiments for which there exists no explanation based on classical concepts. However, if a series of assumptions is made, for instance the de Broglie hypothesis that $p = h/\lambda$, a quantitative description of the mechanism of each phenomenon can be provided. What is now required is some sort of unifying theory that will enable us to predict and explain other occurrences on the atomic scale. In 1926 Schrödinger provided a basis for such a theory by discovering an equation for predicting ψ, the wavefunction of a particle, in any particular circumstance. The equation he discovered is named after him; it replaces Newton's laws when atomic-sized particles are being considered. The theory based on Schrödinger's equation is called *wave* or *quantum mechanics*.

We have seen that particles can possess wave-like properties and that the probability waves associated with a beam of particles can be described in terms of a wavefunction, ψ, given by Eq. (1.29). This expression includes the kinetic energy, T, associated with a particle. In general, however, a particle can also possess potential energy. For instance, it might be an electron moving in a solid; the electron not only has kinetic energy but it also moves in the field due to the lattice and thus has a space-dependent potential energy. In general, the total energy, E, of a particle is therefore

$$E = \hbar\omega = T + V \tag{1.31}$$

where V denotes the potential energy. The one-dimensional wavefunction in the more general case then becomes

$$\psi = A_0 \exp[-j(Et - px)/\hbar] \tag{1.32}$$

What equation does this generalized wavefunction satisfy? We would expect it to be some differential equation comparable to a one-dimensional wave equation, say

$$\frac{\partial^2 H}{\partial x^2} = \epsilon\mu \frac{\partial^2 H}{\partial t^2} \tag{1.33}$$

the solution of which

$$H = H_0 \exp[-j(\omega t - \beta x)] \tag{1.34}$$

gives in this instance the magnetic field of a plane wave propagating in a medium with permittivity ϵ and permeability μ.

Let us try to find a wave equation similar to (1.33) but which has ψ given by Eq. (1.32) as its solution. First, differentiate (1.32) with respect to t

$$\frac{\partial\psi}{\partial t} = -\frac{j}{\hbar}E\psi = -\frac{j}{\hbar}(V + \tfrac{1}{2}mv^2)\psi \tag{1.35}$$

Also, let us differentiate ψ with respect to x, twice

$$\frac{\partial^2\psi}{\partial x^2} = -\frac{p^2}{\hbar^2}\psi = -\frac{m^2v^2}{\hbar^2}\psi \tag{1.36}$$

We may now compare the equations to give, from (1.35)

$$-\tfrac{1}{2}mv^2\psi = -\mathrm{j}\hbar\frac{\partial\psi}{\partial t} + V\psi$$

and from (1.36)

$$-\tfrac{1}{2}mv^2\psi = \frac{1}{2m}\hbar^2\frac{\partial^2\psi}{\partial x^2}$$

Rearranging the equations gives

$$\frac{\partial^2\psi}{\partial x^2} - \frac{2m}{\hbar^2}V\psi + \mathrm{j}\frac{2m}{\hbar}\frac{\partial\psi}{\partial t} = 0 \tag{1.37}$$

This equation is called the one-dimensional time-dependent Schrödinger wave equation. It governs the behaviour in one dimension of all particles. Notice that we have not derived this equation rigorously, since for example the relationships $p = h/\lambda$ and $E = hf$ have been assumed in writing the wavefunctions, which was our starting point. The above steps are only an argument to demonstrate the plausibility of Schrödinger's equation. In fact, there is no proof of the equation. This situation is directly comparable to the lack of proof for Newton's laws. Agreement with experiment is the only check as to its validity; it has been found to be correct when applied to a wide number of circumstances concerning microscopic particles, particularly in its relativistic form. A further test is that, in the classical limit of laboratory-sized experiments, Schrödinger's equation must provide results that agree with those derivable from Newton's laws. It can be shown that this is so and that Eq. (1.37) is quite general and reduces to Newton's laws of motion for large-sized objects. Schrödinger's equation must always be used in preference to Newton's laws, however, when considering the interaction of atomic-sized particles.

If three-dimensional motion is allowed, the wavefunction becomes a function of three space coordinates and time, $\psi(x,y,z,t)$, and is a solution of the three-dimensional time-dependent Schrödinger equation

$$\frac{\partial^2\psi}{\partial x^2} + \frac{\partial^2\psi}{\partial y^2} + \frac{\partial^2\psi}{\partial z^2} - \frac{2m}{\hbar^2}V\psi + \mathrm{j}\frac{2m}{\hbar}\frac{\partial\psi}{\partial t} = 0 \tag{1.38}$$

It should be noted that, although this equation or the simple one-dimensional version will be used predominantly to predict the wavefunction of electrons, it is applicable to *any* particle, provided the appropriate mass and potential energy are included.

For a large class of problems in which the total particle energy is constant, the Schrödinger equation can be simplified by separating out the time- and position-dependent parts. This can be achieved by the standard separation-of-variables procedure. Accordingly we consider the one-dimensional equation for simplicity and assume solutions of the form

$$\psi = \Psi(x)\Gamma(t) \tag{1.39}$$

where Ψ and Γ are respectively functions of position and time only.

We substitute this in Eq. (1.37) to obtain

$$\frac{\hbar^2}{2m}\frac{1}{\Psi}\frac{d^2\Psi}{dx^2} - V = -j\frac{\hbar}{\Gamma}\frac{d\Gamma}{dt} \tag{1.40}$$

Now the left-hand side is a function of space coordinates only, provided V is time-independent, and the right-hand expression is a function of time only. Thus each equation must independently equal some constant, say C. Therefore

$$\frac{d\Gamma}{dt} = \frac{jC}{\hbar}\Gamma$$

or

$$\Gamma(t) = \exp(jCt/\hbar)$$

Comparison of this expression with the time-dependent part of the one-dimensional wavefunction, given in (1.32), indicates that the constant in the expression is equal to $-E$. Thus, if the energy is constant, the time-dependent part of the wavefunction is

$$\Gamma(t) = \exp(-jEt/\hbar) \tag{1.41}$$

and

$$\psi = \Psi(x)\exp(-jEt/\hbar) \tag{1.42}$$

The left-hand side of Eq. (1.40) is then equal to $-E$, to give

$$\frac{d^2\Psi}{dx^2} + \frac{2m}{\hbar^2}(E - V)\Psi = 0 \tag{1.43}$$

This is the one-dimensional time-independent Schrödinger equation. It may be used to find the space-dependent part of the wavefunction whenever the system energy is constant, for example, when considering bound particles or a constant current of particles. In all other situations, for example, when discussing a varying current of particles, the more general time-dependent version of the equation must be employed. For most problems dealing with particles of constant energy, it is usually sufficient to solve the three-dimensional version of Eq. (1.43) to find the dependence of the wavefunction on position $\Psi(x,y,z)$, but if the complete wavefunction $\psi(x,y,z,t)$ is required then $\Psi(x, y, z)$ must be multiplied by $\Gamma(t)$, as indicated by Eq. (1.42).

1.8 Interpretation of the wavefunction ψ

We have shown that it is possible to describe the wave characteristics of a particle in terms of a wavefunction ψ, but we have not yet discussed precisely what property of the particle is behaving in a wave-like manner. There is little difficulty in this respect with other wave types; for example, it is the electric and magnetic field vectors that are oscillating in a radio wave and for sound waves the variable parameter is pressure. However, the physical significance of the wavefunction is not so readily apparent. Since $\psi(x,y,z,t)$ is a function of space and time coordinates, we might expect that it represents the position of a particle at some time t. However, we will see later that in general it is impossible to locate a particle exactly in space without there being any uncertainty as to its position. We can only consider the probability of a particle being at a particular point in space. A further complication is that, since ψ is a solution of Schrödinger's equation, it is usually a complex quantity.

Max Born, in 1926, overcame the difficulty of being unable to attach physical significance to ψ itself by showing that the square of its absolute magnitude, $|\psi|^2$, is proportional to the probability of a particle being in unit volume of space, centred at the point where ψ is evaluated, at time t. Thus, although the exact position of a particle at a particular time cannot be predicted, it is possible to find its most probable location. It follows that $|\psi|^2 \Delta V$ is proportional to the probability that a particle will be found in the volume element ΔV. For example, the probability of finding a particle in the range $x \rightarrow x + dx$, $y \rightarrow y + dy$ and $z \rightarrow z + dz$ is proportional to

$$|\psi(x,y,z)|^2 \, dx \, dy \, dz = \psi\psi^* \, dx \, dy \, dz \qquad (1.44)$$

where ψ^* is the complex conjugate of the wavefunction. Although the direct physical significance of particle waves is not clear, if we solve Schrödinger's equation in particular circumstances and obtain a wavefunction ψ, the *probability density*, $|\psi|^2$, can be used to predict accurately what the spatial distribution of particles will be at some time, t, provided that a sufficiently large number of experiments has been performed.

If a particle exists at all, it is certainly located somewhere in space, and since the probability of its being in an elemental volume is proportional to $|\psi|^2 \, dx \, dy \, dz$, it is convenient to choose the constant of proportionality such that the integral of the probability density over all space equals unity or

$$\iiint_{-\infty}^{+\infty} \psi\psi^* \, dx \, dy \, dz = 1 \qquad (1.45)$$

A wavefunction that satisfies this condition is said to be *normalized*. Whenever wavefunctions are normalized, $|\psi|^2 \Delta V$ *equals* the probability that a particle will be found in a volume ΔV.

Before solving Schrödinger's equation in any particular set of circumstances, it is first necessary to know what boundary conditions must be set on ψ. Since answers to physical problems are obtainable from wavefunctions, they must be well behaved in a mathematical sense. First, ψ must be a continuous and single-valued function of position. Supposing for the moment that this were not so, $\psi\psi^*$ would be discontinuous also, as indicated for the one-dimensional case shown in Fig. 1.12(a). This would imply that the

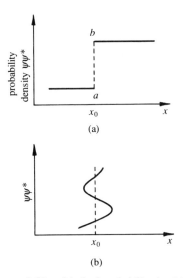

Fig. 1.12 (a) Discontinuous and (b) multivalued probability densities, which are not allowed.

probability of finding a particle is dependent on the direction from which the discontinuity is approached. Coming to x_0 from the left the probability density is a and from the right it is b, which in turn indicates that particles are instantaneously created or destroyed at x_0, which is clearly not allowable. Hence ψ must be continuous.

Consider now the possibility of ψ and hence $\psi\psi^*$ being multivalued, as shown in Fig. 1.12(b). This would imply that, at some position, x_0, there are several probabilities of finding a particle, which is obviously not physically possible. Hence ψ must be single-valued. By similar reasoning, it is possible to show that the spatial derivatives of ψ, $\partial\psi/\partial x$, $\partial\psi/\partial y$ and $\partial\psi/\partial z$ must be continuous and single-valued across any boundary.

1.9 The uncertainty principle

In 1927, Heisenberg published the uncertainty principle, which was subsequently named after him. He showed that, if an attempt is made to measure certain pairs of variables concerned with physical systems, then a lack of precision with which the two variables can be specified simultaneously becomes apparent. Such pairs of variables are momentum and position, and energy and time. Suppose, for example, in a particular experiment, that the energy of a particle can be measured to some accuracy ΔE and the time at which the measurement is taken is known to some accuracy Δt; then a classical theory would indicate that the precision to which these parameters can be measured is limited only by the experimental apparatus and technique. The uncertainty principle shows, however, that if the particle's energy is determined very accurately, so that ΔE is small, there is a proportional increase in the lack of precision in the time measurement and Δt increases. This can be made to sound plausible by the following argument. We know that a single particle can be represented in one dimension by a wavepacket, as shown in Fig. 1.13(a). There is evidently some lack of precision in locating the

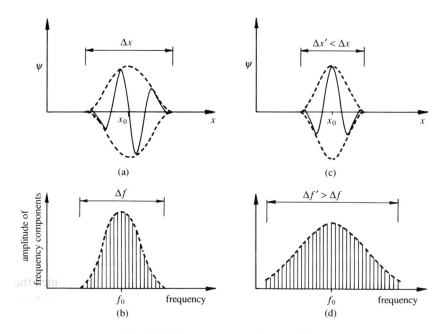

Fig. 1.13   Frequency spectra of wavepackets.

wavepacket and hence the particle, since it is spread over some distance Δx in space. Since the particle can be assumed to be travelling at some velocity v, the time of arrival of the wavepacket at some particular location can only be measured to an accuracy Δt, related to Δx by

$$\Delta x = v \Delta t$$

We have discussed previously how the wavepacket can be synthesized from many waves, each of different frequency. The amplitudes of the various component waves can be obtained by Fourier analysis to give the frequency spectrum of the wavepacket as shown in Fig. 1.13(b). This indicates that the spread in frequency of the component waves is of order Δf, and since $E = hf$, or $\Delta E = h\Delta f$, there is necessarily a corresponding uncertainty in the particle energy.

If, now, the accuracy in the location of x is increased somewhat, Δx and hence Δt decrease and the wavepacket is shorter, as shown in Fig. 1.13(c). This shortening can be achieved only by the addition of further frequency components, with a corresponding increase in the width of the frequency spectrum, as in Fig. 1.13(d). This leads, as we have seen, to an increased ΔE and a greater uncertainty as to the precise value of the energy of the particle. Thus, as Δt is decreased, ΔE is increased and vice versa. If the frequency spectra are obtained in a more quantitative way, it can be shown that the best that can be done, no matter what the experimental apparatus, is that the product of the uncertainties to which each is known is equal to Planck's constant, or

$$\Delta E \, \Delta t \geqslant h \qquad (1.46)$$

Notice that there is no fundamental restriction on the accuracy with which either quantity can be determined individually, only on their product.

Now, the energy of a free particle is $E = hf = \frac{1}{2}mv^2$, from which we obtain

$$\Delta E = h\Delta f = mv\Delta v = v\Delta p$$

Thus, the momentum spectrum for a wavepacket shows similar characteristics to the frequency spectrum; for a short wavepacket, Δx is small but the width of the momentum spectrum Δp is large, and if Δx is made bigger there is a corresponding decrease in Δp. Again, it can be shown that the product of the accuracies for a simultaneous measurement of momentum and position must be greater than $\hbar/2$, or

$$\Delta p \, \Delta x \geqslant \hbar/2 \qquad (1.47)$$

We might, for example, try to locate the exact position of a particle, thus specifying $\Delta x = 0$, but this would be possible only if all knowledge of its momentum were sacrificed since to satisfy Eq. (1.47) Δp must become infinite.

Fortunately, the uncertainties implicit in Heisenberg's principle need not be too restricting for normal laboratory-scale experiments since h is very small, of order 10^{-34} J s. However, the limitations to accuracy, as given in expressions (1.46) and (1.47), become critical for atomic-sized particles, when the magnitudes of the experimental variables can become minute.

1.10 Beams of particles and potential barriers

We shall now consider the interaction of beams of particles with potential barriers of various types. These problems are not only interesting in their own right but also serve as an introduction to discussions of the solution of Schrödinger's equation for particles confined in space, which begin in the next chapter. Interactions in one dimension only will be considered, mostly for mathematical convenience, but the solutions obtained are directly applicable to the motion of particles in devices with large dimensions transverse to the current flow.

First, consider a beam of particles travelling in the x direction with energy E, impinging on a potential barrier at $x=0$, of height $V_2 - V_1$, such that $V_1 < E < V_2$, as shown in Fig. 1.14.

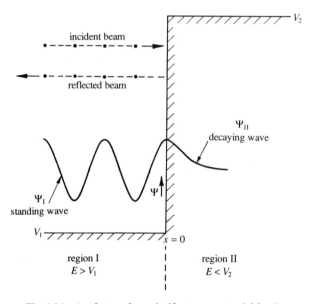

Fig. 1.14 **An electron beam incident on a potential barrier.**

Classically, we should expect particles in region I, where $E > V_1$, and none in region II, where $E < V_2$. We have deliberately chosen to investigate the motion of a beam of particles since, as we have seen, this can be represented by the simple wavefunction.

$$\psi = A_0 \exp[-j(Et - px)/\hbar]$$

Strictly speaking, each particle in the beam should be represented by a wavepacket. However, for a small spread in phase constant, $\Delta\beta$, the reflection of a packet is similar to the reflection of an infinite wave.

Let us apply Schrödinger's stationary-state equation to regions I and II, since E is constant. In region I

$$\frac{d^2\Psi}{dx^2} + \frac{2m}{\hbar^2}(E - V_1)\Psi = 0 \tag{1.48}$$

which, if we let

$$\frac{2m}{\hbar^2}(E - V_1) = \beta^2 \tag{1.49}$$

becomes

$$\frac{d^2\Psi}{dx^2} + \beta^2\Psi = 0 \tag{1.50}$$

Solutions of this equation are of the form

$$\Psi_1 = A\exp(j\beta x) + B\exp(-j\beta x) \tag{1.51}$$

The complete wavefunction for the region, including time dependence, is obtained by multiplying Ψ_1 by $\exp(-jEt/\hbar)$, giving

$$\psi_1 = A\exp\left[-j\left(\frac{Et}{\hbar} - \beta x\right)\right] + B\exp\left[-j\left(\frac{Et}{\hbar} + \beta x\right)\right] \tag{1.52}$$

The first term represents an incident probability wave travelling in the positive x direction and the second term represents the wave reflected by the barrier, travelling in the negative x direction.

In region II, $(E - V_2)$ is negative and we let

$$\frac{2m}{\hbar^2}(E - V_2) = -\alpha^2 \tag{1.53}$$

so that, in this region, the stationary-state equation becomes

$$\frac{d^2\Psi}{dx^2} - \alpha^2\,\Psi = 0 \tag{1.54}$$

This has a general solution

$$\Psi_{II} = Ce^{-\alpha x} + De^{\alpha x} \tag{1.55}$$

Arguing on physical grounds, we would not expect Ψ_{II} to become infinite at large x, so D evidently must be zero, which gives

$$\Psi_{II} = Ce^{-\alpha x} \tag{1.56}$$

We now find relationships between the magnitudes A, B and C by applying the boundary conditions on ψ at the barrier, $x = 0$. That ψ and hence Ψ are continuous at the boundary gives

$$\Psi_I|_{x=0} = \Psi_{II}|_{x=0}$$

which, from Eqs (1.51) and (1.56), gives

$$A + B = C \tag{1.57}$$

Also, for continuity of the derivative of Ψ,

$$(\partial\Psi_{\mathrm{I}}/\partial x)|_{x=0}=(\partial\Psi_{\mathrm{II}}/\partial x)|_{x=0}$$

or

$$j\beta A-j\beta B=-\alpha C \qquad (1.58)$$

Equations (1.57) and (1.58) give

$$A=\tfrac{1}{2}C\,(1-\alpha/j\beta)$$
$$B=\tfrac{1}{2}C\,(1+\alpha/j\beta) \qquad (1.59)$$

Notice that the amplitude of incident and reflected waves are identical, i.e.

$$|A|=|B|$$

but, because there is a phase difference between the forward- and backward-travelling waves, the total Ψ wave or sum of incident and reflected waves is a standing wave. A further point of interest is that the solution Ψ_{II} given in Eq. (1.56), which applies to the right of the barrier, represents a wave whose amplitude decays exponentially with increasing x. Both of these situations are illustrated diagrammatically in Fig. 1.14. It will be seen that there is a finite probability that the incident beam penetrates some distance into the classically forbidden region, since $|\Psi_{\mathrm{II}}|^2$ is greater than zero there. However, if α is large, i.e. $V_2 \gg E$, few particles are found very far inside the boundary.

Now let us turn our attention to the situation where the barrier is not sufficiently high to cause complete reflection of the incident beam. In these circumstances, there is only partial reflection and part of the beam is transmitted through the barrier, as shown in Fig. 1.15. If, as before, we let

$$\beta_{1,2}=\frac{2m}{\hbar^2}(E-V_{1,2}) \qquad (1.60)$$

then the solutions of Schrödinger's equation in the two regions are

$$\Psi_{\mathrm{I}}=A\,\exp(j\beta_1\,x)+B\,\exp(-j\beta_1\,x)$$
$$\Psi_{\mathrm{II}}=C\,\exp(j\beta_2\,x) \qquad (1.61)$$

if it is assumed that there is no reflected wave in region II. Matching Ψ and $\partial\Psi/\partial x$ at the boundary as before we have, from (1.61)

$$A+B=C$$
$$\beta_1(A-B)=\beta_2 C \qquad (1.62)$$

C can be eliminated from these equations to give

$$\frac{B}{A}=\frac{1-\beta_2/\beta_1}{1+\beta_2/\beta_1} \qquad (1.63)$$

which is used to find a reflection coefficient:

$$\frac{\text{density of particles reflected}}{\text{density of particles incident}} = \frac{|\Psi_{\text{ref}}|^2}{|\Psi_{\text{inc}}|^2} = \frac{BB^*}{AA^*} = \frac{B^2}{A^2}$$

$$= \left(\frac{1 - [(E - V_2)/(E - V_1)]^{1/2}}{1 + [(E - V_2)/(E - V_1)]^{1/2}}\right)^2 \quad (1.64)$$

where we have substituted for the β's from Eq. (1.60).

Eliminating B from (1.62) gives the relative amplitude of the transmitted wave:

$$\frac{C}{A} = \frac{2}{1 + \beta_2/\beta_1} = \frac{2}{1 + [(E - V_2)/(E - V_1)]^{1/2}} \quad (1.65)$$

We thus find the surprising result that the amplitude of the transmitted wave is bigger than that of the incident wave, or the probable density of particles in the transmitted electron beam is greater than that in the incident beam! This may be explained by noting that the transmitted particles are moving more slowly than the incident particles. Thus, although $C > A$, the *rate* of flow of particles in the transmitted beam is less than that in the incident beam. Notice also that

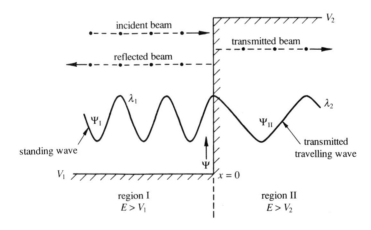

Fig. 1.15   An electron beam incident on a potential step.

since $\beta_2 < \beta_1$, the wavelength of the transmitted Ψ wave is greater than that for the incident wave, as shown in Fig. 1.15.

Our final example in this series is the interaction of a constant-energy beam of particles incident on a classically impenetrable potential barrier of finite width, d, as shown in Fig. 1.16. Applying Schrödinger's equation to the three

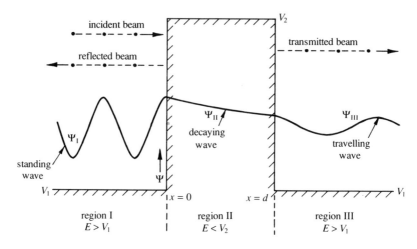

Fig. 1.16 **Partial transmission of an electron beam through a narrow potential barrier.**

regions gives wavefunctions of the form:

$$\Psi_1 = A \exp(j\beta x) + B \exp(-j\beta x)$$
$$\Psi_{II} = C \exp(-\alpha x) + D \exp(\alpha x)$$
$$\Psi_{III} = F \exp(j\beta x)$$

(1.66)

where

$$\beta^2 = \frac{2m}{\hbar^2}(E - V_1) \qquad \text{and} \qquad \alpha^2 = \frac{2m}{\hbar^2}(V_2 - E)$$

(1.67)

as before. Notice that, in order to satisfy the boundary conditions, a highly attenuated reflected wave, magnitude D, has been included in the barrier region; also, it has been assumed that no reflection occurs in region III. We can now apply the usual boundary conditions at $x = 0$ and $x = d$ and hence find the relative values of the wavefunctions at various positions, as shown in Fig. 1.16. If, in particular, B, C and D are eliminated, the amplitude of the Ψ wave in region III in terms of the amplitude of the incident wave in region I can be obtained, thus

$$F = A \exp(-j\beta d)[\cosh(\alpha d) + \tfrac{1}{2}(\alpha/\beta - \beta/\alpha) \sinh(\alpha d)]^{-1}$$

(1.68)

Now, the probability of a particle passing through the barrier, $P_{\text{I-III}}$, is proportional to the ratio of the absolute square of ψ in region III to the absolute square of ψ in the incident region, or

$$P_{\text{I-III}} = |F|^2/|A|^2$$

(1.69)

which can be evaluated using Eq. (1.68).

In most practical cases, αd is large because of high attenuation of the

Ψ wave in the barrier region, and the hyperbolic functions in (1.68) can be simplified to give

$$P_{1-\mathrm{III}}=\frac{|F|^2}{|A|^2}\simeq\frac{\exp(-2\alpha d)}{1+\tfrac{1}{4}(\alpha/\beta-\beta/\alpha)^2}$$

We can now substitute values for α and β as defined in Eq.(1.67) to give

$$P_{1-\mathrm{III}}\approx\exp\{-2[2m(V_2-E)]^{1/2}\,d/\hbar\} \tag{1.70}$$

Thus, if the barrier is sufficiently thin, i.e. d is small, there is a small but finite probability that particles can penetrate the barrier, even though classically this would not seem possible. The particle is said to have *tunnelled* through the barrier. The effect has important physical applications, as will be discovered later when, for example, tunnel diodes and electron emission from a cathode are discussed.

Notice that the probability of tunnelling falls off exponentially with increasing barrier height and thickness. As an example of the sort of figures involved, consider a 1 A electron beam approaching a barrier 1 V high $(V_2-E=1.6\times10^{-19}\,\text{J})$ and 2 nm wide. The probability of tunnelling, using (1.70), is then of order e^{-20} and a current of about 10^{-9} A tunnels through the barrier. However, if the barrier thickness is reduced to 0.1 nm, the tunnelling current is increased to about 0.3 A!

Problems

1. An electron in an atom drops from an excited state to the ground state in a time normally lasting about $10^{-8}\,\text{s}$. If the energy emitted is 1 eV, find (a) the relative uncertainty in the energy, (b) the relative uncertainty in the frequency of the emitted radiation and (c) the number of wavelengths in and the length of the wavepacket of emitted radiation.

Ans. $4\times10^{-7}, 4\times10^{-7}, 2\times10^6, 3\,\text{m}$

2. What is the inherent uncertainty in the velocity of an electron confined in a crystalline solid of volume 10 mm cube?

Ans. $0.07\,\text{m s}^{-1}$

3. Show that the probability of reflection of a beam of electrons of energy E incident on an abrupt potential energy step of height V_0 is determined by a reflection coefficient r given by

$$r=\left(\frac{1-(1-V_0/E)^{1/2}}{1+(1-V_0/E)^{1/2}}\right)^2$$

Also show that the magnitude of the reflection coefficient is independent of whether the electrons are incident from the high or low potential energy side. How is its phase affected?

4. Find the probability of transmission of a 1 MeV proton through a 4 MeV high, 10^{-14} m thick rectangular potential energy barrier $(m_p/m = 1836)$.

  Ans. 5×10^{-4}

2 The electronic structure of atoms

2.1 Introduction

In the previous chapter we showed that microscopic particles can be described in terms of waves and wavepackets and used such a description to discuss the interaction of particles with potential barriers of various shapes. In this section we confine our attention to particles that are constrained to remain localized in some part of space. We shall see that under these conditions the particle cannot have a continuous spectrum of possible energies and the system is such as to allow the particle to have only discrete values of energy; nor can the wavefunction of a confined particle take up any arbitrary value, and only a set of discrete wavefunctions is allowable.

The properties of particles confined to a finite region of space, or *bound*, can be studied by considering them to be trapped in a potential well with very steep sides. The shape of such containers with which we are able to deal is artificial in the extreme for reasons of mathematical simplicity and bears little resemblance to physical reality. However, the results obtained have general characteristics that are most relevant to the description of electrons bound to parent atoms and form a basis for the categorization of atoms in terms of their electronic structure.

2.2 A particle in a one-dimensional potential well

Consider the situation depicted in Fig. 2.1. A particle of mass m and total energy E moves only in the x direction and is constrained to remain in a region of length d by potential barriers at $x = 0$ and $x = d$. The barriers are made artificially steep and high and are infinite in extent so that there is no possibility of the particle surmounting them and escaping. We assume that the potential energy of the particle inside the well is zero and apply Schrödinger's time-independent equation, Eq. (1.43), since the total energy E remains constant. Hence, for $0 \leqslant x \leqslant d$, $V = 0$ and

$$\frac{d^2 \Psi}{dx^2} + \frac{2m}{\hbar^2} E\Psi = 0 \qquad (2.1)$$

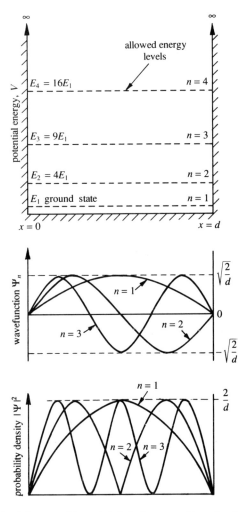

Fig. 2.1 Particle trapped in an infinitely deep, one-dimensional potential well.

We have seen that the general solution of this equation, which gives the wavefunction of the particle inside the well, is

$$\Psi = A e^{j\beta x} + B e^{-j\beta x} \tag{2.2}$$

where

$$\beta^2 = (2m/\hbar^2)E$$

Now, since there is no possibility of the particle penetrating the containing walls and so gaining infinite potential energy, $\Psi\Psi^*$ and hence Ψ must be zero outside the well for $0 \geqslant x \geqslant d$. Thus for continuity of Ψ at the boundaries, the value of Ψ inside the well, given by (2.2), must become zero at $x=0$ and $x=d$. Applying these boundary conditions to Eq. (2.2) gives

$$B = -A$$

and

$$0 = A(e^{j\beta d} - e^{-j\beta d})$$

or

$$\sin(\beta d) = 0$$

Hence

$$\beta d = [(2mE)^{1/2}/\hbar] \; d = n\pi \qquad \text{where } n = 1, 2, 3, \ldots \qquad (2.3)$$

These conditions can now be substituted into Eq. (2.2) to give a general expression for the wavefunctions for the particle in the well:

$$\Psi = C \sin(n\pi x/d) \qquad (2.4)$$

where C is a newly defined constant. We see that the two travelling-wave components of Ψ, as represented by the two parts of Eq. (2.2), are of equal magnitude but move in opposite directions and combine to produce a standing wave in the usual way.

The constant C is a normalizing constant whose value can be obtained by arguing that the probability that the particle is located *somewhere* in the well must be unity. Since the probability that it is located in length dx is $\Psi\Psi^*dx$, then

$$\int_0^d \Psi\Psi^* dx = 1$$

or

$$\int_0^d C^2 \sin^2(n\pi x/d) \, dx = 1$$

from which we find that $C = (2/d)^{1/2}$ and the normalized solution for Ψ becomes

$$\Psi = (2/d)^{1/2} \sin(n\pi x/d) \qquad n = 1, 2, 3, \ldots \qquad (2.5)$$

We see that the wavefunction for the bound particle is one of a set of discrete values, each corresponding to a different value of the integer n. This general conclusion is not confined to this particular example but applies to all bound particles. Each allowed value of the wavefunction, for instance for each value of the integer in Eq. (2.5), is called an *eigenfunction*, which can be loosely translated from the German as 'particular function'.

Returning now to Eq. (2.3), it is evident that the total energy of the bound particle also has only a discrete set of allowed values given by

$$E = \frac{\hbar^2 n^2 \pi^2}{2md^2} = \frac{n^2 h^2}{8md^2} \qquad n = 1,2,3,\ldots \qquad (2.6)$$

Thus, the total energy of the particle in the well has particular allowed values; corresponding to the various integers; the energy is quantized and each particular energy level is called an *eigenvalue*. A set of possible energy levels is shown in Fig. 2.1. Notice that energy levels intermediate to those shown are forbidden and also that a particle in the lowest energy state or ground state, E_1, has a non-zero kinetic energy. Both of these general results are applicable to all bound particles and are at variance with classical mechanical ideas.

This result might have been obtained in a simpler, more intuitive way, using the de Broglie wavelength of the electron, $\lambda = h/p$, directly. Assuming that an electronic wave is reflected at either extremity of the well, since an electron is not allowed to penetrate outside the infinitely high potential walls, a standing wave will be established, analogous to the behaviour of electromagnetic waves in a cavity resonator or a vibrating string with fixed nodes. It follows that the well must contain an integral multiple of half electronic wavelengths of similar form to that shown in Fig. 2.1. Hence

$$d = n\lambda/2 = (n/2)\, h/p$$

which can be substituted in the expression for electron energy

$$E = p^2/2m$$

to give

$$E = \left(\frac{nh}{2d}\right)^2 \frac{1}{2m} = \frac{n^2 h^2}{8md^2}$$

as before.

The probability per unit length that the particle is located at some particular position in the well, x, can be found in the usual way by forming $|\Psi|^2$ from the value of Ψ given in Eq. (2.5). The eigenfunctions, Ψ_n, are plotted in Fig. 2.1 together with the corresponding probabilities of location per unit length, $|\Psi_n|^2$. Notice that the wavelength, λ, of the standing wave of Ψ is quantized, only having discrete values given by $\lambda = 2\pi/\beta = 2d/n$, and that n is the number of loops in the pattern. This result can be compared to the electrical resonances that occur in a transmission line short-circuited at each end; such resonances occur when the wavelength is equal to twice the line length divided by an integer. The analogy is valid and the results similar, since Schrödinger's equation is of the same form as the electromagnetic wave equation and the boundary conditions are similar in each case.

The diagrams of $|\Psi|^2$ indicate that a particle in the ground state is most probably located at the centre of the well. For higher energy states, for example E_2, the particle may be located at some position removed from the centre of the well and there is zero probability of the particle being located at certain other positions. These conclusions do not agree with the classical picture of the particle being reflected elastically from the walls of the well, which results in a constant probability amplitude since the particle may be located anywhere in the well with equal probability. However, at higher energy levels, for large values of n, the agreement is better since the quantum probability oscillates rapidly with position about a mean value, which is the classical probability, and its average value taken over a short length is equal to the classically anticipated one.

The arguments for the infinitely deep well can be extended either theoretically or on a more intuitive basis, as we shall do here, to consider the case of a particle trapped in a one-dimensional potential well of finite depth. Although this problem is slightly more realistic in that it more closely resembles the situation of an electron bound to an atom, it is still somewhat artificial. The problem is illustrated in Fig. 2.2. Consider first the classically

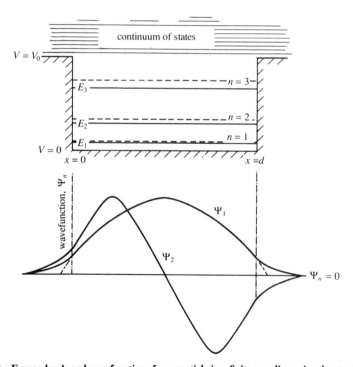

Fig. 2.2 Energy levels and wavefunctions for a particle in a finite one-dimensional potential well.

bound particle whose total energy, E, is less than V_0, the depth of the potential well. As we have seen earlier, there will be a small but finite probability that such particles can penetrate some way through the boundaries of the well before being reflected. The wavefunctions outside the well are attenuated and have the characteristic exponentially decaying form, much the same as that depicted in Fig. 1.14. The wavefunction inside the well no longer fall to zero at the walls since there has to be smooth matching of wavefunctions across each boundary. Thus, although the wavefunctions have a shape generally similar to those for the infinite well, they are slightly modified in that they are no longer purely sinusoidal and their wavelength is increased. This leads to a set of eigenvalues slightly lower in value than the corresponding set for the infinite well, as shown in Fig. 2.2.

When the energy of the particle is greater than the depth of the well, $E > V_0$, then the particle is no longer classically bound and it can take up any level in a continuum of possible energy levels, as shown in the figure.

2.3 The hydrogen atom

We now turn our attention to a quantum-mechanical solution of the hydrogen atom. Not only is this a useful example of the application of Schrödinger's equation to a particle trapped in a more physically realistic potential well, but the results also explain much about the electronic structure of atoms, which lays the foundations for a logical classification of the elements.

The Bohr theory of the hydrogen atom, given in Sec. 1.4, is unacceptable on several counts. First, it assumes that the electron moves in a given orbit, which implies that its position is known precisely at any time, which violates the uncertainty principle. Secondly, the theory arbitrarily assumes that the angular momentum of the orbiting electron is quantized and has values given by Eq. (1.13). Finally, the Bohr theory cannot be extended to treat atoms with more than one outer electron. These limitations are not present in a quantum-mechanical treatment and solutions can be obtained, in principle at least, for more complex atoms and molecules.

We assume that the hydrogen atom consists of a central nucleus of charge $+e$ surrounded by an electron. The nucleus is assumed fixed because of its relatively heavy mass, but to relax this condition would make only a slight quantitative difference to the result. The potential energy of an electron located at a distance r from the nucleus was derived in Sec. 1.4 and is

$$V = -e^2/(4\pi\epsilon_0 r) \tag{2.7}$$

This situation is shown diagrammatically in Fig. 2.3(a). The electron is trapped in the potential well created by the field of attraction set up by the positive nucleus; it is bound to the nucleus. The problem is more difficult than those encountered previously in that the contour of the well is not so simple mathematically and, also, the geometry is now three-dimensional. It is convenient to define the problem in spherical polar coordinates, as shown in Fig. 2.3(b), because of a possible spherical symmetry in the solution. The three-dimensional time-independent Schrödinger equation given in Eq. (1.38) is now written in the new coordinate system for the electron in the well and the

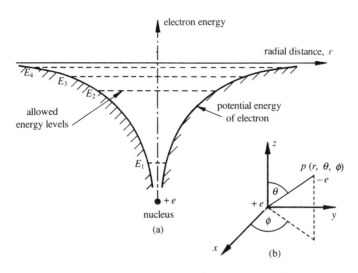

Fig. 2.3 (a) Section through the potential well of a hydrogen atom; (b) spherical polar coordinates.

potential energy function given in Eq. (2.7) can be included at the same time to give

$$\frac{1}{r^2}\frac{\partial}{\partial r}\left(r^2\frac{\partial\Psi}{\partial r}\right)+\frac{1}{r^2\sin\theta}\frac{\partial}{\partial\theta}\left(\sin\theta\frac{\partial\Psi}{\partial\theta}\right)+\frac{1}{r^2\sin^2\theta}\frac{\partial^2\Psi}{\partial\phi^2}$$
$$+\frac{2m}{\hbar^2}\left(E+\frac{e^2}{4\pi\epsilon_0 r}\right)\Psi=0 \tag{2.8}$$

Usually, this equation is solved completely by the separation-of-variables technique, but in order to simplify the problem we shall look for solutions that are independent of the angular coordinates θ and ϕ. The wavefunctions that are solutions of Eq. (2.8) under these particular conditions will have spherical symmetry and will depend on the variable r only. Therefore, we let $\partial/\partial\theta$ and $\partial/\partial\phi$ be zero and carry out the differentiation implicit in the first term, to reduce Eq. (2.8) to

$$\frac{d^2\Psi}{dr^2}+\frac{2}{r}\frac{d\Psi}{dr}+\frac{2m}{\hbar^2}\left(E+\frac{e^2}{4\pi\epsilon_0 r}\right)\Psi=0 \tag{2.9}$$

There is a set of spherically symmetrical or *radial* wavefunctions that satisfy this equation, but the simplest is of the form

$$\Psi_1=A\exp(-r/r_0) \tag{2.10}$$

where r_0 is a constant and A is the usual normalizing constant. Once again, A is evaluated by arguing that the probability of locating the electron somewhere in space must be unity. Since the probability density is $\Psi\Psi^*$ and the volume of a spherical shell of radius r and thickness dr is $4\pi r^2\,dr$, the probability that the electron is located inside such a shell is $\Psi\Psi^*4\pi r^2\,dr$. Hence, using (2.10), we have

$$\int_0^\infty \Psi\Psi^*4\pi r^2\,dr = 4\pi A^2 \int_0^\infty r^2 \exp(-2r/r_0)\,dr = 1$$

which can be integrated by parts to give

$$A = \pi^{-1/2}r_0^{-3/2}$$

and

$$\Psi_1 = \pi^{-1/2}r^{-3/2}\exp(-r/r_0) \tag{2.11}$$

We can find the value of the constant r_0 by substituting this wavefunction into Eq. (2.9), which gives

$$\frac{1}{r_0^2} - \frac{2}{rr_0} + \frac{2m}{\hbar^2}\left(E + \frac{e^2}{4\pi\epsilon_0 r}\right) = 0 \tag{2.12}$$

For this equation to be true for all values of r, the terms containing r must equate to zero, or

$$\frac{2}{rr_0} = \frac{2me^2}{4\pi\hbar^2\epsilon_0 r}$$

which gives

$$r_0 = \frac{4\pi\hbar^2\epsilon_0}{me^2} = \frac{\hbar^2\epsilon_0}{\pi e^2 m} \tag{2.13}$$

Comparing this result with Eq. (1.14) we see that the constant r_0 is numerically equal to the radius of the lowest-energy orbit of the Bohr atom; we will return to a discussion of the physical significance of this result later.

Meanwhile, if we now consider Eq. (2.12) with the sum of the r-dependent terms equal to zero, the total energy of the electron in the lowest energy state is

$$E_1 = -\frac{\hbar^2}{2mr_0^2} = -\frac{me^4}{8\epsilon_0^2 h^2} \tag{2.14}$$

where Eq. (2.13) has been used to evaluate the constant r_0.

A more complete solution of the wave equation so as to include higher-order radial wavefunctions Ψ_2, Ψ_3, etc., leads to the following general expression for the allowed energy levels:

$$E_n = \frac{E_1}{n^2} = -\frac{1}{n^2}\frac{me^4}{8\epsilon_0^2 h^2} \tag{2.15}$$

Again we see that the result of confining the electron to some localized portion of space is to quantize its possible energy levels, as shown in Fig. 2.3. The integer n is called the *principal quantum number*. We shall see later that this is only one of several such numbers that are usually necessary to define the energy of a particular electron in an atom.

We may now use our knowledge of the wavefunction of the electron in the ground state, Ψ_1, to discuss the geometry of the hydrogen atom. We have shown that if the probability that an electron is located in a spherical shell of radius r and thickness dr is dP_r, then

$$dP_r = |\Psi_1|^2 4\pi r^2 \, dr = (4r^2/r_0^3) \exp(-2r/r_0) \, dr \qquad (2.16)$$

and the probability per unit radius for an electron in the ground state to be located at radius r is

$$dP_r/dr = (4r^2/r_0^3) \exp(-2r/r_0) \qquad (2.17)$$

This probability is plotted in Fig. 2.4(a). It can be shown by differentiation that the maximum value of dP_r/dr occurs when $r = r_0$ as shown.

Further, if we let $\rho_1(r)$ be the charge density due to the electron in the ground state, then

$$\rho_1(r) = |\Psi_1|^2 (-e) = -(e/\pi r_0^3) \exp(-2r/r_0) \qquad (2.18)$$

which is shown in Fig. 2.4(b). The charge contained in a spherical shell of radius r and thickness dr is then q_r, where

$$q_r/dr = -(4er^2/r_0^3) \exp(-2r/r_0) \qquad (2.19)$$

This expression is very similar to that in (2.17) and has a maximum value at $r = r_0$ again, as shown in Fig. 2.3(c). Hence a quantum-mechanical interpretation of the hydrogen atom indicates that the electron can no longer be considered as having a fixed orbit; it can be located at any distance from the nucleus, but when in the ground state its most probable location is at the Bohr radius. The charge associated with the electron is smeared out but there is a maximum in the probable charge distribution, again at the Bohr radius.

We have discussed so far only those solutions of the three-dimensional Schrödinger equation that depend on radial distance from the nucleus. A further series of wavefunctions is possible if this assumption of spherical symmetry is relaxed. When the wavefunction has angular dependence on θ and ϕ and neither $\partial/\partial\theta$ nor $\partial/\partial\phi$ is zero, Eq. (2.8) is usually solved by the method of separation of variables. This technique assumes a solution

$$\Psi(r, \theta, \phi) = f_r(r) f_\theta(\theta) f_\phi(\phi) \qquad (2.20)$$

where the functions f_r, f_θ, f_ϕ are only dependent on r, θ and ϕ, respectively. When this trial wavefunction is substituted into Eq. (2.8), the three-dimensional equation in r, θ and ϕ is subdivided into three separate differential equations, one in r only, one in θ only, and one in ϕ only.

The solution for the differential equation in r gives only the radial wave equations and a *principal quantum number*, $n = 1,2,3,\ldots$, which defines the total energy of an electron in a particular state, as we have already discussed.

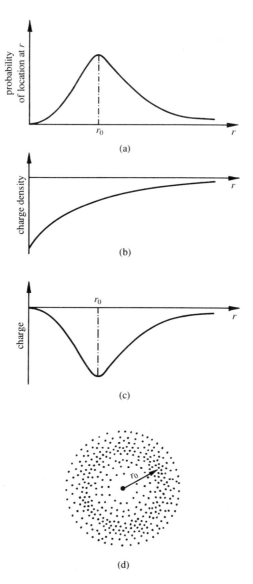

Fig. 2.4 **(a)–(c) Geometry and charge configuration of the hydrogen atom in the ground state; (d) a two-dimensional representation of the smeared-out electronic charge.**

A further set of *quantum numbers*, $l = 0, 1, 2, \ldots, (n-1)$, arises from a solution of the separated differential equation in θ. Such *azimuthal quantum numbers* are associated with the angular momentum of an electron, which is itself quantized.

Finally, that part of Schrödinger's equation which is separated to provide an equation in ϕ can be solved to give a third set of possible quantum numbers, m, which vary from $m = -l$, including $m = 0$, to $m = +l$. These *magnetic orbital quantum numbers* are associated with the fact that an electron in an orbital constitutes a rotating charge and hence an electric current, which has associated magnetic field and magnetic moment. The orientation of this inherent magnetic moment with an externally applied magnetic field is quantized and the quantum number, m, arises because of the discrete number of possible orientations.

We see that in general there exists a set of three possible quantum numbers, n, l and m, and that a particular combination of these is necessary to specify an individual quantum state. In addition, a further quantum number is necessary to define completely a particular quantum state, which is not apparent from the solution to the three-dimensional Schrödinger equation. This was first introduced in an arbitrary manner to explain fine details of atomic spectra. The electron is assumed to spin about its axis in an either clockwise or anticlockwise direction. There are only two possible ways in which the inherent angular momentum vector due to the spin can be oriented with respect to an applied magnetic field. A *spin quantum number*, s, which can have only two values, accounts for this quantization. It has been shown more recently that the assumption of spin need not be introduced so arbitrarily since it arises directly as a solution of a more generalized form of the Schrödinger equation.

2.4 The exclusion principle and the periodic table of elements

We have seen that the particular quantum state of an electronic orbital can be specified by a set of quantum numbers, (n, l, m, s). Such numbers completely define the wavefunctions for a given electron and are usually quoted instead of the wavefunction because they are less cumbersome.

The Pauli exclusion principle provides a method of classifying atoms according to their electronic structure. It states that in a multi-electronic system, which in general can be an atom, a molecule, or a complete crystal, no more than one electron can exist in any one quantum state. For the particular case of an atom, the principle implies that no two electrons can be described by an identical set of quantum numbers. Physically what this means is that no more than two electrons can have the same distribution in space and even then they must have opposite spins. There is no proof of the exclusion statement but its validity is supported by an abundance of experimental evidence.

Consider now the electronic structure of different atoms containing an increasing number of electrons. Electrons will tend to fill the lowest available energy levels first. A consequence of the exclusion principle is that each additional electron must have a different set of quantum numbers and possess a higher energy than the preceding electron. The energy levels are thus filled progressively and additional electrons always have quantum numbers that correspond to the lowest possible unoccupied energy state. A periodic table of elements based on their electronic structure can thus be constructed using this procedure in conjunction with the relationships between the various quantum numbers discussed in the previous section. Thus, for $n = 1$, l and m must both be zero and two electronic states exist, corresponding to the two spin quantum numbers. But for $n = 2$, l can equal 0 or 1, and for $l = 1$, m may have values -1, 0, $+1$; the various combinations of (n, l, m) are thus $(2, 0, 0)$, $(2, 1, -1)$, $(2, 1, 0)$, $(2, 1, 1)$ and each of these is associated with two possible states because of spin, making a total of eight possible states with principle quantum number $n = 2$. This process can be repeated for $n = 3$, $l = 0$, 1, 2, and so on, and the periodic table shown in Table 2.1 results.

Table 2.1 Electronic structure of the lighter elements.

Element	Principal quantum number, n	Azimuthal quantum number, $l = 0, 1, \ldots, n-1$	Magnetic quantum number, $m = -l, \ldots, +l$	Spectroscopic designation
H	1	0	0	$1s$
He	1	0	0	$1s^2$
Li	2	0	0	$1s^2 2s$
Be	2	0	0	$1s^2 2s^2$
B	2	1	-1	$1s^2 2s^2 2p$
C	2	1	-1	$1s^2 2s^2 2p^2$
N	2	1	0	$1s^2 2s^2 2p^3$
O	2	1	0	$1s^2 2s^2 2p^4$
F	2	1	1	$1s^2 2s^2 2p^5$
Ne	2	1	1	$1s^2 2s^2 2p^6$
Na	3	0	0	$1s^2 2s^2 2p^6 3s$
Mg	3	0	0	$1s^2 2s^2 2p^6 3s^2$
Al	3	1	-1	$1s^2 2s^2 2p^6 3s^2 3p$
Si	3	1	-1	$1s^2 2s^2 2p^6 3s^2 3p^2$
P	3	1	0	$1s^2 2s^2 2p^6 3s^2 3p^3$
etc.				

Electrons that have the same principal quantum number, n, are said to be in the same shell. It is evident from the table that the maximum number of electrons per shell is $2n^2$. Within a shell, each state corresponding to a particular integer value of l, the azimuthal quantum number, is given a letter designation:

l	State or subshell
0	s
1	p
2	d
3	f
etc.	

(This peculiar nomenclature arises from the early spectroscopic identification of lines corresponding to various electronic transitions, namely, sharp, principal, diffuse, fundamental, etc.) Thus, in the electronic classification of the lighter elements, given in the last column of Table 2.1, the integers refer to the principal quantum number of each shell, the letters correspond to the value of the azimuthal quantum number, and the indices give the number of electrons in such subshells.

It should be noted that the periodic table does not progress continuously in such a numerically logical manner since for a group of the heavier elements the energy of some electrons in an outer shell is lower than that in an inner subshell and so these levels are filled before the inner subshells are fully occupied.

Problems

1. Compute the three lowest energy levels for electrons trapped in a one-dimensional well of length 0.1 nm.

Ans. 6×10^{-18}, 2.4×10^{-17}, 5.4×10^{-17} J

2. An electron is confined in a one-dimensional potential energy well of length 0.3 nm. Find (a) the kinetic energy of the electron when in the ground state and (b) the spectral frequency resulting from a transition from the next highest state to the ground level.

Ans. 16.6 eV, 3×10^{15} Hz

3 Collections of particles in gases and solids

3.1 Introduction

We have so far discussed only properties of single, isolated particles. As soon as we turn our attention to whole collections of many such interacting particles, for example to consider molecules in a gas or electrons in a solid, it is no longer practicable to describe the detailed dynamics of the system by the application of Newton's laws to each particle. The detailed properties of individual microscopic components could be specified only at a particular time and would change continuously. Further, because of the vast numbers of particles involved in a typical gas or solid, a microscopic treatment would be too cumbersome to be readily solvable or physically meaningful.

Fortunately, we are usually more interested in the macroscopic or bulk behaviour of large numbers of particles, to predict such properties as currents, pressures, and so on. A statistical treatment that describes the average rather than detailed properties of a typical component of the complete assembly of particles is most useful, since the behaviour of the group of particles can then be deduced directly.

In order to determine the average value of, say, the velocity of a molecule in a gas or the energy of an electron in a solid, it is necessary to know first of all how the velocities or energies are distributed throughout the collection of particles. The type of statistics developed to describe such distributions depends not only on the type of particle present but also on the possible interactions between them. For instance, in a neutral gas, molecules can interchange energy by collisions but there is no restriction on the energy of individual molecules. On the other hand, when electrons interact in a solid their allowed energies are restricted by the Pauli exclusion principle, which permits only one electron to occupy a particular quantum state, and a different type of statistics applies.

We shall initially consider collections of neutral particles and develop a statistical method of treatment that not only will be useful in describing some of the properties of gases and plasmas but also is applicable to many other electronic materials. We shall then study the statistics of collections of interacting particles that obey the exclusion principle, since this is necessary for later discussion of the electrical and thermal properties of solids.

3.2 Assemblies of classical particles—ideal gases

Let us consider an ideal gas of neutral molecules containing N molecules per cubic metre. Individual molecules have a random motion and undergo many collisions with other molecules and container walls, which result in changes in the magnitude and direction of their velocity. There is no restriction on the velocity of a particle at a particular time; it may be zero at one instant or relatively high at another, and the energy of each molecule is independent of that of the others. We wish to find how the various velocities and hence energies are distributed between individual molecules.

At any instant, the position of a molecule can be specified by a set of coordinates, (x, y, z) say, and its velocity v can also be resolved into components, (v_x, v_y, v_z), in the x, y and z directions, where

$$v^2 = v_x^2 + v_y^2 + v_z^2 \tag{3.1}$$

Hence, the velocity of each particle can be represented by a vector on a three-dimensional graph with axes v_x, v_y and v_z. Such a *velocity space* is illustrated in Fig. 3.1 A few typical velocity vectors are shown and the dots

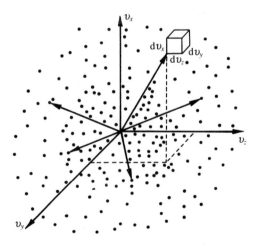

Fig. 3.1 **Particles in velocity space.**

represent the tips of the velocity vectors of the remaining molecules. The problem of finding the distribution of velocities among the particles is reduced to finding the density of such dots in velocity space. This dot density will have spherical symmetry, since there are no preferred velocity directions, and it can be written $P(v^2)$. Thus, considering the elemental volume shown in Fig. 3.1, the number of molecules per unit volume, dN_{xyz}, with velocity components in the range $v_x \to v_x + dv_x$, $v_y \to v_y + dv_y$ and $v_z \to v_z + dv_z$ is

$$dN_{xyz} = P(v^2)\,dv_x\,dv_y\,dv_z \tag{3.2}$$

Alternatively, the number of molecules per unit volume with *speeds* ranging from $v \to v + dv$, dN_v, is equal to the number of vector tips (points in Fig. 3.1) that lie in a spherical shell of radius v and thickness dv in velocity space, or

$$dN_v = P(v^2)4\pi v^2\,dv \tag{3.3}$$

The loosely termed dot density, $P(v^2)$, is called the distribution function for speeds; it gives the number density in velocity space of molecules with a certain speed, v. Notice that in all cases the number of particles in a given range equals the distribution function multiplied by the size of the range.

Since each particle must be represented somewhere in velocity space, the integral of the number in a given range over all velocity space must equal the total number of particles per unit volume. For example

$$\int_0^\infty dN_v = \int_0^\infty P(v^2)4\pi v^2\,dv = N \tag{3.4}$$

3.2.1 The distribution function, $P(v^2)$

The method we use to determine $P(v^2)$ and hence the distribution of velocities in a neutral gas is not very rigorous but has the merit of not being mathematically complex.

Consider two molecules, each of mass M, with velocities v_1 and v_2, which suffer an elastic collision that modifies their velocities to v_3 and v_4 respectively, as shown in Fig. 3.2(a). Since energy is assumed to be conserved in the collision

$$v_1^2 + v_2^2 = v_3^2 + v_4^2 \tag{3.5}$$

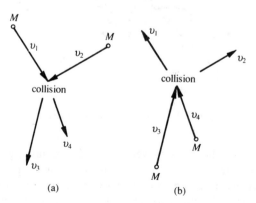

Fig. 3.2   Elastic collisions between gas molecules.

Now, the number of molecules having velocity v_1 is proportional to the distribution function evaluated at this velocity, $P(v_1^2)$. Similarly, the number of molecules with velocity v_2 is proportional to $P(v_2^2)$. Thus, the likelihood of two such particles colliding is proportional to the number in each class, and

$$\text{collison probability} \propto P(v_1^2) P(v_2^2) \tag{3.6}$$

The probability of reverse collisions occurring between molecules with velocities v_3 and v_4 to produce particles with velocities v_1 and v_2 as depicted in Fig. 3.2(b) is, by similar argument, proportional to $P(v_3^2)P(v_4^2)$. Thus, provided the geometry remains constant and the system is in equilibrium, the probability of collisions of either type must be the same, and

$$P(v_1^2)P(v_2^2) = P(v_3^2)P(v_4^2) \tag{3.7}$$

Now the only type of solution that satisfies Eqs (3.5) and (3.7) simultaneously is

$$P(v^2) = A \, \exp(-\beta v^2) \tag{3.8}$$

which can be verified by substitution.

Equation (3.8) is the general form of the distribution function for speeds, and all that remains is to find the values of the constants A and β. First, the expression for $P(v^2)$ can be substituted in the normalizing equation (3.4) to obtain a relationship between A and β as follows:

$$N = 4\pi A \int_0^\infty \exp(-\beta v^2) v^2 \, dv \tag{3.9}$$

It may be verified from tables of definite integrals that

$$\int_0^\infty \exp(-\beta v^2) v^2 \, dv = \pi^{1/2}/(4\beta^{3/2})$$

which can be substituted in Eq. (3.9) to give

$$A = (\beta/\pi)^{3/2} N \tag{3.10}$$

We now require a further relationship so that A and β can be obtained explicitly. We shall make use of the fact that the temperature of the gas, T, is defined in such a way that the mean particle energy per degree of freedom is $\frac{1}{2}kT$. Thus, since there are N molecules per cubic metre in the gas, each having three degrees of freedom, the total energy per unit volume is $\frac{3}{2}NkT$. Now, the number of particles in the speed range between v and $v+dv$ is given by Eq. (3.3), where $P(v^2)$ is now given by Eq. (3.8), and each particle in the range has energy $\frac{1}{2}Mv^2$. Thus, the total energy per unit volume is

$$\int_0^\infty (\tfrac{1}{2}Mv^2) \, [A \, \exp(-\beta v^2)] \, 4\pi v^2 \, dv = \tfrac{3}{2}NkT \tag{3.11}$$

We again turn to tables of definite integrals to confirm that

$$\int_0^\infty \exp(-\beta v^2) v^4 \, dv = 3\pi^{1/2}/8\beta^{5/2} \tag{3.12}$$

Making use of this fact in Eq. (3.11) and eliminating first A and then β from the resulting equation and Eq. (3.10) gives

$$\beta = M/(2kT) \tag{3.13}$$

and

$$A = N\left(\frac{M}{2\pi kT}\right)^{3/2} \tag{3.14}$$

which can be substituted in Eq. (3.7) to give

$$P(v^2) = N\left(\frac{M}{2\pi kT}\right)^{3/2} \exp\left(-\frac{Mv^2}{2kT}\right) \tag{3.15}$$

3.2.2 Maxwell–Boltzmann distribution function

The number of molecules in the speed range v to $v + dv$ is given by Eq. (3.3). We can now be more explicit and include the expression for $P(v^2)$ given in Eq. (3.15), to obtain

$$dN_v = 4\pi N\left(\frac{M}{2\pi kT}\right)^{3/2} \exp\left(-\frac{Mv^2}{2kT}\right) v^2 \, dv \tag{3.16}$$

This can be simplified by writing

$$dN_v = Nf(v) \, dv \tag{3.17}$$

where

$$f(v) = 4\pi\left(\frac{M}{2\pi kT}\right)^{3/2} \exp\left(-\frac{Mv^2}{2kT}\right) v^2 \tag{3.18}$$

The function $f(v)$ is called the normalized Maxwell–Boltzmann distribution function for speeds. It is evident from the defining Eq. (3.17) that $f(v)$ is the fraction of the total number of molecules per unit volume in a given speed range, per unit range of speed.

A graph of $f(v)$ versus v is shown in Fig. 3.3(a). Areas under such a graph represent fractional numbers of molecules in given speed ranges. For example, the area of the region of length dv shown is $f(v)dv$, which, by referring to Eq. (3.17), is seen to be equal to dN_v/N, the fraction of the total number of molecules per unit volume in the speed range dv. As a further example, the fraction of the total number of molecules with speeds less than v' shown on Fig. 3.3(a) is equal to the area under the curve to the left of v'.

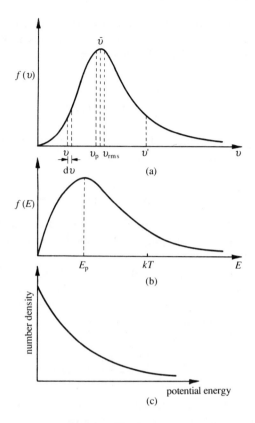

Fig. 3.3 (a) and (b) Maxwell–Boltzmann distributions for speed and energy; (c) Boltzmann distribution of particles in a field of force.

3.2.3 Use of distribution functions to calculate average values

We have determined the distribution of speeds among the constituent molecules of an ideal gas, in terms of a normalized distribution function $f(v)$, and are now able to use this to evaluate average values of speeds and velocities, which in turn can describe bulk properties of the gas, as discussed earlier. It should be pointed out that the averaging procedures developed are quite general and can be applied to *any* distribution function, even though we shall be specifically concerned with the speed distribution function.

What is the most probable speed of a molecule, v_p, in a gas obeying Maxwell–Boltzmann statistics? The defining Eq. (3.17) indicates that this occurs when $f(v)$ is maximum, as shown in Fig. 3.3(a). Hence v_p can be found by differentiation of Eq. (3.18), which gives

$$v_p = (2kT/M)^{1/2} \tag{3.19}$$

We next turn our attention to the average speed of a molecule in the gas, $\bar{v}$. This may be obtained by finding the number of molecules with a given speed, multiplying by the speed, summing all such contributions, and dividing by the total number of molecules. Thus, by definition,

$$\bar{v} = \frac{\int_0^\infty dN_v \, v}{N} = \int_0^\infty v f(v) \, dv \tag{3.20}$$

For the particular case of a Maxwell–Boltzmann distribution, Eq. (3.18) can be substituted in this expression and the integration carried out to give

$$\bar{v} = 2\left(\frac{2kT}{\pi m}\right)^{1/2} = \frac{2}{\pi^{1/2}} v_{\mathrm{p}} \tag{3.21}$$

Thus, although $\bar{v}$ and v_{p} are nearly equal, $\bar{v}$ is slightly higer than v_{p}, as shown in Fig. 3.3(a), because in the averaging process the higher speeds are weighted more heavily than the lower.

The mean-square speed and root-mean-square (r.m.s.) speed, v_{rms}, are often required, for instance, for calculating the average kinetic energy of gas molecules. We can use the same averaging process as described previously to define a mean-square speed; then

$$\overline{v^2} = \frac{\int_0^\infty dN_v \, v^2}{N} = \int_0^\infty v^2 f(v) \, dv \tag{3.22}$$

This can be evaluated for the distribution function defined by Eq. (3.18), again making use of tables of definite integrals, to give

$$\overline{v^2} = 3kT/M \tag{3.23}$$

and the root-mean-square speed is then

$$v_{\mathrm{rms}} = (\overline{v^2})^{1/2} = (3kT/M)^{1/2} = (\sqrt{\tfrac{3}{2}}) v_{\mathrm{p}} \tag{3.24}$$

as shown on Fig. 3.3(a).

It will be noticed in passing that Eq. (3.23) can be rearranged to show that the average value of the kinetic energy of a molecule, $M\overline{v^2}/2$, is equal to $3kT/2$, which is consistent with our earlier assumption.

3.2.4 Energy distribution function

We now study how the energy of molecules is distributed throughout the ensemble. We make use of the fact that the translational kinetic energy of a particular molecule, designated E, is given by

$$E = Mv^2/2 \tag{3.25}$$

We consider energies in the range between E and $E + dE$ and require an expression for the number of molecules per unit volume with energies in this range dN_E. Differentiating Eq. (3.25) gives

$$dv = \frac{dE}{Mv} = \frac{dE}{M}\left(\frac{M}{2E}\right)^{1/2} = \frac{dE}{(2EM)^{1/2}} \qquad (3.26)$$

Equations (3.25) and (3.26) are substituted in Eq. (3.16) to give

$$dN_E = 4\pi N \left(\frac{M}{2\pi kT}\right)^{3/2} \exp\left(-\frac{E}{kT}\right) \frac{2E}{M} \frac{dE}{(2EM)^{1/2}}$$

or

$$dN_E = Nf(E)\,dE \qquad (3.27)$$

where $f(E)$, the distribution function for energies, is given by

$$f(E) = \frac{2}{\pi^{1/2}} \left(\frac{1}{kT}\right)^{3/2} E^{1/2} \exp\left(-\frac{E}{kT}\right) \qquad (3.28)$$

The distribution function, $f(E)$, as defined by Eq. (3.27), gives the fraction of the total number of molecules per unit volume in the unit energy range centred on energy E. It can be seen from Eq. (3.28) that the shape of the $f(E)$ versus E curve is dominated by the $E^{1/2}$ term at low energies, rises to a maximum value, and then has an exponentially decaying tail at higher energies when the $\exp(-E/kT)$ term takes over, as shown in Fig. 3.3(b). It is left as an exercise to show that the most probable energy, which occurs when $f(E)$ is maximum, is

$$E_p = kT/2$$

3.2.5 Boltzmann distribution

We have assumed previously that the gas molecules are not acted on by any external forces. Under these conditions, the energy E refers to the kinetic energy (KE) of a molecule; also, the density of the gas does not vary from point to point.

We now relax this assumption and consider the assembly of molecules to be in a field that causes each molecule to experience a force, which, athough it may vary spatially, is identical for each constituent. For example, the molecular gas may be in a gravitational field or the ensemble may be electrons under the influence of an electric field. In either case, a more general theory shows that Eq. (3.28) is still applicable, provided E is the total energy and includes a potential energy (PE) term. The distribution function is then

$$f(E) \propto \exp(-E/kT) \propto \exp[-(KE+PE)/kT] \qquad (3.29)$$

This expression is quite general and applies to any assembly of particles that interact infrequently. For example, consider a collection of electrons that originally has a Maxwellian distribution, say. When it is subjected to an electric field, applied such that the potential energy, eV, varies with position, the electron distribution everywhere still remains Maxwellian but the energy distribution function has an additional multiplying factor, $\exp(-eV/kT)$, which affects the local electron density. The electron densities n_1 and n_2, at two points where the potential energies are eV_1 and eV_2, respectively, are then related by

$$n_2/n_1 = \exp[-e(V_2 - V_1)/kT] \tag{3.30}$$

Hence, the electron density decreases exponentially with increasing potential energy. Such a relationship between particle density and potential energy is called a *Boltzmann distribution* and is of the form shown in Fig. 3.3(c). It has general applicability to many systems other than the ones mentioned.

3.3 Collections of particles that obey the exclusion principle

We have discussed assemblies of particles in which each constituent is independent of any other in the sense that it can take up any energy value; in other words, several molecules can exist in the same quantum state. We now turn our attention to collections of particles, electrons in particular, that interact quantum mechanically with each other in such a way that the occupancy of a particular state is restricted by the Pauli exclusion principle, i.e. no two electrons are allowed to occupy the same state. The ensemble must again be treated statistically, but the additional restriction leads to distribution functions of a form different from those encountered previously. The distribution function to be derived is most important, since it is applicable to free electrons in metals and semiconductors and allows many electrical phenomena in such materials to be understood, which otherwise could not be explained using classical concepts.

The arguments leading to a distribution function are similar to those followed previously, but this time, since we are dealing with quantum-mechanical systems, the uncertainty principle suggests that, rather than assign definite energy values to each particle, the probability that an energy state is occupied should be considered. Accordingly, let us consider a particular state, E_1, and let the probability of this level being occupied be $p(E_1)$. As a consequence, the probability that there are electrons at two different levels, E_1 and E_2, *simultaneously* is $p(E_1)p(E_2)$. Now, suppose two such electrons interact in such a way that they are transferred to two other energy levels, E_3 and E_4. The precise nature of the interaction will be discussed in later chapters. Energy will be assumed to be conserved by this process, and

$$E_1 + E_2 = E_3 + E_4 \tag{3.31}$$

The exclusion principle implies that such an interaction is permissible only if vacant energy levels exist at E_3 and E_4. Now, the probability that level E_3 is *not* occupied is $1 - p(E_3)$ and the probability that states E_3 and E_4 are vacant simultaneously is $[1-p(E_3)][1-p(E_4)]$. Thus, the probability that two electrons in states E_1 and E_2 interact and are transferred to states E_3 and E_4 is

$$p(E_1)p(E_2)[1-p(E_3)][1-p(E_4)]$$

Since the system is assumed to be in thermal equilibrium, the probability of the reverse process, that is, two electrons in the E_3 and E_4 states interact and are transferred to E_1 and E_2 states, must be the same, and hence

$$p(E_1)p(E_2)[1-p(E_3)][1-p(E_4)]=p(E_3)p(E_4)[1-p(E_1)][1-p(E_2)]$$

or

$$\left(\frac{1}{p(E_1)}-1\right)\left(\frac{1}{p(E_2)}-1\right)=\left(\frac{1}{p(E_3)}-1\right)\left(\frac{1}{p(E_4)}-1\right) \qquad (3.32)$$

The solution that satisfies both this equation and the energy conservation condition, Eq. (3.31), is of the form

$$[1/p(E)]-1=A\,\exp(\beta E) \qquad (3.33)$$

where A and β are constants. That this function simultaneously satisfies both equations can be verified by substitution.

Equation (3.33) can be rearranged to give

$$p(E)=1/[1+A\exp(\beta E)] \qquad (3.34)$$

For high-energy states, the exponential term dominates the denominator and

$$p(E)\simeq A\,\exp(-\beta E) \qquad (3.35)$$

Thus, at high energies the distribution approaches the Boltzmann distribution of Eq. (3.29) and hence the constant β can be identified with $1/kT$. That the two distribution functions become almost identical for high-energy states, even though one system (Boltzmann) allows multiple occupancy of states and that at present being discussed does not, is plausible on physical grounds. At high energies the number of electrons distributed over many available states is small and there are many more energy levels than electrons to occupy them. Under these conditions there is little chance of two or more electrons occupying the same state and whether the exclusion principle is included in the statistics or not becomes irrelevant to the form of the distribution function. Equation (3.34) therefore becomes

$$p(E)=1/[1+A\exp(E/kT)] \qquad (3.36)$$

Finally, it is customary to assume, without any loss of generality, that the constant A is redefined such that

$$A=\exp(-E_F/kT)$$

where the new constant, E_F, is called the *Fermi energy*; the physical significance of this constant energy will become apparent later. Meanwhile, Eq. (3.36) becomes

$$p(E) = \frac{1}{1 + \exp\left[(E - E_F)/kT\right]} \tag{3.37}$$

This is the *Fermi–Dirac function*; it gives, for any ensemble obeying the exclusion principle, the probability that a particular state E is occupied. That the form of the function is consistent with the exclusion principle is apparent from (3.37); $p(E)$ can never exceed unity since the exponential term is always positive; hence the probability of occupancy of a particular state cannot exceed unity and no more than one electron per quantum state is allowed.

The nature of the probability function is best appreciated by plotting $p(E)$ versus E at various temperatures, as in Fig. 3.4. At 0 K the exponential in Eq. (3.37) has a value of 0 or ∞, depending on the sign of $(E - E_F)$. Thus, if $E < E_F$, the probability function is equal to unity and for $E > E_F$, the function equals zero, as shown in Fig. 3.4(a). Physically, what this implies is that at 0 K

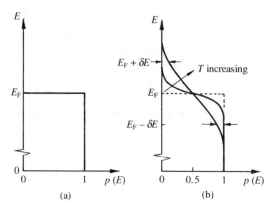

Fig. 3.4   Fermi–Dirac probability function (a) at 0 K and (b) at T K.

all available states up to an energy level E_F are filled, whereas all levels above E_F are empty, i.e. no electrons can possibly have energy greater than E_F. This serves as a preliminary definition of E_F and explains why it is alternatively called the *Fermi brim*.

When the temperature is increased to T K, there is a possibility that electrons that were originally at levels near to the maximum possible energy at 0 K, namely the Fermi level E_F, can now occupy states for which E is greater than E_F. However, energy cannot be transferred to those electrons in low-energy states because there are no unoccupied energy levels immediately above for them to move into and the probability function at these energies remains substantially as it was for $T = 0$ K. These arguments are substantiated by the shape of the probability function for $T > 0$ K, as shown in Fig. 3.4(b). It will be seen that at low temperatures there is a small but finite probability that electrons will occupy available states for which $E > E_F$ but the probability rapidly decreases with increasing energy. As the temperature is increased, this *tail* of the probability function becomes more pronounced and the probability of occupancy of higher energy states is correspondingly increased.

Notice that the probability that an electron occupies the Fermi level, i.e. $E = E_F$, is always 0.5, independent of the temperature. Another interesting fact is that the probability function is symmetrical about the value $(E = E_F, p(E) = 0.5)$. This can be shown as follows. Consider an energy level that is δE *above* the Fermi level, as shown in Fig. 3.4(b). The probability that the level is occupied is, using Eq. 3.37

$$p(E_F + \delta E) = \frac{1}{1 + \exp[(E_F + \delta E - E_F)/kT]} = \frac{1}{1 + \exp(\delta E/kT)} \qquad (3.38)$$

Now consider the level that is δE *below* E_F. The probability that this level is *not* occupied is

$$1 - p(E_F - \delta E) = 1 - \frac{1}{1 + \exp(-\delta E/kT)} = \frac{1}{1 + \exp(\delta E/kT)} \qquad (3.39)$$

Comparison of Eqs (3.38) and (3.39) demonstrates the symmetry of the probability about E_F, since the probability that a level at a particular energy increment above E_F is occupied equals the probability that a level below E_F by the same energy increment is empty.

Finally, it should be pointed out that in many solids E_F is typically of the order of a few electronvolts and at ordinary temperature kT is only a fraction of an electronvolt. Thus, since the probability function goes from nearly unity to nearly zero over an energy range near E_F of only a few kT, the ordinate of the probability curve in Fig 3.4(b) has been artificially compressed so that details of the variation of $p(E)$ near to $E = E_F$ are not lost.

Problems

1. For a hypothetical gas the number of molecules per unit volume dN_v in the speed range from v to $v + dv$ is given by

$$dN_v = Kv\, dv \qquad v_0 > v > 0$$
$$dN_v = 0 \qquad v > v_0$$

where K is a constant. The total number of molecules per unit volume is N. Draw a graph of the distribution function and find the constant K in terms of v_0 and N. Compute the average, r.m.s. and most probable speeds in terms of v_0.

    Ans. $2N/v_0^2$, $2v_0/3$, $v_0/\sqrt{2}$, v_0

2. A hypothetical gas with N molecules per cubic metre has a speed distribution function

$$f(v) = Cv^2 \qquad \text{for } v_0 > v > 0$$
$$f(v) = 0 \qquad \text{for } v > v_0$$

Find the mean-square fluctuation of the speeds, which is defined as the mean-square speed minus the square of the mean speed.

Ans. $0.04v_0^2$

3. Show that the Maxwell–Boltzmann distribution function for the speeds of molecules in a gas can be written in terms of a most probable speed v_p, thus

$$f(v) = \frac{4}{\sqrt{\pi}} \frac{v^2}{v_p^3} \exp\left(-\frac{v^2}{v_p^2}\right)$$

Use this expression to find the mean speed and the r.m.s. speed in terms of v_p. Assume

$$\int_0^\infty \exp(-ax^2) x^{2k+1} \, dx = \frac{k!}{2a^{k+1}}$$

and

$$\int_0^\infty \exp(-ax^2) x^{2k} \, dx = \frac{1, 3, 5 \dots (2k-1)}{2^{k+1}} \left(\frac{\pi}{a^{2k+1}}\right)^{1/2}$$

Ans. $2v_p/\sqrt{\pi}$, $(\sqrt{\tfrac{3}{2}})v_p$.

4. In a Maxwellian gas the number of particles colliding with unit surface area of its container per second is

$$\frac{N}{4} \left(\frac{8kT}{\pi M}\right)$$

Caesium atoms are contained within a furnace at temperature T K. There is also a hot tungsten wire (radius r, length l) inside the furnace and caesium atoms striking this wire became singly ionized. The resulting ions are collected on a nearby negative electrode. Show that the ion current I to this electrode is given by

$$I = erlp(2\pi/kTM)^{1/2}$$

where p is the vapour pressure of the caesium at temperature T and M the mass of the caesium atom. Can you suggest a practical use for this device?

5. A gas possesses a Maxwellian velocity distribution. Show that the fraction of molecules in a given volume that possess a velocity component v_x whose magnitude is greater than some selected value v_{0x} is given by

$$\exp(\tfrac{1}{2}-\pi^{-1/2})\int_0^{mv_{0x}^2/2kT}\exp\left(\frac{-mv_{0x}^2}{2kT}\right)d\left(\frac{mv_{0x}^2}{2kT}\right)^{1/2}$$

The definite integral $\int_0^\infty \exp(-\beta s^2)ds=\tfrac{1}{2}(\pi/\beta)^{1/2}$ will be required.

6. Show that the most probable energy of a molecule in a Maxwellian gas is $kT/2$.

7. Show that the number of molecules, N_0, in a Maxwellian gas whose energies lie between zero and E_0, where $E_0 \ll kT$, is given approximately by

$$\frac{N_0}{N}=\frac{4}{3\sqrt{\pi}}\left(\frac{E_0}{kT}\right)^{3/2}$$

Hence calculate approximately the percentage of molecules whose energies are less than 1 per cent of kT.

8. At $T=0$ K the electron energy levels in a metal are all occupied for $E<E_F$ and are empty for $E>E_F$. The energy distribution is then of the form

$$\Delta N/N=CE^{1/2}\Delta E \qquad \text{for } E<E_F$$
$$\Delta N/N=0 \qquad \text{for } E>E_F$$

where C is a constant. Find (a) the average electron energy under these conditions and (b) the percentage of the total number of electrons with energies between $0.1E_F$ and $0.2E_F$.

Ans. $0.6E_F$, 5.8 per cent.

4 Conduction in metals

4.1 Introduction

In this chapter we shall be considering the conduction of electricity in good conductors, typically metals. We shall see that in such materials the valence electrons are no longer associated with any one particular ion core but are free to wander about the lattice under the influence of external forces. The metal is then considered simply as a 'container' of free electrons, which are only trapped within the boundaries of the metal. Earlier arguments about bound particles would suggest that confining the electrons in this manner leads to a set of discrete energy eigenvalues, much the same as for electrons trapped in the one-dimensional well of Sec. 2.2. Further, since electrons in the metal will be subject to the exclusion principle, Fermi rather than Maxwell–Boltzmann statistics will be applicable to them.

4.2 A simple model of a conductor

The potential energy of an electron located at some distance r from the nucleus of a single, isolated metal atom will be of the general form shown in Fig. 2.3(a). Consequently, there will be a set of allowed electron energy levels associated with the atom, each specified by a particular set of quantum numbers, similar to that shown in Fig. 2.3(a). Electrons are trapped within the potential energy well and are thus bound to the atom; not even the outermost valence electrons usually have sufficient energy to escape.

When such atoms are incorporated in the lattice of a metal, the potential energy distribution between neighbouring atoms is different from that of the individual atoms, as shown in Fig. 4.1. Since potential energy is a scalar quantity, the potential energies of the individual atoms add and the net energy profile is depressed, i.e. is made more negative, as shown in the figure. This causes some of the less tightly bound electrons, lying in the outermost levels of the individual atoms, for example those in E_4 of Fig. 4.1, to have energies higher than the binding energies of atoms in the lattice. Such valence electrons can no longer be associated with a particular atom and are free to move about the lattice in the vicinity of any ion core. Notice that electrons occupying the lower energy states are unaffected by the atoms being incorporated into the

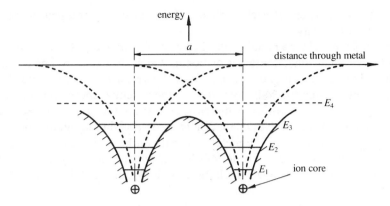

Fig. 4.1 Potential energy and energy levels of neighbouring atoms in a metal lattice.

metal and remain bound to the parent nucleus. We can estimate the number density of free electrons by assuming, conservatively, that each atom provides only one such electron. A typical lattice constant of 0.1 nm then suggests a number density of $[(10^{-10})^{-1}]^3 = 10^{30}$ atoms per cubic metre and a similar minimum number of free electrons in the same volume.

The situation is somewhat different near the surface of the metal. Consider a section of the metal lattice shown in Fig. 4.2(a). The nearest atom to a particular surface obviously has no neighbouring atom outside the metal and the potential energy is not depressed as in the interior of the metal. Hence, potential barriers exist at each surface, as shown, which are normally unsurmountable by electrons. Thus, although electrons are considered to be free to wander unimpeded about the inside of the metal, they are reflected by

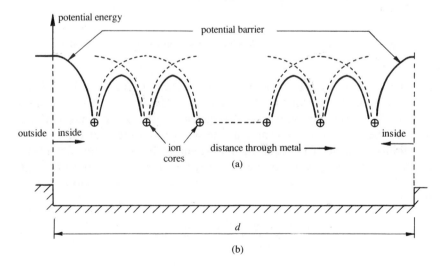

Fig. 4.2 (a) Section through a metal lattice; (b) potential box model.

the potential barriers at each surface and are effectively trapped inside the material in this way.

A simple model for a metal might then be as shown in Fig. 4.2(b). It consists of a three-dimensional container or box in which electrons are free to move without hindrance, but under normal conditions electrons are prevented from penetrating outside the walls by assuming the potential energy outside to be very high. Such a box will have dimensions that are the same as those of the metal being considered. Notice that there is some slight ambiguity about the position of the surface of the metal. All that can be said is that a surface is located within a distance of order of one lattice constant away from the last nucleus in a section through the lattice, as shown in Fig. 4.2(a).

4.3 Electrons trapped in a three-dimensional potential box

If the free-electron model of a metal is assumed, the electron energy states will be the same as those for electrons trapped in a three-dimensional potential box. Consider a box with dimensions shown in Fig. 4.3. Electrons are free to

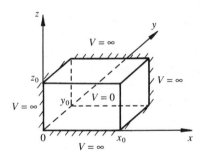

Fig. 4.3 **Dimensions of a three-dimensional potential box model of a conductor.**

move inside the box since the potential energy there, V, is assumed to be zero. Everywhere else V is assumed to be infinite, in order to confine the electrons to the box. By reasoning similar to that given for the one-dimensional well problem (Sec. 2.2), Ψ must vanish at the walls to ensure that there is zero probability of an electron acquiring infinite energy by penetrating outside the box. Schrödinger's equation applied to electrons with constant total energy, E, inside the box, where $V=0$, and stated in the coordinate system shown is, from Eq. (1.43),

$$\frac{\partial^2 \Psi}{\partial x^2} + \frac{\partial^2 \Psi}{\partial y^2} + \frac{\partial^2 \Psi}{\partial z^2} + \frac{2m}{\hbar^2} E \Psi = 0 \tag{4.1}$$

This equation can be solved by the separation-of-variables technique by assuming

$$\Psi = f_x(x) f_y(y) f_z(z) \tag{4.2}$$

where the f's are functions only of the variable used as a subscript. This trial solution can be substituted in Eq. (4.1) to give

$$\frac{1}{f_x}\frac{d^2 f_x}{dx^2} + \frac{1}{f_y}\frac{d^2 f_y}{dy^2} + \frac{1}{f_z}\frac{d^2 f_z}{dz^2} = -\frac{2mE}{\hbar^2} \tag{4.3}$$

Then, by the usual arguments, since E is a constant, making the right-hand side of the equation constant, each of the left-hand terms must individually equal constants, C_1^2, C_2^2, C_3^2, say, and

$$\frac{d^2 f_x}{dx^2} = C_1^2 f_x \qquad \frac{d^2 f_y}{dy^2} = C_2^2 f_y \qquad \frac{d^2 f_z}{dz^2} = C_3^2 f_z \tag{4.4}$$

As has been established already, the boundary conditions on the wave-functions require that $\Psi = 0$ at the walls of the box. Hence

$$\Psi = 0 \qquad \text{at } x = 0 \text{ and } x = x_0$$
$$y = 0 \text{ and } y = y_0 \tag{4.5}$$
$$z = 0 \text{ and } z = z_0$$

Possible solutions to Schrödinger's equation, given by Eq. (4.4) and also satisfying the boundary conditions (4.5), are then

$$f_x = A \sin(n_x \pi x / x_0)$$
$$f_y = B \sin(n_y \pi y / y_0) \tag{4.6}$$
$$f_z = C \sin(n_z \pi z / z_0)$$

where the n's are quantum numbers, which independently have integer values $1, 2, 3, 4, \ldots$, and A, B, C are newly defined constants. That these functions are solutions of Eq. (4.4) and also satisfy conditions (4.5) can be verified by substitution.

The complete expression for the wavefunctions can now be obtained by substituting the x-, y- and z-dependent parts back into Eq. (4.3). This expression is then normalized, using much the same technique as for the one-dimensional potential well in Sec. 2.2, to give finally

$$\Psi_{n_x n_y n_z} = \left(\frac{2}{x_0}\right)^{1/2} \sin\left(\frac{n_x \pi x}{x_0}\right) \left(\frac{2}{y_0}\right)^{1/2} \sin\left(\frac{n_y \pi y}{y_0}\right) \left(\frac{2}{z_0}\right)^{1/2} \sin\left(\frac{n_z \pi z}{z_0}\right) \tag{4.7}$$

Notice that this wavefunction is the product of three one-dimensional wavefunctions of the type encountered in Sec. 2.2. (This result may come as no surprise to readers familiar with electromagnetic theory, since as the one-dimensional potential well is analogous to a short-circuited resonant transmission line, the three-dimensional potential box has its electromagnetic equivalent in a short-circuited resonant cavity.)

The wavefunction expression of Eq. (4.7) is next substituted back into Schrödinger's equation to obtain the permitted quantized energy levels of electrons in the box, to give

$$E_{n_x n_y n_z} = \frac{\hbar^2}{2m}\left[\left(\frac{n_x \pi}{x_0}\right)^2 + \left(\frac{n_y \pi}{y_0}\right)^2 + \left(\frac{n_z \pi}{z_0}\right)^2\right]^2 \qquad (4.8)$$

where n_x, n_y, n_z are three quantum numbers that specify a particular electronic energy state.

The problem is simplified if the containing volume is considered to be a cube of side d. The expression for the energy eigenvalues given by Eq. (4.8) then simplifies to

$$E = \frac{h^2}{8md^2}(n_x^2 + n_y^2 + n_z^2) \qquad (4.9)$$

Finally, if we write

$$n^2 = n_x^2 + n_y^2 + n_z^2 \qquad (4.10)$$

the expression for eigenvalues of electrons in a cubical potential box is

$$E = \frac{h^2}{8md^2}n^2 \qquad (4.11)$$

which is of the same form as that obtained for the one-dimensional potential well (Eq. (2.6)).

4.4 Maximum number of possible energy states

It is instructive to find the maximum number of possible energy states for which n, as defined by Eq. (4.10), is less than some maximum value, n_F say. The discussion and result will be most pertinent to later consideration of the distribution of electron energies on a metal.

A particular value of n, which has components n_x, n_y and n_z, all positive integers, can be visualized in n space as shown in Fig. 4.4(a). In order to count the number of possible combinations of (n_x, n_y, n_z) up to the maximum value n_F, which will then automatically specify the possible energy levels via Eq. (4.9), it is easiest first of all to assume n_x fixed at some value, n_{x0} say, and consider a plane that is a section through n space as shown in Fig. 4.4(b). Each point in the diagram represents a particular combination of quantum numbers n_y and n_z. Notice that any area on this plane is numerically equal to the number of possible combinations of n's enclosed by the area. For example, the section indicated by broken lines is 12 units of area in extent and encloses 12 combinations of n_y and n_z. Now, for the fixed quantum number n_{x0}, the other quantum numbers are related by

$$n^2 - n_{x0}^2 = n_y^2 + n_z^2$$

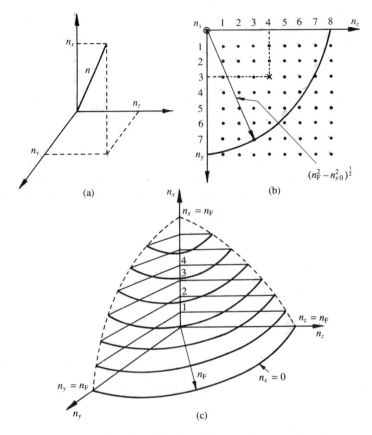

Fig. 4.4 Diagrams illustrating the evaluation of the maximum number of energy states of electrons trapped in a potential box.

and thus the maximum number of possible combinations of n_y and n_z for $n \leqslant n_F$ must lie within a circle centred at the origin and of radius $(n_F^2 - n_{x0}^2)^{1/2}$, as shown; by our previous argument this number of combinations must be numerically equal to the area of the circle in the first quadrant.

If n_x is now allowed to assume other values the total number of possible combinations of (n_x, n_y, n_z) is included in a series of slices through n space, each corresponding to different integer values of n_x, as shown diagrammatically in Fig. 4.4(c). It follows that the total number of possible combinations of the quantum numbers for $n \leqslant n_F$ is numerically equal to that volume of a sphere of radius n_F that is contained in the first quadrant of n space. This latter condition is necessary since the n's are always positive integers. Hence

$$\text{total number of combinations of } (n_x, n_y, n_z) = \tfrac{1}{8}(\tfrac{4}{3}\pi n_F^3) = \pi n_F^3/6 \qquad (4.12)$$

Now, in general, each electron energy state is characterized by four quantum numbers; in addition to n_x, n_y and n_z there is a spin quantum number, which can have either of two values. Thus, there are two spin states for any particular combination of the quantum numbers n, which gives, from Eq. (4.12),

$$\text{total number of possible energy states} = 2(\pi n_F^3/6) = \pi n_F^3/3 \qquad (4.13)$$

This gives the total number of available energy levels. These are not in general necessarily all occupied by electrons. We shall assume, however, that in some circumstances, the physical significance of which will be discussed later, all states are occupied up to some maximum energy state characterized by n_F. If the number density of electrons in the box is N, we can then equate the total number of electrons to the total number of available states, since no more than one electron can occupy a particular state, by the exclusion principle. Hence

$$Nd^3 = \pi n_F^3/3$$

or

$$n_F = (3N/\pi)^{1/3} d \qquad (4.14)$$

The energy of the highest occupied state can then be obtained by substituting this value of n_F into Eq. (4.11) to give

$$E_{F0} = \frac{h^2}{8m}\left(\frac{3N}{\pi}\right)^{2/3} \qquad (4.15)$$

The subscripts to E anticipate later discussions of the physical implications of the result, when this maximum energy will be seen to correspond to the Fermi energy of a metal at 0 K.

4.5 The energy distribution of electrons in a metal

In order to find out the manner in which energy is distributed among the free electrons in a metal, it is first necessary to determine the energy distribution of allowed energy levels that are available for occupation by the electrons. In other words, the number of available energy states lying in a range of energies, say between E and $E + dE$, is required.

It is found to be convenient to define a function $S(E)$, called the *density distribution of available states* or simply the *density of states*, which is defined in such a way that $S(E) dE$ is the number of available states per unit volume in the energy range considered; the problem is then modified to that of finding the form of $S(E)$.

The relationship between the energy of a state, E, and its quantum number designation, n, for a metal cube of side d is given in Eq. (4.11). In the energy range dE there will be a corresponding range of quantum numbers, dn say. We can evaluate the number of states in the range by finding the volume that the range dn occupies in n space, as was explained in the previous section. Thus, considering a spherical shell in n space of radius n and thickness dn, as shown

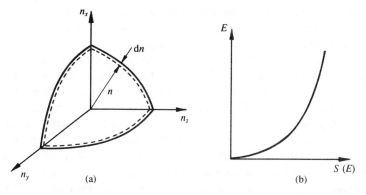

Fig. 4.5 (a) **Spherical shell used to evaluate** $S(E)$; (b) **relationship between** $S(E)$ **and energy** E.

in Fig. 4.5(a), the number of states with quantum numbers between n and $n+dn$ is numerically equal to the volume of the shell in the first octant, multiplied by 2; the additional factor accounts for the two possible spin states corresponding to each value of n. Thus the number of available states in the range is

$$2(4\pi n^2 \, dn)/8 = \pi n^2 \, dn \qquad (4.16)$$

which by definition of the density-of-states function, $S(E)$, is equal to

$$S(E) \, dE \, d^3$$

Hence

$$S(E) = \frac{\pi n^2}{d^3} \frac{dn}{dE} \qquad (4.17)$$

Equation (4.11) can be differentiated to give

$$n^2 \frac{dn}{dE} = \frac{(8\sqrt{2})m^{3/2} d^3 E^{1/2}}{h^3}$$

which can be substituted in Eq. (4.17) to provide the following expression for the density of available states:

$$S(E) = \frac{(8\sqrt{2})\pi m^{3/2}}{h^3} E^{1/2} \qquad (4.18)$$

To reiterate, the total number of *available* energy levels per unit volume in a given energy range, dE, is obtained by multiplying this distribution function, $S(E)$, by the size of the range. Thus an alternative definition of $S(E)$ is that it is the number of available states per unit volume, per unit of energy centred at E. Notice that Eq. (4.18) is independent of d; the expression for $S(E)$ is quite general and is independent of dimensions.

The relationship between the density of states function $S(E)$ and energy E is shown in Fig. 4.5(b). It will be seen that the number of available energy levels increases parabolically with increasing energy.

However, in general, not all the available energy states are filled, since, for example, it is extremely unlikely that an electron can gain sufficient energy to occupy one of the relatively very high levels. What determines whether a particular energy level, E say, is filled or not is the probability that an electron can possess energy E. We have seen that for particles that obey the exclusion principle, as electrons in a metal do, such a probabilty is given by the Fermi–Dirac function, $p(E)$. Hence, the number of electrons per unit volume that are in a given energy range depends not only on the number of available states in the range but also on the probability that electrons can acquire sufficient energy to occupy the states, or

number of electrons per unit volume with energies between E and $E+dE$
=(number of available states per unit volume in the range E to $E+dE$)
×(probability that a state of energy E is occupied)

and

$$N(E)\,dE = S(E)\,dE\,p(E)$$

where $N(E)$ is the number of electrons per unit volume per unit of energy centred at E. Thus, the number density of electrons in a unit energy range is obtainable from the distribution function of available states and the probability function since

$$\boxed{N(E)=S(E)p(E)} \qquad (4.19)$$

The number density of electrons as a function of energy can thus be deduced from Eq. (4.19) and Figs 3.4 and 4.5(b), and is plotted at 0 K and some higher temperature in Fig. 4.6. At 0 K, $N(E)$ increases parabolically with E, following

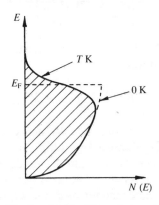

Fig. 4.6 **Number density of free electrons in a metal.**

the $S(E)$ curve and all levels are filled up to the Fermi level, E_F; all levels above the Fermi level are empty. At higher temperature some electrons in the levels near to the Fermi level can gain sufficient energy to have energies greater than E_F. Thus, as the temperature is increased, the number of electrons just below the Fermi level decreases and there is a corresponding increase in the number of electrons in the high-energy tail with energies greater than E_F.

4.6 The Fermi level in a metal

If the number of free electrons per unit volume in a metal, n, is known, it is then possible in principle to calculate the value of the Fermi energy, E_F. This can be done by normalizing, by summing the number density per unit energy, $N(E)$, over all possible energies and equating to n, thus

$$n = \int_0^\infty N(E)\,dE \qquad (4.20)$$

Writing $N(E)$ in terms of the density-of-states function, $S(E)$, and the probability function, $p(E)$, using Eq. (4.19), and including the more specific expressions of Eqs (4.18) and (3.37) give

$$n = \int_0^\infty S(E)p(E)\,dE = \frac{(8\sqrt{2})\pi m^{3/2}}{h^3}\int_0^\infty \frac{E^{1/2}\,dE}{1+\exp[(E-E_F)/kT]} \qquad (4.21)$$

This integral is difficult to evaluate except when $T=0$ K; then $p(E)$ equals unity for all $E \leqslant E_{F0}$ and is zero elsewhere, and

$$n = \frac{(8\sqrt{2})\pi m^{3/2}}{h^3}\int_0^{E_{F0}} E^{1/2}\,dE$$

The Fermi energy at 0 K is thus given by

$$E_{F0} = \frac{h^2}{8m}\left(\frac{3n}{\pi}\right)^{2/3} = 3.65 \times 10^{-19}\,n^{2/3}\,\text{eV} \qquad (4.22)$$

a result that was anticipated in Eq. (4.15).

Typical values for the Fermi energy at 0 K as calculated from Eq. (4.22) and a knowledge of n are: for silver, $E_{F0} = 5.5$ eV; for copper, $E_{F0} = 7.0$ eV; and for aluminium, $E_{F0} = 11.7$ eV. Thus, the Fermi energy for a good conductor is of the order of a few electronvolts. This emphasizes an essential difference between a classical gas and the electron-gas model of a metal at 0 K; in the former all particles have zero energy while in the latter all electron energies up to the Fermi energy are possible.

At temperatures other than 0 K, the Fermi energy can be obtained from Eq. (4.21) by numerical integration. At room temperature it can be shown that a reasonably good approximation for E_F is

$$E_F = E_{F0}\left[1 - \frac{\pi^2}{12}\left(\frac{kT}{E_{F0}}\right)^2\right] \tag{4.23}$$

This equation shows that, whereas E_F decreases with increasing temperature, since kT is usually much smaller than E_{F0}, E_F is not far removed from E_{F0} and is fairly insensitive to temperature changes.

4.7 Conduction processes in metals

We shall first examine electrical conduction in terms of the free-electron model of a metal. Although there are drawbacks to the treatment, which will be discussed subsequently, what follows will serve as a simple introduction to the essential features of conduction processes.

Conduction in electrical conductors is governed by a fundamental experimental law, Ohm's law, which in its most general form may be written

$$J = \sigma \mathscr{E} \tag{4.24}$$

where J is the current density, in a material of conductivity σ, produced by the application of an electric field $\mathscr{E}$. If we assume that the current flow is due to the movement of n free electrons per unit volume, each of charge $-e$ and travelling with velocity v, we can write

$$J = -nev \tag{4.25}$$

Now, an electron subjected to an electric field, $\mathscr{E}$, by definition experiences an accelerating force, $-e\mathscr{E}$. Thus, in the absence of any restraining force the free electrons in a metal with an externally applied electric field accelerate progressively, and as a consequence of Eq. (4.25) the current density increases with time. This is clearly at variance with Ohm's law, which requires the current to be constant for a particular applied field. It is evident that, for the two expressions for current density to be compatible, the electron velocity must remain constant for any particular applied field. The constant velocity can be explained in terms of 'collisions' of the electrons with the crystal structure in which they move. The precise mechanism of the collisions will be discussed more fully later. Meanwhile, it is sufficient to say that a free electron can be accelerated from rest by the application of an external electric field, acquires a linearly increasing velocity for a short time, but then undergoes some form of collision that reduces its velocity to zero; the process then repeats itself. As a consequence, the electron acquires a constant average drift velocity in the direction of the accelerating force, which is superimposed on to its random thermal motion. This situation is shown diagrammatically in Fig. 4.7.

The effect of collisions in this case, and, incidentally, for similar processes in semiconductors which are considered in following chapters, is thus to introduce a viscous or frictional force, which inhibits the continual accelera-

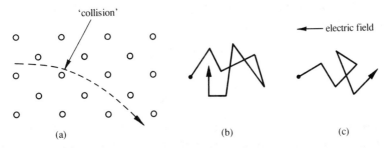

Fig. 4.7 **(a) Random motion of an electron through the crystal lattice; (b) as for (a) but on a larger scale; and (c) drift motion of an electron under the influence of an external electric field.**

tion of the carriers by the field and limits the velocity to some average drift velocity, v_D.

If the average time between collisions is τ_r, then the number of collisions per second is $1/\tau_r$ and the average rate of change of momentum or the frictional force on the carriers due to collision is mv_D/τ_r. The equation of motion of an electron subject to an applied field in the x direction, $\mathscr{E}_x$, is then

$$-e\mathscr{E}_x = m\frac{d}{dt}(v_{Dx}) + \frac{m(v_{Dx})}{\tau_r} \tag{4.26}$$

The solution of this equation is

$$v_{Dx} = \frac{-e\tau_r\mathscr{E}_x}{m}[1 - \exp(-t/\tau_r)] \tag{4.27}$$

which may be verified by substitution. We can now substitute this value of v_{Dx} in Eq. (4.25) to obtain the current density

$$J = n(-e)v_{Dx} = \frac{ne^2\tau_r\mathscr{E}_x}{m}[1 - \exp(-t/\tau_r)] \tag{4.28}$$

We see that the electron drift velocity and the current density both rise exponentially with time to become constant in a time comparable to τ_r.

We can understand the significance of τ_r further if we suppose that after some initial application of the field it is suddenly reduced to zero. Equation (4.26) indicates that the decay of drift velocity from its initial value at $t=0$, v_{D0}, back to the thermal equilibrium state when $v_D=0$ is then governed by the equation

$$v_D = v_{D0}\exp(-t/\tau_r) \tag{4.29}$$

The time constant, τ_r, is often referred to as the electron *relaxation time*, since it controls the exponential way in which electron drift velocity and hence current relax back to zero when the field is suddenly removed. It is typically of order 10^{-14} s and so, for any time after the application of an electric field that is long compared to this, the mean drift velocity and current density are constant, their steady-state values being

$$v_{Dx} = -(e\tau_r/m)\mathscr{E}_x \tag{4.30}$$

and

$$J_x = (ne^2 \tau_r/m)\mathscr{E}_x \tag{4.31}$$

The minus sign in the former equation indicates that the electrons drift in the negative x direction, in the opposite direction to the field. This corresponds to a conventional current flow in the opposite direction, so J_x remains positive.

Equations (4.30) and (4.31) can be somewhat simplified if we assume that τ_r is independent of $\mathscr{E}_x$, which is usually permissible. We then notice that the drift velocity of electrons is directly proportional to the applied field. The constant of proportionality, usually designated μ, is called the *mobility*. Thus

$$v_D = -\mu \mathscr{E}_x \tag{4.32}$$

where the electron mobility

$$\mu = e\tau_r/m \tag{4.33}$$

The mobility is thus defined as the incremental average electron velocity per unit of electric field.

We can now rewrite the current equation in terms of the mobility, and Eq. (4.31) becomes

$$J_x = ne\mu\mathscr{E}_x \tag{4.34}$$

Comparing this equation with Eq. (4.24), it is seen that the conductivity of the metal, σ, can be expressed as

$$\sigma = ne\mu = ne^2 \tau_r/m \tag{4.35}$$

This expression is quite general and holds for any conduction process, provided μ, n and m are specified for the particular process.

The discussion so far has excluded any mention of the distribution of allowed electron energy levels or the exclusion principle, which must apply to electrons in a metal. A graphical representation of the statistical distribution of the energies of conduction electrons will now be considered, since it gives more physical insight into the conduction process.

Each conduction electron occupies a particular energy state, which has an associated velocity that can be represented as a point in the three-dimensional velocity space shown in Fig. 4.8(a). In the absence of an applied electric field, the electron velocities are random; for every group of electrons travelling with a particular velocity, there will be a similar number travelling with the same speed but in the opposite direction, and the distribution in velocity space will have spherical symmetry. At zero temperature the distribution will be most compact and all levels will be occupied out to a velocity, v_F, corresponding to

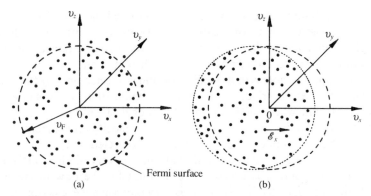

Fig. 4.8 Distribution of electrons in velocity space: (a) with no applied field and (b) with field $\mathscr{E}_x$ applied.

the Fermi energy. Moreover, since $E_F = \frac{1}{2}mv_F^2$, the outermost boundary containing all possible velocities will be a sphere, shown as a broken circle in Fig. 4.8(a). Such a boundary is called a *Fermi surface*, even when it is not spherical as it is in this diagram.

At more elevated temperatures the boundary of the distribution in velocity space becomes more diffuse since a small proportion of electrons now have velocities greater than v_F. This 'fuzzing' of the edges of the sphere is not very pronounced because the corresponding range of energies is only a few kT, which is generally much less than E_F.

When an electric field $\mathscr{E}_x$ is applied as shown in Fig. 4.8(b), all electrons in the distribution are subjected to a force $e\mathscr{E}_x$ in the $-x$ direction. Only those electrons near to the Fermi level can move, since only they have unoccupied energy levels immediately adjacent into which to move. Suppose for the moment that all electrons suffer a randomizing collision simultaneously at time $t = 0$ and that at this instant the electrons have the equilibrium velocity distribution shown in Fig. 4.8(a). At some time t later the distribution as a whole moves in the negative x direction by a velocity increment $e\mathscr{E}_x t/m$. After a further series of collisions the distribution will tend to revert to its equilibrium position. Of course, in a more realistic situation, collisions occur in a more random manner; some electrons are returning to their equilibrium position in velocity space after a collision while others are being accelerated by the field. Hence, on average, application of an electric field causes the entire equilibrium velocity distribution to be shifted slightly by an amount $e\mathscr{E}_x \tau_r/m$ in the opposite direction to the field as shown in Fig. 4.8(b). The velocity distribution is no longer symmetrical about the origin and the precise cancellation of electron velocity components in the direction of the field does not occur. There is now a slightly preferred electron velocity direction in the opposite direction to the field, which results in a drift velocity

$$v_D = -e\mathscr{E}_x \tau_r/m$$

Thus, it is the difference between the equilibrium and the shifted steady-state distributions that accounts for the drift velocity and resulting current flow on the application of a field. Of course, as soon as the field is removed, the steady distribution relaxes to the symmetrical equilibrium form.

A further point to note is that the distribution is affected only by the applied field near to the Fermi surface and so the most significant relaxation time is not the average for the whole distribution but that of electrons near to E_F, τ_{rF} say. Equations (4.20) and (4.31), which give the drift velocity and current density, are seen to be still applicable provided τ_r is interpreted as τ_{rF}.

Finally, it should be emphasized that only a very slight shift for the equilibrium distribution is necessary to account for the flow of current in a metal, and that depicted in Fig. 4.8(b) is very much exaggerated for the sake of clarity.

The free-electron model has been used successfully to explain many aspects of electronic conduction in metals, particularly when Fermi statistics is employed. There are, however, some details of the properties of electrical conductors that cannot be accounted for quantitatively by the simple model. For example, in order that the electron velocity be limited to a drift velocity, it has been found necessary to postulate some sort of collision mechanism followed by energy randomization. Early theories assumed that the collisions were between electrons and ion cores, which occupy most of the volume of a metal. That such a theory is unacceptable can be seen by estimating the mean free path or average distance between collisions, $\bar{l}$. For electrons near the Fermi level

$$\bar{l}_F = \tau_{rF} \, v_F \tag{4.36}$$

where v_F is the velocity of an electron with the Fermi energy. Since the Fermi energy is relatively insensitive to temperature, v_F is given approximately by

$$v_F \approx (2E_{F0} \, e/m)^{1/2}$$

If, for example, we consider copper, $E_{F0} \approx 7 \, \text{eV}$ and $v_F \approx 10^6 \, \text{m s}^{-1}$. The relaxation time for copper can be estimated using Eq. (4.35); if a measured conductivity of $6 \times 10^7 \, \text{S m}^{-1}$ and a free-electron density of 10^{29} are assumed, then τ_{rF} is of order $10^{-14} \, \text{s}$. Hence the mean free path from Eq. (4.36) is of the order of tens of nanometres. This is very much longer than the lattice constant, which is of order $0.1 \, \text{nm}$. Clearly, collisions are not occurring between electrons and metallic ions since this would inevitably lead to mean free paths of the same order as the lattice constant. A more accurate description of the nature of the collision process will be deferred until the next chapter.

Another property of conductors for which the simple model does not account quantitatively is the temperature dependence of resistivity. It is well known experimentally that the resistance of a metal increases almost linearly with temperature. Thus, since

$$\sigma = \frac{ne^2 \, \tau_{rF}}{m} = \frac{ne^2 \, \bar{l}_F}{m v_F}$$

and since v_F is almost independent of temperature, l_F would be expected to decrease almost linearly with increasing temperature; this is at variance with a free-electron theory, which suggests a $T^{-1/2}$ dependence of the mean free path and the conductivity. The discrepancy will be accounted for in the next chapter.

Problems

1. A particular metal contains 10^{28} free electrons per cubic metre. Find the number density of electrons in the energy interval 2.795 to 2.805 eV at $T = 300$ K.

   Ans. $8 \times 10^6 \, \text{m}^{-3}$

2. The Fermi level in copper at 0 K is 7.0 eV. Estimate the number of free electrons per unit volume in copper at this temperature.

   Ans. $8.4 \times 10^{28} \, \text{m}^{-3}$

3. Calculate the Fermi energy at 0 K in copper given that there is one conduction electron per atom, that the density of copper is 8920 kg m^{-3} and its atomic weight is 63.54.

   Ans. 7.06 eV

4. Use the equation of motion of an electron in a metal under the influence of an electric field $\mathscr{E}$, Eq. (4.26), to show that if an alternating field $\mathscr{E}_0 \exp(j\omega t)$ is applied, the effective conductivity of a metal may be written

$$\sigma = \sigma_0/(1 + j\omega\tau_r)$$

where σ_0 is the low-frequency conductivity. [*Hint*: Write $C \exp(j\omega t)$ as a solution of the equation, where C is to be found.] What do you infer from the result?

5 Electrons in solids—
an introduction to band
theory

5.1 Introduction

The conduction theory presented in the previous chapter assumes that many free electrons are available within the body of the material, which behave as classical particles. In a metal the free valence electrons are shared by all atoms in the solid; hence there is a tendency for the periodic potential of the crystal lattice as seen by the conduction electrons to be smeared out and to appear almost constant. This accounts for the success of the free-electron model in explaining most, if not all, of the conduction phenomena in metals. However, for materials with different crystal structures, for example in the important case of covalent bonded solids such as some semiconductors, valence electrons are located much nearer to the parent atoms and cannot be associated with the entire collection of atoms as in a metal. The free-electron model fails for such materials since the potential seen by valence electrons can no longer be regarded as constant since it varies rapidly, particularly near to ion cores in the lattice.

A quantum-mechanical model that overcomes this difficulty assumes that the conduction electrons, as well as being subject to the restriction of the exclusion principle as before, are not entirely free but move in the perfectly periodic potential of a crystal. Such a distribution of potential arises because of the regular spacing of ion cores in the lattice and its perodicity is equal to the lattice constant. We shall see that in this situation the energy of electrons can be situated only in allowed bands, which are separated by forbidden energy regions. Within a particular allowed band, electrons behave in much the same way as free electrons; they can again interact with externally applied fields to produce conduction effects but the interaction parameters have to be modified to account for the presence of the lattice.

The so-called band theory of solids, which is developed from the periodic potential model, has been most successful in explaining some of the anomalies predicted by the free-electron model and also can account for the differing electrical properties of conductors, semiconductors and insulators. What determines the conduction properties of a particular material is whether the electronic states within an allowed energy band are empty or full.

The more complete model also accounts for apparent changes in effective electron mass with position in an energy band. Further, it will be shown that the properties of a material with an almost filled band are identical to those of a material containing a few positive charge carriers in an otherwise empty band; this is a quantum-mechanical justification of the concept of a *hole*, which will be used extensively when discussing semiconductors later.

Finally, departures from the assumed perfect periodicity of potential will be shown to account for resistive effects in a practical material.

5.2 Allowed energy bands of electrons in solids

5.2.1 General concepts

It was shown in Chapter 2 that the electrons in an isolated atom are only allowed to possess discrete values of energy. The exclusion principle also stipulates that each energy level, which is defined by a set of three quantum numbers, can only be occupied by at most two electrons, provided they have opposite spins.

When atoms are packed closely together in a solid such that the electronic orbitals of neighbouring atoms tend to overlap, the allowed electron energy levels are modified from those of the individual constituent atoms. Consider, for example, two identical atoms that are gradually brought together. As the outermost orbitals overlap, electrons that originally had the same energy in the isolated atoms have their energies slightly modified so that the exclusion principle is not violated for the two-atom system; each allowed energy level is split into two closely spaced levels. If the atoms are brought still closer together such that the electrons in inner orbitals of each atom interact, the lower energy levels split in a similar manner.

Energy level splitting of atoms in close proximity can be explained in terms of a simple quantum-mechanical model, as follows. Consider, for example, two hydrogen atoms separated by an initially large distance, r. The electronic energy structure of each atom will be as shown in Fig. 2.3(a) and for the two atoms is as shown in Fig. 5.1(a). To a first approximation each atom can be represented by a one-dimensional rectangular potential well of width δ, as shown in Fig. 5.1(b). We have seen that the wavefunctions of bound electrons in such a potential well are nearly sinusoidal (or exactly sinusoidal if the well is infinitely deep) and that the corresponding energies are discrete eigenvalues. For example, the possible wavefunctions of the lowest energy state ($n = 1$) are as shown in Fig. 5.1(c). Notice that the negative ψ solution, shown as a broken curve, is included, since it is equally possible because it is only the quantity $|\psi|^2$ that has any physical significance. The general expression for the energy eigenvalues of a trapped electron is given in Eq. (2.6); hence the lowest energy

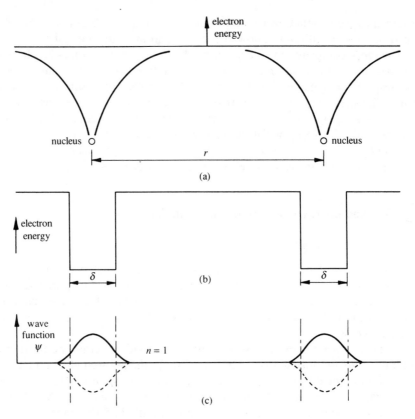

Fig. 5.1 (a) Potential energy of electrons in two isolated hydrogen atoms; (b) one-dimensional potential well equivalent; (c) wavefunction of trapped electrons in the lowest energy state.

state for an individual well is

$$E_1 = h^2/(8m\delta^2) \tag{5.1}$$

Now, consider the atoms when they are brought more closely together such that $r > \delta$, as shown in Fig. 5.2(a). The boundary conditions stipulate that both the wavefunction, ψ, and its gradient, $\partial\psi/\partial x$, must always remain continuous. It would appear that there are two possible configurations for the wavefunction of the complete system, as shown, one in which ψ is symmetrical about the centre line dividing the atoms and one in which ψ is antisymmetrical. When the atoms are brought even nearer, such that $r = \delta$, the wavefunctions of the lowest-energy electrons merge into the symmetrical and antisymmetrical forms shown in Fig. 5.2(b) The electron energies corresponding to each wavefunction of the complete system can again be obtained from Eq. (2.6) and are

$$E_{1,\text{sym}} = \frac{h^2}{8m(2\delta)^2} = \frac{1}{4}\frac{h^2}{8m\delta^2} \tag{5.2}$$

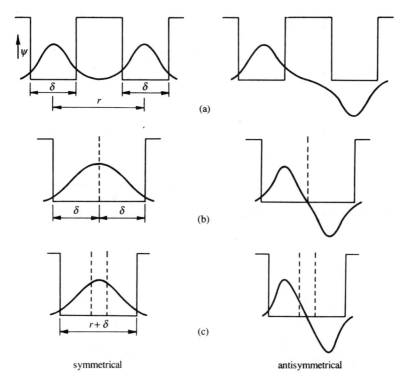

symmetrical   antisymmetrical

Fig. 5.2 Possible wavefunctions for the lowest energy state of a two-hydrogen-atom system as the separation is changed.

and

$$E_{1,\text{antisym}} = \frac{2^2 h^2}{8m(2\delta)^2} = \frac{h^2}{8m\delta^2} \tag{5.3}$$

For even closer separation such that $r < \delta$, the possible wavefunctions for the system are shown in Fig. 5.2(c) and the corresponding electron energy values are

$$E_{1,\text{sym}} = \frac{h^2}{8m(r+\delta)^2} \tag{5.4}$$

and

$$E_{1,\text{antisym}} = \frac{4h^2}{8m(r+\delta)^2} \tag{5.5}$$

These possible energy states of the system for various spacings r are collected together in Fig. 5.3. It will be seen that as soon as the spacing between the atoms is such that $r \simeq \delta$, two definite energy states exist for the system where for the individual atoms there was only one.

It may be helpful to consider the electrical circuit analogue of closely

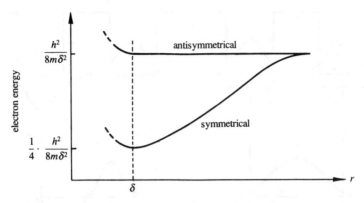

Fig. 5.3   Energy of lowest states in a system of two hydrogen atoms separated by distance r.

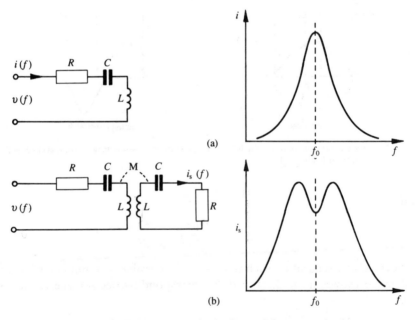

Fig. 5.4   (a) Series resonant circuit; (b) coupled resonant circuits.

coupled resonant circuits. The simple series LRC circuit depicted in Fig. 5.4(a) has a single current maximum when the frequency of the applied sinusoidal voltage is equal to the resonant frequency, f_0, as shown. When two such circuits are tightly coupled via mutual inductance as shown in Fig. 5.4(b), two current maxima occur at frequencies displaced slightly to either side of f_0. Such frequency splitting can be compared to energy level splitting in atomic systems, particularly when it is remembered that energy and frequency are related quantities in quantum-mechanical systems.

Now let us consider briefly the case of three atoms brought into close proximity. For a particular separation there will now be three possible configurations for the wavefunctions, as shown in Fig. 5.5(a). The corresponding energies associated with the particular wavefunctions shown are as depicted in Fig. 5.5(b). Again, we see that each level of an individual atom is split into the same number of levels as the number of atoms in the system.

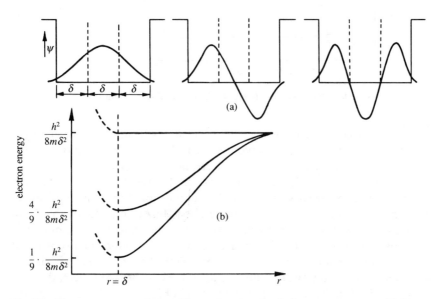

Fig. 5.5 **Three-atom system: (a) possible wavefunctions for the lowest energy states; (b) electron energy as a function of atomic separation.**

The extension of this argument is straightforward and it is reasonable to expect that in a system of n interacting atoms each discrete energy level of an individual atom is split into n closely spaced levels as the atoms are brought together. A system containing seven atoms is shown diagrammatically in Fig. 5.6; notice that the higher energy levels split at larger separations. This is because the electrons at these levels are on average further from the nucleus and interact with neighbouring atoms more readily.

Of course, in a more realistic system such as a solid the number of interacting atoms is much higher than seven; a typical figure may be, for example, 10^{22}. Also the total width of each band of allowed energy levels is of order 1 eV and depends not on the number of atoms grouped together but on their interatomic spacing. Since in this instance 10^{22} discrete levels have to be accommodated in an energy range that is only 1 eV in extent, individual levels in a band are of necessity very closely spaced together; the allowed energy levels within a band are therefore said to be *quasi-continuous*. Summarizing,

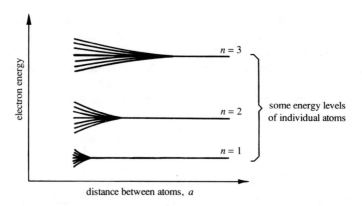

Fig. 5.6 Electron energy levels of a hypothetical seven-atom system.

the allowed electron energies in a solid occur in bands in which the energy is *almost* continuously variable, separated by forbidden energy regions, which correspond to energy levels that the electrons cannot attain.

Let us now be more specific and consider the useful example of carbon, which possesses properties similar to those of the more common semiconductors. The electronic structure of a single carbon atom is seen from Table 2.1 to be $1s^2 2s^2 2p^2$; that is, its inner principal shell is filled but there are only four electrons in its outer shell and there are four vacancies in the outer subshell. If we first of all consider a 'gas' of such atoms with the interatomic spacing or *lattice constant*, a, so large that no interaction occurs between them, then the energy levels of each atom are as shown at (i) in Fig. 5.7. As the carbon atoms are brought into closer proximity (i.e. a is reduced), level splitting occurs as described, which results in bands of allowed energies, as for example at (ii). For even closer spacings as at (iii), the bands can overlap. Eventually, as a is reduced still further, the energies of the outer-shell electrons can lie in one of two bands, separated by a forbidden gap, as at (iv).

Of couse, it is not possible to vary the interatomic spacing, continuously as we have assumed for convenience, since for a particular crystalline solid the lattice constant is fixed. The band structure of a particular allotropic form of carbon will then correspond to a vertical slice through Fig. 5.7; for example, carbon in the diamond form has a band structure similar to that at (iv). We shall see later that it is the magnitude and form of the band structure of a particular solid that completely specify the electrical conducting properties peculiar to it. Before doing so we will consider a more quantitative approach to the investigation of the band structure of solids.

5.2.2 Mathematical model of a solid

In an ideal solid, the ion cores of the crystal are spaced with perfect regularity and the potential experienced by an electron in the solid, V, is periodic in

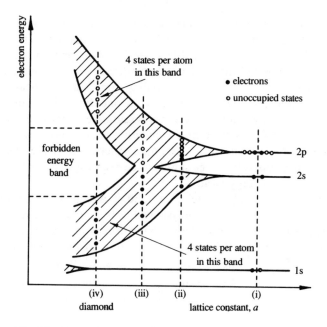

Fig. 5.7 **Energy bands for carbon with varying interatomic spacing.**

space; in any given direction, V repeats itself after distances equal to the lattice constant in that particular direction, a say. Thus

$$V(x) = V(x+a) = V(x+2a) = \ldots$$

Of course, the precise nature of $V(x)$ is complex and the solution of Schrödinger's equation including such a function is difficult. The problem is considerably eased if a simpler mathematical model of the solid is assumed in which the potential in a given direction, as seen by electrons in higher-lying energy bands, changes abruptly from some value V_0 to zero with a periodicity a, as shown in Fig. 5.8. The model, which was proposed by Kronig and Penney,

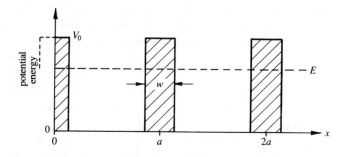

Fig. 5.8 **One-dimensional idealized model of an array of atoms in a solid.**

thus consists of a regular one-dimensional array of square-well potentials. Such a variation in potential is roughly similar to that of a linear array of atoms, as shown for example in Fig 4.2(a), but the approximation is most crude; the chief merit of the model is that it predicts qualitatively the effects that are seen in real solids while retaining some mathematical simplicity. Furthermore, the discussion is limited to a one-dimensional model; while many of the results are qualitatively representative of those obtainable for a three-dimensional model, complete generalization is not always possible. Note also that the regions of low and high potential in the model alternate periodically; this property will be shown to be responsible for the allowed and forbidden electron energy bands.

Even when the simple model of a solid is assumed, the mathematical treatment is involved and the analysis will only be outlined. An analogous electrical system will be discussed later that will explain some of the properties of the model. Applying Schrödinger's equation to the potential variation shown in Fig. 5.8, we have that, inside the wells, $V = 0$ and

$$\frac{\partial^2 \Psi}{\partial x^2} + \beta^2 \Psi = 0 \tag{5.6}$$

where

$$\beta^2 = 2mE/\hbar^2$$

and in the barriers, where $V_0 > E$,

$$\frac{\partial^2 \Psi}{\partial x^2} - \alpha^2 \Psi = 0 \tag{5.7}$$

where

$$\alpha^2 = 2m(V_0 - E)/\hbar^2$$

Equations (5.6) and (5.7) can be solved using the appropriate boundary conditions. It is usual at this stage, however, to make a further simplification to the model so that the problem becomes more tractable. This consists of letting the width of the barriers, w, go to zero and their height to infinity, in such a way that the 'strength' of the barrier, wV_0, remains constant; in other words, the potential is considered to be a periodic delta function. One type of solution that satisfies Eqs (5.6) and (5.7) is then found to be of the form

$$\Psi = U_k(x)e^{jkx} \tag{5.8}$$

When the normal exponential time dependence is included, the exponential part of the solution represents a plane wave of wavelength $\lambda = 2\pi/k$, which is travelling in the positive or negative x direction, depending on the sign of k. The factor $U_k(x)$, called a Bloch function, is a periodic function that varies with the same periodicity as the lattice, a. Thus the solution to Schrödinger's equation consists of travelling waves that are modulated periodically in space.

When the usual boundary conditions of continuity of Ψ and $\partial\Psi/\partial x$ are applied, it is found that Eq. (5.8) is only a solution for particular values of electron energy, E, which satisfy

$$\left(\frac{mawV_0}{\hbar^2}\right)\frac{\sin(\beta a)}{\beta a}+\cos(\beta a)=\cos(ka) \qquad (5.9)$$

where β is as defined in Eq. (5.6). Notice that the left-hand side is a function of the electron energy and the strength of the potential barriers, whereas the right-hand side consists of a wavelength term.

Notice also that whereas the $\cos(ka)$ term lies in the range between -1 and $+1$, there is no such limitation on the left-hand side of the equation, which can assume values outside this range, depending on the value of wV_0; this is illustrated graphically in Fig. 5.9(a). Whenever the left-hand side is greater than 1 or less than -1, no travelling-wave solution of the type described by Eq. (5.8) exists for that particular value of electron energy, E. Such values of E lie in the forbidden bands.

The relationship between E and the wavenumber, k, for the travelling-wave solution in allowed bands can be derived from Eq. (5.9) if a particular value of wV_0 is assumed, and is typically as shown in Fig. 5.9(a). The resulting band structure of allowed electron energy bands separated by forbidden energy gaps is also included in the diagram.

It will be noticed from Eq. (5.9) and Fig. 5.9(a) that if the barrier strength is increased, i.e. wV_0 is made larger, the allowed bands become much narrower and the forbidden bands are correspondingly widened. At the other extreme, as wV_0 is reduced to zero, Eq. (5.9) reduces to $\beta=k$ and hence from Eq. (5.6)

$$\beta=k=2\pi/\lambda=(2mE)^{1/2}/\hbar$$

$$E=h^2/(2m\lambda^2) \qquad (5.10)$$

This energy expression is identical to that for the free electron, Eq. (4.11); the result is not surprising, since as wV_0 becomes zero the potential barriers are removed and electrons can move freely inside the solid.

Considering Eq. (5.9) again, and the solution in the form of an $E-k$ diagram, Fig. 5.9(b), the value of k is not uniquely determined, e.g. if on the right-hand side of the equation k is replaced by $k+2\pi n/a$, where n is an integer, the right-hand side remains the same. This implies that each plot of $E-k$ in the various bands can be shifted left or right by an integral multiple of $2\pi/a$. It is often convenient to use this concept and to define a *reduced wavevector* limited to the region

$$\pi/a\geqslant k\geqslant-\pi/a$$

as shown by the broken curves in Fig. 5.9(b), in which the curves marked a and b are shifted by $-2\pi/a$ and $2\pi/a$ along the axis to appear as a' and b' in the reduced zone; similar shifts of $2\pi n/a$ in k values transform curves c, d, e and f to c', d', e' and f' in the reduced zone.

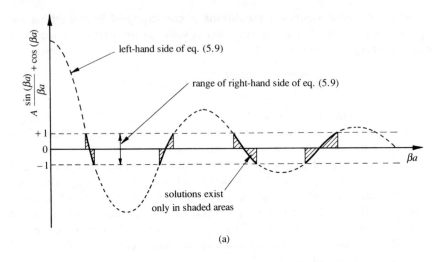

(a)

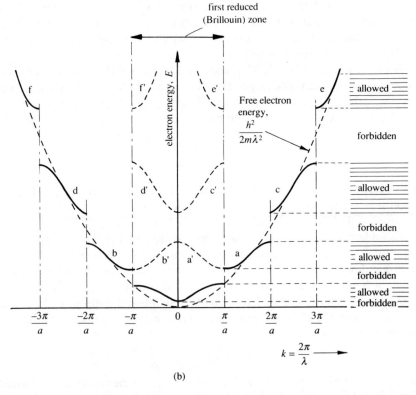

(b)

Fig. 5.9 **(a) Range of possible solutions of Eq. (5.9). (b) Electronic energy as a function of wavelength and the energy band structure of a hypothetical solid.** ----**reduced zone representation.**

Let us now try to gain some physical insight into the striking behaviour shown in the E–k diagram whenever $k = n\pi/a$, by employing the following electric circuit analogy. Consider the coaxial transmission line shown in Fig. 5.10(a); the voltage, V, at any point x on the line is given by

$$\frac{d^2 V}{dx^2} + \omega^2 \mu_0 \epsilon_r \epsilon_0 V = 0 \tag{5.11}$$

If the line is air-filled but periodically loaded at intervals of a with dielectric discs of relative permittivity ϵ_r and thickness w, as shown, then the voltage equation becomes similar in form to Eqs (5.6) and (5.7), provided $\omega^2 \mu_0 \epsilon_0 \equiv \beta^2$

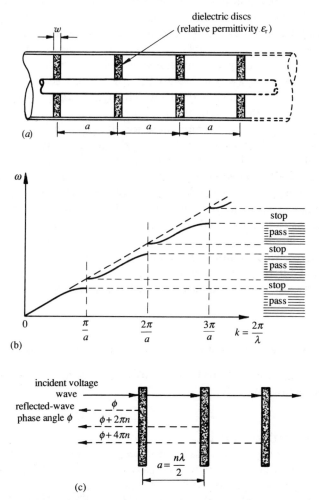

Fig. 5.10 (a) A periodically loaded transmission line; (b) ω–k diagram of the loaded line; (c) total reflection occurring when $ka = n\pi$.

and $\omega^2 \mu_0 \epsilon_r \epsilon_0 \equiv -\alpha^2$. Hence, in the circuit analogy, voltage V is equivalent to electron energy E and frequency bands are analogous to electron energy bands. The ω–k diagram for the loaded line can be obtained by solving Eq. (5.11) using the appropriate boundary conditions and results in series of pass bands of frequency and stop bands that occur when $ka = \pi, 2\pi, 3\pi, n\pi$, as shown in Fig. 5.10(b). The stop bands arise because when $ka = n\pi$ the discs are spaced an integral number of half-wavelengths apart and reflections from successive discs add in phase as shown diagrammatically in Fig. 5.10(c). Then even if individual reflections are weak, their combined effect is to produce total reflection; hence, for this condition, no travelling-wave solution exists for the voltage and only standing-wave solutions are possible.

The situation in the stop bands is analogous to what occurs at the forbidden energy levels in a solid; at some critical wavelength (or k value) the partial reflections of travelling electron waves from successive potential barriers add constructively to produce a reflected wave of the same amplitude as the incident wave and only a standing-wave solution occurs. This may be confirmed by putting $k = \pm n\pi/a$ into Eq. (5.8); then, since $U_k(x)$ is periodic with periodicity a, Ψ is also periodic, which suggests that Ψ has a standing-wave form around these particular k values.

We have seen that it is impossible for an electron to possess energy corresponding to that in a forbidden band. However, an interesting consequence of the analysis is that, within the allowed energy bands, travelling-wave solutions exist that are unattenuated, since $\alpha = 0$; this implies that there is no electron scattering in the uniform lattice of a perfect crystal and within an allowed band an electron can move in a completely unrestricted manner. This statement has to be reconciled with the fact that in a practical solid we have seen that, as a consequence of Ohm's law, the electron must be subjected to a viscous force, which inhibits its continual acceleration. Further discussion of the essential differences between ideal and practical solids, which can resolve this difficulty, will be deferred until after we have discussed other consequences of the E–k diagram.

5.3 The velocity and effective mass of electrons in a solid

Let us first consider a free electron; its kinetic energy E and momentum p are related parabolically since

$$p = mv \qquad \text{and} \qquad E = p^2/2m \qquad (5.12)$$

which is identical to Eq. (5.10).

However, electrons in a solid are not free; they move under the combined influence of an external field plus that of a periodic potential due to atom cores in the lattice. As a result, the electron energy is no longer continuous and the energy–momentum relationship, since $p = \hbar k$, will be similar in shape to that shown in Fig. 5.9(b), as depicted in Fig. 5.11(a). Now, an electron moving

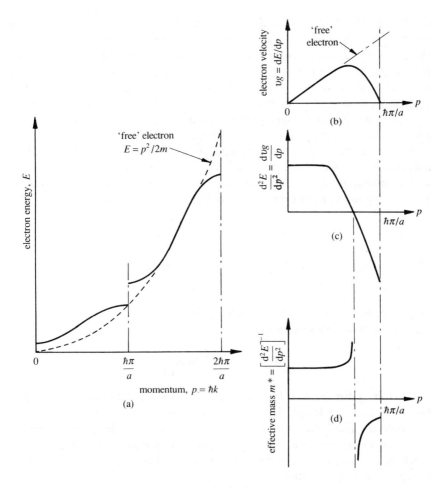

Fig. 5.11 Variation of energy, velocity and effective mass of an electron in a solid.

through the lattice can be represented by a wavepacket of plane waves grouped around some value of k, each wave component being of the form

$$\exp[-j(Et/\hbar - kx)]$$

The group velocity, v_g, of the wavepacket, which we have seen is identical to the electron velocity is, from Eq. (1.30)

$$v_g = \frac{\partial \omega}{\partial k} = \frac{\partial (E/\hbar)}{\partial k} = \frac{1}{\hbar}\frac{\partial E}{\partial k} = \frac{\partial E}{\partial p} \tag{5.13}$$

Thus, the velocity of an electron, which is represented by a packet of waves centred near to a particular value of $k = k_0$, say, is proportional to the slope of the $E-p$ or the $E-k$ curve evaluated at k_0. The slope of the $E-p$ curve in Fig. 5.11(a) is shown in Fig. 5.11(b) and gives the relative electron velocity over the lowest energy band. Notice that the electron velocity falls to zero at each band edge; this is in keeping with our finding that electronic wavefunctions become standing waves at the top and bottom of a band, i.e. $v_g = 0$ there. It is evident that because of interaction with the lattice the momentum of an electron in a solid is no longer simply related to either its energy or its velocity and the classical equation (5.12) is no longer applicable.

Now, consider an electronic wavepacket moving in a crystal lattice under the influence of an externally applied uniform electric field. If the electron has an instantaneous velocity, v_g, and moves a distance δx in the direction of an accelerating force, F, in time δt, it acquires energy, δE, where

$$\delta E = F\delta x = Fv_g \delta t = \frac{F\delta E}{\hbar\, \delta k}\delta t$$

which, in the limit of small increments in k, can be rearranged to give

$$dk/dt = F/\hbar \qquad (5.14)$$

Digressing for a moment we see that for a classically free electron this equation reduces to Newton's second law

$$F = \frac{d}{dt}(\hbar k) = \frac{dp}{dt} = m\frac{dv}{dt} \qquad (5.15)$$

But this is not the case for the electron in a solid. This is not to say that Newton's laws no longer hold, but is a consequence of the fact that the externally applied force is not the only force acting on the electrons; as we have seen, forces associated with the periodic lattice potential are also present. The acceleration of an electronic wavepacket in a solid is equal to the time rate of change of its velocity, thus

$$\text{acceleration} = \frac{dv_g}{dt} = \frac{d}{dt}\left(\frac{dE}{dp}\right) = \frac{d^2 E}{dt\,dp}$$

or, using Eq. (5.14)

$$\frac{dv_g}{dt} = \frac{dp}{dt}\frac{d^2 E}{dp^2} = F\frac{d^2 E}{dp^2}$$

which can be rearranged to give

$$F = \left(\frac{d^2 E}{dp^2}\right)^{-1}\frac{dv_g}{dt} \qquad (5.16)$$

Comparing this equation with the classical equation of motion for a particle, Eq. (5.15), we see that the quantity $(d^2 E/dp^2)^{-1}$ is equivalent to the mass of the free electron. Thus if, for an electron moving in the periodic lattice of a solid, we define an *effective mass*, m^*, where

$$m^* = \left(\frac{d^2 E}{dp^2}\right)^{-1} = \hbar^2 \left(\frac{d^2 E}{dk^2}\right)^{-1} \tag{5.17}$$

then

$$F = m^* dv_g/dt \tag{5.18}$$

By this means it is possible to treat electrons in a solid in a semiclassical manner since quantum-mechanical interactions are included in the effective mass term; an electron of mass m, when placed in a crystal lattice, responds to applied fields as if it were of mass m^*, interaction with the lattice being responsible for the difference between m and m^*. That it is possible using the device of an effective mass to treat an electron in a solid as a classical particle should not be allowed to mask the fact that the electron–lattice interaction is essentially quantum-mechanical. This is emphasized by the fact that m^* can vary over a range from a few per cent of m to much greater than m, which cannot be explained by classical arguments. A further point is that m^* is not a constant but is a function of energy. We can see how it varies typically over an energy band by noting the definition implicit in Eq. (5.17) and forming the reciprocal of the second derivative of energy with respect to momentum; these steps are shown graphically in Figs 5.11(c) and (d). It will be seen that m^* can vary appreciably with position in the band; at the low-energy edges of the band electrons have positive effective mass but at the top end of a band their effective mass can, surprisingly, become negative!

The changing sign of the effective mass of an electron can be explained physically, as follows. Suppose an electron is situated at a point a on the $E-p$ diagram of Fig. 5.12. If an electric field, $\mathscr{E}$, is impressed in the direction shown, the electron will accelerate and moves to the right on the diagram to some point b, where both its energy and its velocity have increased; this conventional behaviour corresponds to a positive effective mass. Now, consider an

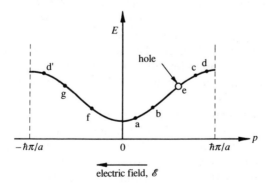

Fig. 5.12 $E-p$ diagram of a hypothetical solid.

electron at the upper end of a band, at c say. When the field is applied again the electron moves to d, where its energy is increased but its velocity has decreased (see for example Fig. 5.11(b)). The electron appears to have been decelerated by the previously accelerating force and since Eq. (5.18) applies always this can be accounted for by the electron having a negative effective mass. It will be apparent from Fig. 5.11 that negative effective masses occur whenever the E–p curve is concave downwards and that the electron mass is positive whenever the curve is concave upwards. Now, the direction of acceleration of an electron is determined by the sign of both its effective mass and its charge, so an alternative way of accounting for the properties of an electron with negative mass is to consider it as a particle with positive mass but having a *positive* charge e, since acceleration $= -e\mathscr{E}/(-m^*) = +e\mathscr{E}/(+m^*)$. Thus, when electrons at the top of a band are acted on by an applied field the resulting currents correspond to the movement of electrons with a positive charge $+e$ and a positive effective mass m^*.

In most materials, at least one band is only partly filled. For example, we have seen that, in a metal, conduction electrons occupy all levels up to around the Fermi energy, E_F. If, as is often the case with other materials, a band is almost entirely filled, it is often convenient to discuss its properties in terms of the relatively few unfilled states rather than those occupied by the many remaining electrons. An unfilled energy state providing a vacancy that can be occupied by an electron wavepacket is called a *hole*. Consider the hole shown diagrammatically in Fig. 5.12 at e. Because of the symmetry of the E–p diagram an electron with positive momentum, say at b, will be cancelled by another electron, in this case at f, with the same magnitude of momentum but oppositely directed. This cancellation of momentum by pairs applies to all electrons except the one at g, which is at the corresponding energy level to the hole; this electron has a negative velocity, v_g (from Eq. (5.13)), and so the hole at e must have a corresponding positive velocity, which can be accounted for by attributing a positive charge to the hole. Hence an electron moving in one direction results in a current flow that is equivalent to a hole or vacancy moving in the opposite direction. In general, the effective mass of a hole is again dependent on its location in the band and the sign of the mass can be found by arguments similar to those used previously. However, since holes are often located at the top of a band, if they are designated as positive charge and moreover move in the opposite direction to electrons, in this case they are accelerated in the same direction as the field, and they must usually possess a positive effective mass.

We see that conduction in a nearly filled band can be accounted for by considering the motion of a small number of positive particles of charge $+e$, possessing positive effective mass, m_h^*, which are called holes and correspond to the number of unoccupied electron states in the band. This concept is vital to an understanding of the details of conduction processes in semiconductors, which will be discussed later.

5.4 Conductors, semiconductors and insulators

The band structure of a solid is a convenient method of classifying its conduction properties. Of course, electrical engineering materials can readily be characterized experimentally by means of their conductivities, but the band theory explains the essential differences between materials with widely differing conductivities.

The conduction process in any material is dependent on the availability of charge carriers. Clearly, if a given energy band is unoccupied it can make no contribution to electronic conduction. What is not quite so obvious is the fact that there can be no net conduction effects if all the bands are completely full either. Consider, for example, a completely full band with E–p diagram as in Fig. 5.12 (disregarding the hole now). We have seen that because of symmetry of the graph there can be no net electron momentum when no external field is applied; obviously, no current flows when the field is zero. When the field, $\mathscr{E}$, is applied, electrons at a or b, say, are accelerated by the field and their momentum is increased. However, an electron at d, say, can have its momentum so increased that it reaches the boundary of the band edge, is reflected and reappears at d' with oppositely directed momentum. Thus, since all levels remain filled before and after the application of the field, and since the distribution of electron momenta is unaltered, there is still no net flow of current.

In order for conduction to occur there must be empty available states in a particular band. Then, when an electric field is applied, as shown in Fig. 5.13,

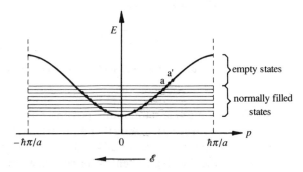

Fig. 5.13 **Partially filled energy band with an applied field.**

electrons at levels just below the empty ones can gain energy from the field and move into the available levels, for example from a to a', and all other electrons move to the right in momentum space, as shown. This results in a net electron momentum in the opposite direction to the field that is no longer zero and an electron current flows.

In a good conductor, the essential requirement of many carriers being available in a partially filled energy band is achieved by the two outermost bands, one of which is completely empty and one full, overlapping; this situation is demonstrated diagrammatically in Fig. 5.7, where the section at (iii) represents the band structure of a metal. It is conventional to draw the band structure of a particular material with fixed lattice constant as shown in Fig. 5.14. While the extremities of each band are, in fact, dependent on crystal

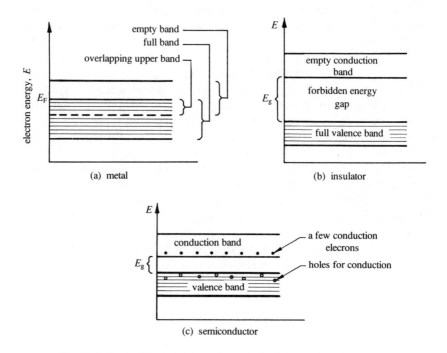

Fig. 5.14 Typical band structure of metals, insulators and semiconductors.

orientation, usually the maximum and minimum possible values for the band edges, regardless of direction, are chosen. A further point is that the abscissa on such diagrams has no significance. It is usually only necessary to show the outermost *filled* band and the next highest empty band, since lower bands are usually completely occupied and play no part in the conduction process. The band diagram of a metal is as shown in Fig. 5.14(a), then. All levels are filled in the band up to some level approaching E_F, above which there are many empty states. Only electrons near the Fermi level participate in conduction and they behave as if they had an effective mass, m^*, evaluated at $E = E_F$. Since E is not located near a band edge, m^* is nearly equal to m, as shown in Fig. 5.9(b); this accounts for the success of the free-electron model of a conductor.

At the other conductivity extreme, insulators are characterized by a band structure consisting of a completely full band, the *valence* band, separated from an empty band, the *conduction* band, by a wide forbidden energy gap of several electronvolts, as shown in Fig. 5.14(b). At all ordinary temperatures, the statistical probability of electrons from the full band gaining sufficient energy to surmount the energy gap and becoming available for conduction in the conduction band is slight. This very limited number of free conduction electrons at all but very elevated temperatures accounts for the high resistivity of insulators.

Semiconductors, as their name suggests, have conduction properties that are intermediate between those of metals and insulators. They have a band structure as shown in Fig. 5.14(c), which is basically similar to that of insulators, except that the gap energies are very much smaller, being typically of order 1 eV. Since the gap is appreciably smaller than for insulators, it is statistically more probable at ordinary temperatures for electrons in the otherwise full valence band to be elevated across the forbidden gap to the empty conduction band, where they are available for conduction. An essential difference between conduction in metals and conduction in semiconductors is that when electrons in a semiconductor gain sufficient energy to occupy the conduction band they automatically create vacancies in the valence band due to their absence. Thus, additional current flow is possible due to charge motion in the now partially empty valence band; such currents can be described in terms of the motion of holes, as discussed in the previous section. The relatively low conductivities of semiconductors compared with metals is, of course, a consequence of the relatively small number of charge carriers, both electrons and holes, available for conduction.

5.5 Electrical resistance of solids

We have seen that for a perfect solid there is no attenuation of electrons in an allowed energy band, i.e. there are no electron 'collisions' and the solid is transparent to the electron; in this situation the mean free path for collisions is infinite. In the previous discussion of electrical conduction in metals it was found necessary to postulate some sort of collision mechanism to provide a frictional force to account for the terminal drift velocity of electrons, which ensures that Ohm's law is obeyed. Since conduction processes in other solids are similar, with the added complication of hole conduction in some materials, it is again clearly necessary for some kind of collisional damping to be present. An electron in an allowed band can then gain energy from an applied field and move higher in the band, but then can suffer a collision (more usually it is said to be *scattered*), give up its energy in the form of Joule heating of the lattice, and return to lower down in the energy band. The conductivity of the solid can again be expressed using equations similar to Eq. (4.35), except that, since the carriers are no longer necessarily free, their effective mass must be included to account for interaction of carrier and lattice.

But what are electrons colliding with? What are they scattered by? It was shown in Sec. 4.7 that electrons colliding with ion cores cannot account for the electrical resistance of a solid. It is the interaction of electrons with the slightly aperiodic potential fields experienced in real solids that can cause scattering effects and hence account for electrical resistance effects. Such deviations from the perfect periodicity in potential, which was assumed for an ideal solid, can be due to thermal lattice vibrations, lattice defects, or the presence of impurity atoms and boundaries, some or all of which are normally present in practical materials.

The most important scattering process at ordinary temperatures can occur in crystals where impurity atom or structural imperfections are negligible. The departure from periodicity necessary to produce scattering is in this case brought about by thermal vibration of the lattice atoms about their equilibrium position. Such a displacement alters the local potential, hence its regular periodicity, and an electron travelling in this field can have both the magnitude and direction of its momentum altered. Such an event constitutes what we have thought of as a collision and is often called *lattice scattering*. The description is something of a misnomer since it is not the lattice that produces the scattering so much as its thermally induced vibrations. Scattering interactions between carriers and lattice vibrations become more probable the higher the temperature because of the larger amplitude of vibrations. Thus, we see qualitatively that the average time between collisions, or relaxation time, τ_r, and hence the carrier mobility decreases with increasing temperature in materials in which lattice scattering is the dominant mechanism, such as relatively pure or structurally perfect crystals.

A further scattering mechanism is attributable to the presence of impurity atoms in the lattice, which, may be ionized or otherwise, although the former are more important. Such atoms alter the local electrostatic potential and create the necessary aperiodicity in the field to cause *impurity scattering* of the electrons. The effectiveness of the deflection of an electron by an ionized impurity is greater the lower the velocity of the electron; hence impurity scattering tends to dominate in purer crystals at lower temperatures when thermal scattering is weak, as well as being important when the impurity concentration is high.

Other possible scattering mechanisms are due to vacancies, dislocations and other lattice imperfections. It is also conceivable that conduction electrons could be scattered by holes and vice versa but the probability that these processes occur is slight.

Now let us try to estimate more quantitatively the effects that departures from periodicity have on the resistivity of a real solid. Again, it is convenient to revert to the dielectric disc-loaded coaxial line model of a solid discussed in Sec. 5.2.2. For a uniform disc spacing, a, we saw that there was no attenuation in the line, i.e. all reflected waves cancel, unless $ka = n\pi$. If we now assume a small deviation from periodicity δ, such that $\delta \ll a$, it can be shown that the reflection coefficient will be proportional to δ and the back-scattered power to

$\overline{\delta^2}$. Thus the fraction of power reflected, dP, in a length of line dx is given by

$$dP/P = K\overline{\delta^2}dx$$

where K is some constant. Integrating we have

$$P(x) = P_0 \exp(-K\overline{\delta^2}x) \qquad (5.19)$$

where P_0 is the power at $x=0$.

Applying this result to the analogous case of electron waves travelling in the periodic potential of a solid, then if the mean-square departure from periodicity in potential is $\overline{\delta^2}$, it follows by comparison with Eq. (5.19) that

$$\psi^2 = \psi_0^2 \exp(-K\overline{\delta^2}x) \qquad (5.20)$$

In other words, electrons are scattered by the aperiodic potential and the probable electron density falls off with distance as $\exp(-K\overline{\delta^2}x)$. Notice that if $\overline{\delta^2} = 0$ there is no scattering. When δ is finite, however, the electron density falls to $1/e$ of its initial value in a mean free path, $\overline{l}$, and so

$$\overline{l} \propto 1/(K\overline{\delta^2}) \qquad (5.21)$$

Since the principal scattering mechanism is due to thermal vibrations we shall consider the consequences of Eq. (5.21) applied to this case. Ion cores in a realistic solid have a natural frequency of vibration about their equilibrium position; also, as one is displaced, the position of its neighbour is affected and acoustic waves propagate. A simple mechanical analogy is that of a linear chain of masses, M, joined by springs of stiffness C. Each simple oscillator has kinetic energy $Mx^2/2$ and potential energy $Cx^2/2$. Now, the equipartition of energy condition requires that the *mean* energy in each energy state at any particular temperature, T, is $kT/2$. Hence

$$C\overline{x^2}/2 = kT/2 \qquad (5.22)$$

and the mean-square deviation from the equilibrium is proportional to T. Thus, for a practical solid,

$$\overline{\delta^2} \propto T$$

and from Eq. (5.21)

$$\overline{l} \propto T^{-1}$$

It follows from (4.37) that the conductivity, σ, is proportional to $\overline{l}$ and hence

$$\sigma \propto T^{-1}$$

or the resistivity, ρ, is proportional to temperature. This result is in accordance with practice, where

$$\rho = \rho_0(1 + \alpha_T T) \qquad (5.23)$$

where α_T is the temperature coefficient of resistance. The residual resistivity term in the expression is due to lattice defects, which are present even at very low temperatures. Such defects are particularly large in disordered alloys such as nichrome and their effect is to tend to swamp the increase in ρ due to lattice vibrations. Hence such alloys are useful where high resistance combined with a low temperature coefficient of resistance are required.

Now, the energy of the acoustic wave due to thermal vibrations of the lattice is quantized and can only change in units of hf. The quantum of acoustic energy is called a *phonon* (cf. photon of electromagnetic energy). Hence, at very low temperatures, Eq. (5.22) is no longer valid since kT becomes comparable to hf and a phonon description of the interaction must be used. This can account for the departure from linearity of the $\rho-T$ curve of a practical solid at very low temperatures.

Although a phonon is actually an acoustic wave propagating through a solid, it is often convenient to think of the associated quantum of energy, the phonon, as a particle in the solid capable of interacting with other particles. Thus, the mechanism we have discussed can be considered as the collision between electrons and phonons and is indeed sometimes referred to as *electron–phonon collisional scattering*.

It is now evident that the classical collisions that have been postulated to account for resistive effects are realized in a solid by the interaction of electron waves with other waves due to lattice vibrations.

Problems

1. The $E-k$ diagram for an energy band in a particular material is as shown. If

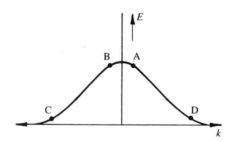

an electric field is applied to the material in the negative k direction (force in the positive direction), find (a) the polarity of the effective masses of the four wavepackets made up of groups of states near A, B, C and D, (b) the direction of the velocity of each of the four wavepackets and (c) the direction of the acceleration of each. What are the physical consequences of these results?

Ans (a)$-$, $-$, $+$, $+$, (b) $-$, $+$, $+$, $-$, (c) $-$, $-$, $+$, $+$

2. The conductivity of a metal having n free electrons per unit volume is given by Eq. (4.37) and the Fermi energy by Eq. (4.22). Consider a metal with a simple-cubic lattice structure of side 0.2 nm and one free conduction electron per atom. Assuming that the mean free path for electron collisions with the lattice is 100 lattice constants, find the relaxation time for an electron with the Fermi energy.

Ans. 2.2×10^{-14} s

6 Semiconductors

6.1 Introduction

We have seen, in Chapter 5, that the electrical properties of a solid are characterized by its band structure. In particular, a semiconductor has two bands of interest (neglecting the bound electrons, since these play no part in any conduction process); these are the valence band and the conduction band, which are separated by a forbidden energy gap, E_g, as depicted in Fig. 6.1. At

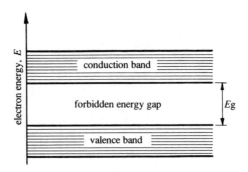

Fig. 6.1   Band structure of a semiconductor.

0 K the valence band is completely full and the conduction band is empty; the semiconductor behaves as an insulator.

It is the possibility at higher temperatures of electrons being transferred across the energy gap into the conduction band, leaving vacant levels in the valence band, that gives the semiconductor its particular conducting properties. In the following section, we shall discuss the properties of semiconducting crystals that give rise to their distinguishing band composition.

6.2 The crystal structure of semiconductors

When atoms, similar or dissimilar, are brought together in close proximity to each other, they may link together to form a stable solid; *bonds* exist between adjacent atoms. There are several mechanisms for bonding, but we shall only concern ourselves with those pertinent to semiconductors. It has been shown that there is a tendency for an atom to form closed outer shells. In order to satisfy this requirement, atoms in solids tend to lose or gain valence electrons, or share them with other neighbouring atoms. It is the manner in which the valence electrons are shared or interchanged that determines the nature of the bond between atoms.

In a semiconducting material the tendency for each atom to form closed outer shells is satisfied by means of a covalent bond; adjacent atoms share electrons, which are located predominantly in the region between the atoms. As a simple example of the mechanism of this type of bonding, let us first consider the hydrogen molecule. This is synthesized from two hydrogen atoms, each of which has a single electron in its vicinity. Both constituent atoms have unfilled shells, since there is a vacancy for one additional electron. As the two atoms are brought into close proximity, the orbitals of each unpaired electron overlap. Provided the electrons have opposite spins, they move together to form an electron pair bond. Electrons with identical spins cannot form bonds since this would violate the Pauli exclusion principle. The two electrons lie predominantly between and are shared by the atoms, in this way completing the outer shell of each, as shown in Fig. 6.2. The term *covalent*

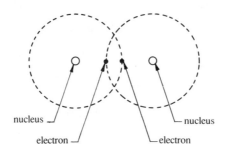

nucleus

electron

nucleus

electron

Fig. 6.2 Covalent bonding of a hydrogen molecule.

is used to describe this particular type of electron pair bond because neighbouring atoms share valence electrons to complete the bond.

The covalent bonding that takes place in semiconductors and is typified in the hydrogen molecule can be derived in a more rigorous and quantitative manner by means of quantum mechanics. Such a theory is a little beyond our scope but its elements can be mentioned briefly. Because the hydrogen molecule is symmetrical, each electron has an identical wavefunction, which can be either symmetric or antisymmetric, as discussed in Chapter 5. When Schrödinger's equation is solved for the case of electrons having opposite spins, the resulting symmetric wavefunctions of both electrons can be evaluated. They predict a relatively high value of $|\psi|^2$ midway between the nuclei; there is a high probability that the binding electrons are located in this region and a charge density exists there.

The elements that form the basic materials used in semiconducting devices all form covalent bonds. Such covalent crystals are characterized by their hardness and brittleness. They are brittle because adjacent atoms must remain in accurate alignment, since the bond is strongly directional and is formed along a line joining the atoms. The hardness is a consequence of the great strength of the paired electron bonding.

The most important elemental semiconducting elements all appear in group IV of the periodic table, as shown in Fig. 6.3. Silicon and germanium are

Valency group	IIIA	IVA	VA	VIA	VIIA
	B 5 2p Boron	C 6 $2p^2$ Carbon			
	Al 13 3p Aluminium	Si 14 $3p^2$ Silicon	P 15 $3p^3$ Phosphorus	S 16 $3p^4$ Sulphur	
	Ga 31 4p Gallium	Ge 32 $4p^2$ Germanium	As 33 $4p^3$ Arsenic	Se 34 $4p^4$ Selenium	
	In 49 5p Indium	Sn 50 $5p^2$ Tin	Sb 51 $5p^3$ Antimony	Te 52 $5p^4$ Tellurium	I 53 $5p^5$ Iodine
			Bi 83 $6p^3$ Bismuth		

Key :

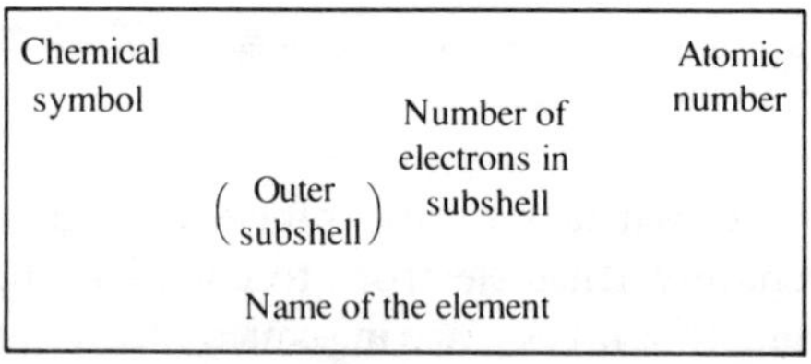

Chemical symbol		Atomic number
	Number of electrons in $\left(\begin{array}{c} \text{Outer} \\ \text{subshell} \end{array} \right)$ subshell	
	Name of the element	

Fig. 6.3 **Part of the periodic table of elements showing those elements of principal interest for application to semiconductor devices.**

of principal interest; the other elements in the group only have semiconducting properties in unusual circumstances.

Elements in adjacent columns, such as boron, phosphorus, arsenic, antimony, bismuth, sulphur, selenium, tellurium and iodine, also can possess some semiconducting properties to a greater or lesser degree. However, we will initially concern ourselves only with group IV elements, C, Si and Ge, since these materials have the most useful electrical properties. Their unifying characteristic is that they each have only four electrons in their outer s and p shells, while there are four vacancies in the outer subshell, facts that can be ascertained from the periodic table (Table 2.1). Such elements are said to be *tetravalent*; we would expect them to have similar bonding mechanisms and chemical properties. Each constituent atom can be represented schematically by the simple model depicted in Fig. 6.4(a).

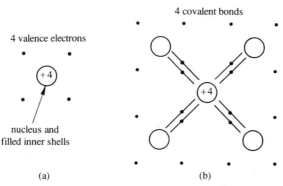

(a) (b)

Fig. 6.4 **Two-dimensional representation of (a) a tetravalent atom and (b) a crystalline solid formed from such atoms.**

Because there is a tendency for the atoms to complete outer shells in the solid state, these materials crystallize into what is known as the *diamond structure*. Since each atom of the solid has four vacancies in its outer shell, covalent bonds exist between the atom and its four nearest neighbours; it shares one of its valence electrons with each of the four neighbours, thus effectively filling its outer shell. This situation is illustrated schematically in Fig. 6.4(b). In order that the four surrounding atoms shown in Fig. 6.4(b) may each in turn have four neighbouring adjacent atoms, the three-dimensional lattice is built up from a basic building block consisting of an atom surrounded by four others, which occupy the corners of a regular tetrahedron, as shown in Fig. 6.5(a). In this diagram the tetrahedron is enclosed in an imaginary unit cube so that the packing of such unit cells in the crystalline solid can be illustrated a little more clearly. The diamond-structure crystal is built up from many such primitive cells in the manner shown in Fig. 6.5(b).

The three-dimensional model of the crystal structure is too cumbersome for most of our discussions; it may be replaced by the two-dimensional representation, shown in Fig. 6.6(a). Although the model is oversimplified and overemphasizes the role of the covalent bonding, it will be adequate for most purposes.

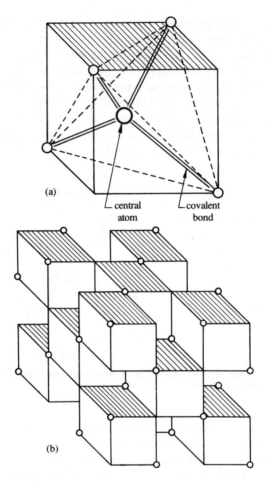

Fig. 6.5 **(a) A unit cell of a tetravalent atom; (b) packing of such atoms to form a crystal with a diamond structure.**

In the next section, conduction processes in elemental semiconductors will be discussed in terms of the model and compared with the band-structure description of conduction.

A further important group of semiconductors are the compounds formed from two elements, one in group III of the periodic table and one in group V, the so-called intermetallic III–V compounds. The most common examples of these compounds are gallium arsenide and indium antimonide. Since one constituent atom has a valency of three and the other five, the bonding in this solid is covalent in character, eight electrons being shared by two neighbouring atoms. Each atom is again surrounded by a tetrahedral arrangement of atoms of the other atomic type; the crystal structure is thus similar to the

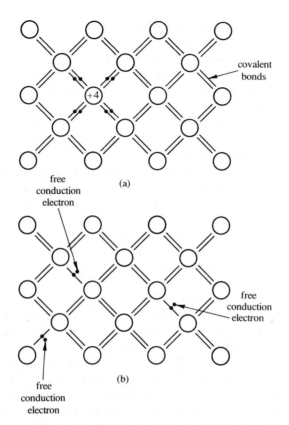

free
conduction
electron

(a)

covalent
bonds

free
conduction
electron

free
conduction
electron

(b)

Fig. 6.6   **A representation of a tetravalent semiconductor crystal: (a) at low temperature and (b) at room temperature.**

diamond structure of Fig. 6.5, with alternate group III and group V atoms. Although preparation of the intermetallic compounds to produce single crystals of sufficiently high quality is more difficult than for elemental semiconductors, they offer additional technical advantages that make this worth while. For instance, the range of semiconducting compounds offers a much wider selection of possible energy gaps than is available with elemental semiconductors and there is a corresponding flexibility in the value of mobility by the choice of a particular III–V compound. We shall see later how both these parameters contribute to influence the electrical properties of a semiconductor.

6.3 Conduction processes in semiconductors

We have seen that in the absence of thermal excitation or any external stimulus the valence band of a semiconductor is filled to capacity and the conduction band is completely empty. In terms of our crystal model, provided the crystal is sufficiently cool and is shielded from, for example, extraneous light sources, all valence electrons take part in covalent bonding and none are free to move through the crystal. If conduction is to occur the valence electrons as a consequence must move to higher energy levels, but this is precluded at low temperatures since there are no available energy levels in the valence band and the empty states in the conduction band can only be reached if the valence electron is given energy in excess of the forbidden gap energy, E_g. It is evident that the covalent bonds can be broken since, for example, it is possible to smash a diamond crystal, but this indicates that energy is required to break the bond. The unifying feature of the bond and crystal representation of a semiconductor is that the minimum energy required to break a covalent bond is equal to the gap energy, E_g.

The covalent bond can be ruptured in several ways, depending on the means whereby energy is imparted to the electrons in the bond. One way is to increase the temperature of the crystal above 0 K. Some of the energy of the resulting lattice vibrations is transferred to the valence electrons. Eventually, as the temperature is increased, sufficient energy is given to an electron in a bond to allow it to break free. This process can be pictured crudely as the shaking loose of electrons by thermal vibration of atoms of the crystal lattice. Fig. 6.6(b) illustrates this situation schematically.

A freed electron can move through the body of the material under the influence of applied fields until it encounters another broken bond, when it is drawn in to complete the bond, or *recombines*, and takes no further part in the conduction process. Thus, by increasing the temperature of a semiconductor above 0 K, its electrical characteristics are changed from those of an insulator to those of a conductor.

When a covalent bond is broken, releasing an electron, a net positive charge exists, which is associated with the broken bond and arises because of the loss of negative electronic charge from an originally neutral environment. This is the crystallographic manifestation of the hole, which was discussed in the previous chapter. It is possible for the positively charged vacancy, or hole, to move under the influence of an applied field, even though the atom whose bond was broken initially remains stationary. The sequence of events is illustrated in Fig. 6.7.

If the electric field is applied in the direction shown, it is possible for one of the electrons in a neighbouring covalent bond to move to fill the vacancy, as in Fig. 6.6(b). This process is repeated in Figs 6.6(c) and (d) and we see that motion of electrons in covalent bonds causes migration of the hole in the direction of the applied field. Thus, as we would expect from previous arguments, the vacant site moves like a fictitious positively charged particle, which has an effective mass, a charge of approximately the same magnitude as, but of opposite sign to, that of an electron, and all the characteristics of a real particle. The concept of a hole is valuable in that it is much simpler to discuss the

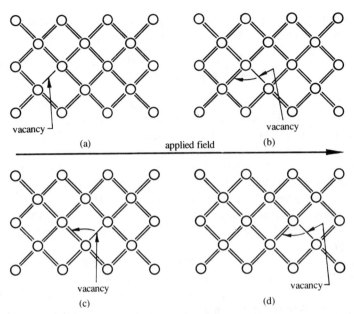

Fig. 6.7 **Movement of a vacancy under the influence of an applied electric field.**

motion of positive charged particles than that of bound electrons moving into vacancies.

Notice that, for every bond broken, one conduction electron is liberated and one hole created. The subsequent motion of the electron is independent of the hole motion and vice versa; both proceed until they recombine with particles of the opposite sign. It is the presence of two charge-carrier types in the semiconductor that causes the essential difference in conduction mechanism between semiconductors and metallic conductors. Because the conduction process described is inherently dependent on the structure of pure crystalline materials, such materials are called *intrinsic* semiconductors.

Attractive though the semiconductor model, typified by Figs 6.6 and 6.7, is, both as a pedagogical device and as an easily visualized pictorial representation of the motion of charged carriers in semiconductors, it has serious limitations, which should always be borne in mind. The limitations come about because of the uncertainties implicit in the uncertainty principle, discussed in Chapter 1. Suppose, for instance, that we estimate the mean thermal velocity of an electron in the semiconductor, v_{th}, by equating its mean kinetic energy, $\frac{1}{2}mv_{th}^2$ to $3kT/2$, by invoking the equipartition-of-energy law. At room temperature, this relation yields a mean thermal velocity of about 10^5 m s^{-1}. Now, for the sake of argument, let us assume that the instantaneous thermal velocity lies in the range of 0.5–1.5 times the mean value. Thus the uncertainty in velocity, Δv, is 10^5 m^{-1}s. The corresponding uncertainty in momentum, Δp, is $m\Delta v$. The Heisenberg relationship then predicts that the uncertainty in particle location, Δx, is in this instance at least

$$\Delta x = \frac{h}{\Delta p} = \frac{6.63 \times 10^{-34} \times 10^{10}}{9.11 \times 10^{-31} \times 10^{5}} \approx 7 \text{ nm}$$

Thus, even when the momentum of the electron is known only very approximately, the electron can still only be located to within a range of about 30 times the lattice constant, if this is assumed to be of the order of 0.2 nm. If the momentum is specified more accurately, then the uncertainty in location becomes even more pronounced. It is clear that the concept of an electron moving between adjacent bonds to effect hole movement can only be pictorial and has no real physical significance. Nevertheless, if the problem is treated in a more rigorous quantum-mechanical way, as in Chapter 5, the motion of a positively charged hole can in most instances correctly represent that of bound electrons.

A further point to emphasize is that, whenever the dynamics of holes and electrons in solids is being considered in a classical way, the mass parameter to be used is their *effective mass*, m_h^* or m_e^*, as discussed in Chapter 5. We saw there that the effective mass of a particle in a solid is dependent on its position in a particular band, but often the mass can be considered constant in practice since electrons and holes are usually located near band edges. Further, the effective masses of holes and electrons in a particular solid are not usually identical.

6.4 Density of carriers in intrinsic semiconductors

Returning to the energy band representation of a semiconductor, we have seen that a broken covalent bond corresponds to an electron being raised in energy so as to occupy the conduction band, leaving a hole in the valence band, as shown in Fig. 6.8. Conduction, as well as being possible because of electrons in the conduction band, can also occur as a result of the vacancies created in the

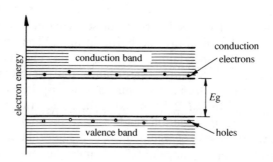

Fig. 6.8. Band structure of an intrinsic semiconductor, illustrating electron–hole pair generation.

valence band, and both electrons and holes contribute to the overall conduction process, as discussed in the previous chapter.

If the material is sufficiently pure, as is assumed for the intrinsic semiconductor, each broken bond generates an electron–hole pair, and electrons and holes thus occur in equal numbers. Thus

$$n = p = n_i \qquad (6.1)$$

where n is the number density of electrons, p is the number density of holes and n_i, the number density of charge carriers of either sign in an intrinsic semiconductor, is called the *intrinsic density*.

Whenever an electron–hole pair is created thermally, the state of excitation of the resulting particles is brief. For instance, the electron may undergo a transition that carries it to another energy state in the conduction band or it may revert to a vacant site in the valence band, i.e. it may *recombine* with a hole. Under thermal equilibrium conditions, the number of conduction electrons per unit volume, n, and the concentration of holes, p, are constant. How do these concentrations vary with the temperature? A simple argument is as follows.

At some particular absolute temperature, T, let the number of electrons in the conduction band be n and the number of holes in the valence band be p. Our initial simplifying assumptions are that the width of energy bands is small compared with the gap energy and that as a consequence all levels in each band can be considered as having the same energy. This situation is illustrated in Fig. 6.9, where we have, for convenience, taken an arbitrary zero-energy reference level at the top of the valence band.

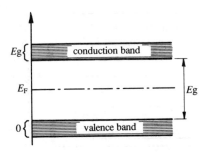

Fig. 6.9 Simplified band structure of an intrinsic semiconductor.

Now the number of electrons in the conduction band is

$$n = NP(E_g)$$

where $P(E_g)$, the probability of an electron having energy E_g, is given by the Fermi–Dirac function, Eq. (3.37), and N is the total number of electrons in both bands. Thus, in this case

$$n = \frac{N}{1 + \exp[(E_g - E_F)/kT]}$$

(6.2)

where E_F is the Fermi level, as usual.

The probability of an electron being in the valence band with zero energy is again obtained from the Fermi-Dirac function, Eq. (3.37), by putting $E = 0$, and the number of electrons in the valence band is thus

$$n_v = \frac{N}{1 + \exp(-E_F/kT)}$$

(6.3)

Now, the total number of electrons in the semiconductor, N, is the sum of those in the conduction band, n, and those in the valence band, n_v. Therefore, using (6.2) and (6.3)

$$N = \frac{N}{1 + \exp[(E_g - E_F)/kT]} + \frac{N}{1 + \exp(-E_F/kT)}$$

which can be simplified to give

$$E_F = E_g/2$$

(6.4)

Thus, in the intrinsic semiconductor, the Fermi level lies midway between the conduction and valence bands. That this conclusion is physically plausible is best seen with the aid of Fig. 6.10. At 0 K all available energy levels are filled, up

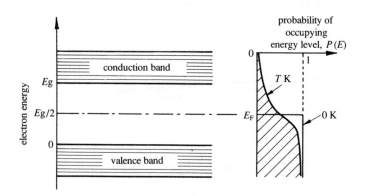

Fig. 6.10 Position of the Fermi level in an intrinsic semiconductor.

to E_F, and since energies in the forbidden gap are excluded, this indicates that the valence band is completely filled and the conduction band is empty, as we would expect. This fact only locates the Fermi level *somewhere* in the forbidden gap but it is the symmetry of the distribution function about E_F that ensures that E_F is located midway between the valence and conduction bands. Thus, if

E_F is in the position shown in Fig. 6.10, the probability that an electron occupies an energy level in the conduction band, at energy $E_g/2$ above the Fermi level, is identical with the probability that an electron *does not* occupy a level in the valence band, at $E_g/2$ below the Fermi level, which is equal to the probability of there being a hole in the valence band. This is in accordance with the physical arguments that the number of electrons in the conduction band equals the number of holes in the valence band.

The number of conduction electrons at any given temperature, T, which also equals the number of holes in the intrinsic material, is given by Eq. (6.2), which, using (6.4), becomes

$$n = \frac{N}{1 + \exp(E_g/2kT)} \tag{6.5}$$

Now, typically, the gap energy in a semiconductor varies from a fraction of an electronvolt to several electronvolts. Thus, at room temperature, when kT is of order 0.025 eV, $E_g \gg kT$ and the exponential part of the denominator of (6.5) dominates. Then

$$n \simeq N\exp(-E_g/2kT) \tag{6.6}$$

Thus, the number of conduction electrons, and hence the number of holes, in an intrinsic semiconductor decreases exponentially with increasing gap energy, which accounts for the lack of charge carriers in insulators with large gap energies. It will also be noted that the number of available charge carriers increases exponentially with increasing temperature.

The foregoing treatment is only very approximate since we have seen that all states in either the conduction or valence bands do not possess the same energy, as has been assumed. A more rigorous analysis must include additional terms that account for the density of available energy states in either band, as well as the probability function, as was discussed in Chapter 4.

The number of conduction-band electrons then becomes

$$n = \int_{\substack{\text{conduction} \\ \text{band}}} S(E)P(E)\,\mathrm{d}E \tag{6.7}$$

where $S(E)$ is the density of available states in the energy range between E and $E + \mathrm{d}E$, and $P(E)$ is the probability that an electron can occupy a state of energy E.

We saw in Chapter 4 that the density-of-states function for free electrons in a metal is given, per unit volume, by

$$S(E) = \frac{(8\sqrt{2})\pi m^{3/2}}{h^3} E^{1/2} \tag{6.8}$$

Now, electrons in the conduction band are not free, but Eq. (6.8) can be used to describe the density of states of electrons in the periodic potential of the crystal lattice if we replace m by an effective mass m_e^*, as we have already discussed. At the same time, we shall again arbitrarily choose the origin of the energy axis to be at the top of the valence band. Under these conditions the density-of-states term becomes

$$S(E) = \frac{(8\sqrt{2})\pi(m_e^*)^{3/2}}{h^3}(E - E_g)^{1/2} = C(E - E_g)^{1/2} \qquad (6.9)$$

where C for the moment is taken as constant.

The number density of conduction electrons using Eqs (6.7) and (6.9) is then given by

$$n = C\int_{E_g}^{\infty} \frac{(E - E_g)^{1/2}\,dE}{1 + \exp[(E - E_F)/kT]} \qquad (6.10)$$

Notice that we have assumed that the upper limit in the integration becomes infinite so as to introduce some mathematical simplification. It does not significantly increase the value of the definite integral because of the very rapid increase of the exponential term in the Fermi factor with increasing energy, which makes contributions from the top of the conduction band relatively unimportant. We can simplify (6.10) by again ignoring 1 in the denominator in comparison with the exponential term, since, as we have seen in our approximate treatment, energies in the conduction band are many times kT above the Fermi level. Thus

$$n \simeq C\int_{E_g}^{\infty}(E - E_g)^{1/2}\exp[-(E - E_F)/kT]\,dE \qquad (6.11)$$

This equation is best solved by making the substitution

$$x = (E - E_g)/kT$$

and evaluating the resulting definite integral using tables of standard integrals. Following this procedure and inserting the value of the constant C we obtain

$$n = 2\left(\frac{2\pi m_e^* kT}{h^2}\right)^{3/2}\exp[-(E_g - E_F)/kT] \qquad (6.12)$$

The non-exponential term only varies relatively slowly with temperature compared to the exponential term and it is convenient to consider it as a pseudo-constant, N_c, and write

$$n = N_c\exp[-(E_g - E_F)/kT] \qquad (6.13)$$

Turning now to the density of holes in the valence band, p, this is computed by a development analogous to the previous one. The essential difference is that the Fermi factor used must now be the one pertinent to holes. This can be obtained by equating the probability of a hole occupying a given energy level to the probability that an electron *does not* occupy that level. Thus the Fermi factor for holes is given by

$$1 - P(E) = 1 - \frac{1}{1 + \exp[(E - E_F)/kT]} = \frac{\exp[-(E_F - E)/kT]}{1 + \exp[-(E_F - E)/kT]} \quad (6.14)$$

Again, we know from our earlier approximate analysis that the energy levels in the valence band lie many kT below the Fermi level, $(E_F - E) \gg kT$, and hence Eq. (6.14) can be expressed approximately as

$$1 - P(E) \simeq \exp[-(E_F - E)/kT] \quad (6.15)$$

The density of available states in the valence band is again taken to be the same as that pertaining to free electrons, with the necessary effective mass and energy zero change, giving

$$S(E) = \frac{(8\sqrt{2})\pi(m_h^*)^{3/2}}{h^3}(-E)^{1/2} \quad (6.16)$$

The density of holes is then obtained by multiplying Eqs (6.15) and (6.16) together as before and integrating over all energies from $-\infty$ to 0. The lower limit of the integration is again chosen for mathematical convenience and is permissible because of the rapid decay of the exponential term of Eq. (6.15) with decreasing energy. We then arrive at

$$p = 2\left(\frac{2\pi m_h^* kT}{h^2}\right)^{3/2} \exp(-E_F/kT) \quad (6.17)$$

Again, it is usual to write the first term as a pseudo-constant with temperature, N_v; then

$$p = N_v \exp(-E_F/kT) \quad (6.18)$$

A most interesting and useful result can be obtained by forming the product of the number densities of holes and electrons, using Eqs (6.13) and (6.18). Thus

$$np = n_i^2 = N_c N_v \exp(-E_g/kT) \quad (6.19)$$

or, for the intrinsic material,

$$n_i = 2\left(\frac{2\pi kT}{h^2}\right)^{3/2} (m_e^* m_h^*)^{3/4} \exp(-E_g/2kT) \quad (6.20)$$

Notice that this expression agrees with the less quantitative one derived earlier in Eq. (6.6), since the temperature dependence is largely controlled by the rapidly varying exponential term. We see that the density of carriers in an intrinsic semiconductor is independent of the Fermi level and for a given material is constant at any given temperature. It is also most important to observe that, since the gap energy does not change with impurity concentration, and since Eq. (6.19) does not contain E_F, this equation, although derived by considering intrinsic semiconduction, is equally valid for semiconductors containing impurities. We shall return to this point later.

We can calculate the position of the Fermi level in our improved model of the intrinsic semiconductor by equating the electron and hole concentrations, using Eqs (6.12) and (6.17). Then

$$(m_e^*)^{3/2}\exp[-(E_g - E_F)/kT] = (m_h^*)^{3/2}\exp(-E_F/kT)$$

which is rearranged to give

$$E_F = \tfrac{1}{2}E_g - \tfrac{3}{4}kT\log(m_e^*/m_h^*) \tag{6.21}$$

Thus, if the effective masses of holes and electrons are identical, the Fermi level is again midway between valence and conduction bands, as in our more approximate model. Otherwise, the Fermi level is temperature-dependent, and E_F increases slightly with temperature if, as is generally the case, $m_h^* > m_e^*$.

A final point to note before leaving this subject is that we have tacitly assumed throughout that the effective masses are constant, whereas we know them to be functions of position in an energy band. Since they are functions of energy they should be included under the integrals over the energy bands. However, this step leads to unnecessary mathematical complication and is usually ignored with only a small error, since, for example, the mass of the electron remains sensibly constant over the narrow lower portion of the conduction band that it usually occupies.

The mathematical derivation of the density of carriers in an intrinsic material via the Fermi and density-of-states functions is illustrated in Fig. 6.11, which may serve as additional clarification of the process.

6.5 Extrinsic or impurity semiconductors

Intrinsic semiconductors as such are comparatively rarely included in semiconductors devices but, because of their purity and predictable electrical

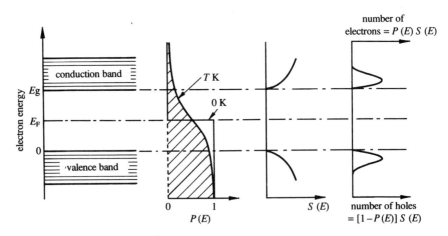

Fig. 6.11 Calculation of the number density of carriers in an intrinsic semiconductor.

characteristics, they often constitute the basic raw material that is modified during the manufacturing process by the addition of controlled amounts of impurity. We shall see that the number of available charge carriers, their type and hence the electrical conductivity of a semiconductor are extremely sensitive to the concentration of certain types of impurity. It is the ability to alter the electrical characteristics of the material at will by the addition of impurities that makes the *extrinsic* semiconductor so important and interesting.

The most significant impurities that are introduced into the semiconductors we have discussed so far, notably silicon and germanium, are those elements which occur at either side of them in the periodic table, in columns III and V. Deliberate addition of controlled quantities of these impurity elements is called *doping*. Usually, only minute quantities of such *dopants* are required, in the range one part in 10^{10} to one part in 10^3. Extrinsic or doped semiconductors are classified into two main categories, according to the type of charge carrier that predominates, n-type and p-type, which are discussed in the following sections.

6.5.1 n-Type semiconductors

If controlled amounts of impurity elements from column V of the periodic table, e.g. phosphorus, arsenic and antimony, are added to an intrinsic elemental semiconductor, the extrinsic material as a result is rich in negative conduction electrons and is said to be n-type. These dopants are all characterized by a valency of five, since they have five electrons in their outer shells. They can be represented diagrammatically as in Fig. 6.12(a). When the intrinsic material is doped with one of the column V dopants, say for example antimony, each antimony atom occupies a site usually occupied by a germanium or silicon atom, since, apart from the difference in number of valence electrons, both parent and impurity atoms are similar in dimensions and electronic structure. The substitutional nature of the dopant atoms has been verified, for example, by X-ray crystallography. The substituted antimony impurity atom, depicted in the two-dimensional model of Fig. 6.12(b), requires only four of its five valence electrons to form covalent bonds with its neighbours and, as a result, it has net positive charge. That the 'extra' electron is not so tightly bound as the other valence electrons can be seen by an extension of the Bohr theory for the hydrogen atom, given in Chapter 1. It is reasonable to assume that the fifth electron rotating round the positively charged impurity atom behaves in a manner similar to that of the electron in the Bohr model. There are, however, two principal differences: first, the electron is influenced by the periodic potential of the crystal lattice and it is thus its effective mass, m^*, rather than its free mass, m, that is important; secondly, if we assume that the radius of the orbit is sufficient to embrace many crystal lattice points, as we shall verify later, the electron may be considered to be moving in a dielectric medium of relative permittivity ϵ_r.

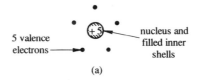

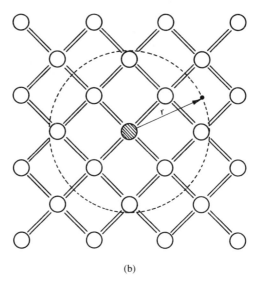

(b)

Fig. 6.12 **(a) Representation of an n-type impurity atom and (b) substitution of an n-type impurity atom into a semiconductor crystal lattice.**

In Chapter 1 the ionization energy for the hydrogen atom was shown to be

$$E = -\frac{m_e e^4}{8\epsilon_0^2 h^2} = -13.6\,\text{eV}$$

By analogy, the binding energy of each extra electron in the doped semiconductor is then

$$E_b = -\frac{13.6}{\epsilon_r^2}\left(\frac{m_e^*}{m_e}\right)\text{eV} \qquad (6.22)$$

In order to test the validity of our assumption regarding the radius of the orbit, we again modify the expression for the radius in atomic hydrogen, obtained in Chapter 1, so as to include the effective mass of the electron and the relative permittivity of the semiconductor, thus

$$r_h = r_{\text{hydrogen}} = \frac{h^2 \epsilon_0}{\pi e^2 m_e} \approx 0.05\,\text{nm}$$

and, similarly,

$$r_n = r_{\substack{\text{n-type} \\ \text{semiconductor}}} = \frac{h^2 \epsilon_r \epsilon_0}{\pi e^2 m_e^*} = \epsilon_r \left(\frac{m_e}{m_e^*} \right) r_h \qquad (6.23)$$

If we include characteristic values of $m_e^* = 0.6 m_e$ for the effective mass of conduction electrons in germanium and $\epsilon_r = 16$, then the radius in germanium is approximately 27 times that in hydrogen, or about 1.4 nm. This radius is sufficiently large to include many hundreds of lattice points within an orbit and our use of the relative permittivity of the bulk material is justified.

We can now substitute typical values in Eq. (6.22) to estimate the binding energy of electrons in n-type semiconductors. For example, for germanium,

$$E_b = -13.6 \times 0.6/16^2 = 0.03 \, \text{eV}$$

In practice, the experimentally determined values for the binding energies of P, As and Sb in germanium are almost identical and are about 0.01 eV, which gives additional support for our simple model.

Because the binding energies are so small, then, even at room temperature, nearly all the impurity atoms lose an electron into the conduction band by thermal ionization. The additional electrons in the conduction band contribute to the conductivity in exactly the same way as those excited thermally from the valence band. The essential difference between this mechanism and the intrinsic process is that the ionized impurities remain as fixed positively charged centres in the lattice and no *holes* are produced. Thus, group V impurities donate extra carrier electrons, without producing additional holes, and are called *donors* or sometimes *n-type dopants*.

In the energy band scheme, each donor atom gives rise to a new isolated donor level, E_D, just below the bottom of the conduction band. Thus, as a consequence of the introduction of n-type dopant, an additional set of highly localized electronic states, with discrete energies situated in the forbidden energy gap, is introduced. At 0 K all donor levels are filled but at room temperature and above they are empty, as shown in Fig. 6.13.

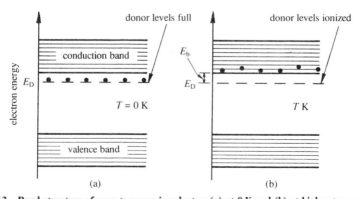

Fig. 6.13 **Band structure of an n-type semiconductor: (a) at 0 K and (b) at higher temperatures.**

Since at room temperature each impurity atom donates an additional charge carrier, even small concentrations of dopant substantially increase the carrier concentration and hence the conductivity. As a simple example, let us compute the electron density in intrinsic germanium and compare this with n-type germanium, doped with phosphorus. If we assume the gap energy $E_g = 0.75\,\text{eV}$, that $kT \approx 0.025$ at room temperature, and that $m_e^* = m_h^* = m$ for simplicity, Eq. (6.20) predicts that $n_i = 10^{19}\,\text{m}^{-3}$. There are about 10^{28} atoms per cubic metre (atom m^{-3}) in germanium; hence if we assume that it is doped with P at the rate of one part in 10^6, then 10^{22} atom m^{-3} of impurity are introduced into the intrinsic material. At room temperature each dopant atom is ionized and contributes one conduction electron; thus 10^{22} conduction electron m^{-3} have been donated. We see that the introduction of impurity atoms at the rate of one part in a million has increased the carrier concentration and hence the conductivity a thousand-fold!

6.5.2 p-Type semiconductors

Semiconductors with a majority charge carrier concentration of holes are called p-type. They can be produced by adding impurities from column III of the periodic table, e.g. boron, aluminium, gallium or indium, to intrinsic silicon or germanium. These p-type impurities are characterized by three valence electrons in their outer shell, Fig. 6.14(a). Each impurity atom again adds substitutionally to the parent crystal lattice but this time there is one electron short for completing the four covalent bonds between it and the four neighbouring atoms, as shown in Fig. 6.14(b). The vacancy thus created by the impurity is not a hole, since it is bound to the atom, but at some temperature above 0 K an electron from a bond of a neighbouring parent atom can fill the vacant electron site, leaving a hole in the valence band for conduction. In accordance with our concept of the hole as a positively charged particle of

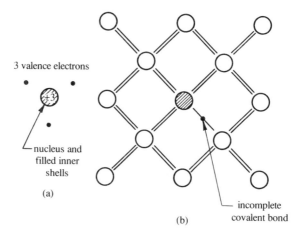

Fig. 6.14 (a) Representation of a p-type impurity atom and (b) substitution of a p-type impurity atom into a semiconductor crystal lattice.

effective mass, m_h^*, and since it may be considered to rotate at a large radius round a fixed negative charge, we can again estimate the ionization energy in a way analogous to that used in the preceding section. The only difference from the result in Eq. (6.22) is that the effective mass of a hole replaces that of an electron. Hence, the energy required by a valence electron for it to fill the vacancy created by an impurity atom and thus create a hole, E_A, is of similar magnitude to the ionization energy of a donor atom and is typically of order 0.01 eV.

Dopants from column III are called *acceptors*, since they accept electrons to create holes for conduction. Such impurities create discrete acceptor levels just above the top of the valence band, separated from it by energy E_A, as shown in Fig. 6.15(a). Again, since E_A is small compared with the thermal energy of an

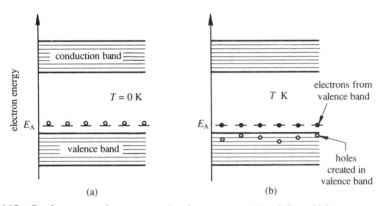

Fig. 6.15   Band structure of a p-type semiconductor: (a) at 0 K and (b) at higher temperatures.

electron at room temperature, nearly all acceptor levels are occupied and each acceptor atom creates a hole in the valence band, as in Fig. 6.15(b). Note that no additional electrons are created by the process but, as before, each acceptor atom of dopant can produce a charge carrier and the available number of holes is very much greater than for the intrinsic material, even for relatively moderate doping concentrations.

We have seen that in extrinsic semiconductors there are often more carriers of one type than of the other. It is convenient to designate those which predominate as *majority carriers* and the remainder as *minority carriers*. By using these terms it is often possible to discuss the general characteristics of doped semiconductors without specifying whether the material is n- or p- type.

6.6 Electron processes in real semiconductors

6.6.1 Direct- and indirect-gap semiconductors

The simplified energy band models of semiconductors developed so far, for example as shown in Fig. 5.14, are sufficient for most purposes. However, in certain situations, particularly when carriers are involved in interband transitions such as in optical and some microwave devices, which will be described later, the simple band model does not contain sufficient information to enable carrier effects to be fully explained. In these circumstances, the more complete energy–momentum (E–p) diagram for the semiconductor concerned, of the type shown in Fig. 5.9(b), must be reverted to. E-p diagrams for real semiconductors are more complicated than the theoretically derived curves shown in Fig. 5.9(b), not only because E is a function of p, but also because the shape of the curve is dependent on the crystal orientation. However, their band diagrams can conveniently be subdivided into two main categories, as typified by the simplified versions for GaAs and Si shown in Fig. 6.16.

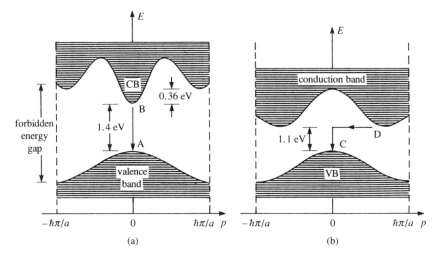

Fig. 6.16   Simplified energy–momentum diagrams for (a) GaAs, a direct–gap semiconductor, and (b) Si, an indirect-gap semiconductor.

In so-called *direct-gap* semiconductors, typified by the band diagram in Fig. 6.16(a), a maximum in the valence band, at A for example, corresponds in the momentum scale to a minimum in the conduction band, at B. In other words, the minimum band gap occurs at $p=0$. Hence an electron can be elevated from the conduction band at A into the valence band at B with the minimum of energy and there is no momentum change necessary during the process. The corresponding *direct* transition of an electron in the other direction, from B to A, together with the emission of a photon and again with no momentum change, is also statistically favourable, which explains, as we shall discover, why such materials are used for example in solid-state lasers. Such favourable direct-gap carrier transitions characterize this group of semiconductors.

In contrast, the group of *indirect-gap* semiconductors, of which Si is the principal member, have a simplified energy band structure, typically as in Fig. 6.16(b). The valence band again is maximum at the origin on the momentum axis, at C, but the minimum in the conduction band is situated towards the edges of the diagram, at D for example. An electron transferred across the minimum energy gap from C to D or vice versa must also experience a simultaneous change in momentum, as shown in the diagram, the momentum change being achieved by interaction with lattice phonon vibrations. Such an *indirect* process is much less likely than the direct transitions involving no momentum changes that occur in GaAs, etc.

6.6.2 Recombination and trapping processes

We have seen that recombination takes place in a semiconductor when an electron is permanently removed from the conduction band to recombine with a hole in the valence band. Although such a process is possible in direct-gap semiconductors such as InSb and GaAs, the probability for direct recombination is very slight in the indirect-gap semiconductors Si and Ge, because momentum must be conserved and this is rarely possible for the colliding free electron and hole. In these materials the conduction-band electron usually returns to the valence band only after several intermediate transitions. For instance, it may first drop to some localized state in the forbidden gap, whence, after some random time interval, it is translated to some empty state in the valence band; in other words, it recombines with a hole from the valence band at the localized level, which is called a recombination centre. The sequence of events is illustrated in Fig. 6.17(a).

Whereas recombination results in the permanent loss of a carrier, *trapping* describes the temporary removal of a carrier to a localized level, again usually located in the forbidden gap, from where it may eventually return to its original band. This process is shown diagrammatically in Fig. 6.17(b). If the trapping is fast, and it can be for as short a time as, say, 10^{-7}s, then little signal

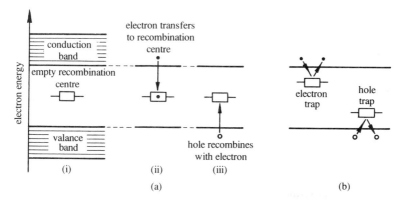

Fig. 6.17 (a) Sequence of events in the recombination process and (b) trapping in semiconductors.

information is lost at relatively low frequencies, but 'slow' traps that can retain the carrier for many seconds completely modify electrical information being processed by a particular device.

Recombination and trapping centres occur at isolated energy levels in the forbidden gap. Such states are associated with some kind of imperfection in the crystal or its surface and are a consequence of lattice defects, especially dislocations, vacancies, or additional atoms in the lattice. Such imperfections can be avoided by modern high-quality crystal growth. Sometimes, however, impurities are added deliberately to create additional trapping centres. For example, gold is often added to Si to remove unwanted holes. Similarly, impurities producing 'slow' traps are introduced into some phosphorescent materials used for clock faces, for example, so that they luminesce long after excitation.

6.7 The density of carriers and the position of the Fermi level in extrinsic semiconductors

The distribution of electrons or holes in allowed energy states is dependent on the position of the Fermi level. If this can be determined for a particular semiconductor, the number of charge carriers of either type available for conduction at any particular temperature is readily determined.

In general, a semiconductor can be doped with both acceptor and donor impurities and in this case it is difficult to obtain an exact analytical expression for E_F, which usually has to be obtained by numerical computation. We will first consider the general case and then relax some of the generality so as to make the problem more tractable and to produce trends in the movement of the Fermi level.

The principle used to determine the Fermi level is not unique but is common to many semiconductor calculations. The Fermi level in a particular crystal automatically adjusts itself so that overall charge neutrality exists in the crystal. Therefore, if a condition can be found when the total negative charge is equal to the total positive charge, the Fermi energy is immediately specified.

In an extrinsic semiconductor, it is not only the holes and electrons that have charge, but ionized impurity atoms are also present and they, too, are charged, as we have seen. The condition for electrical neutrality is thus

$$p + N_d^+ = n + N_a^- \tag{6.24}$$

where N_d^+ is the number density of ionized donor atoms and N_a^- is the number density of ionized acceptor atoms, which is the number of acceptors that have received an electron from the valence band and have thus generated a hole.

If we consider the energy level structure shown in Fig. 6.18, the number of

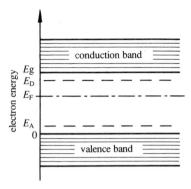

Fig. 6.18 Energy levels in an extrinsic semiconductor.

electrons and holes per unit volume are given by Eqs (6.13) and (6.18), i.e.

$$n = N_c \exp[-(E_g - E_F)/kT]$$
$$p = N_v \exp(-E_F/kT)$$

The concentration of ionized acceptors, N_a^-, is equal to the total acceptor concentration, N_a, multiplied by the probability of finding an electron at the acceptor level, E_A, at some particular temperature T. Thus

$$N_a^- = \frac{N_a}{1 + \exp[(E_A - E_F)/kT]} \qquad (6.25)$$

Similarly, N_d^+, the concentration of ionized donors, is equal to the total number density of donors times the probability of finding a hole at energy E_D, or, alternatively, times one minus the probability of finding an electron at the level. Thus

$$N_d^+ = N_d \left(1 - \frac{1}{1 + \exp[(E_D - E_F)/kT]}\right) = \frac{N_d}{1 + \exp[-(E_D - E_F)/kT]} \qquad (6.26)$$

Substituting the expressions for n, p, N_a^- and N_d^+ into the charge neutrality expression, Eq. (6.24), gives

$$N_v \exp(-E_F/kT) + \frac{N_d}{1 + \exp[-(E_D - E_F)/kT]}$$

$$= N_c \exp[-(E_g - E_F)/kT] + \frac{N_a}{1 + \exp[(E_A - E_F)/kT]} \qquad (6.27)$$

If the gap energy and the various doping concentrations and impurity levels are known, this equation can be used to determine the exact position of the Fermi level and hence the carrier concentrations in a particular crystal sample.

By considering the following special cases and using simplifying assumptions, it is possible to reduce the complexity of Eq. (6.27) and to gain some insight into the approximate location of the Fermi level.

6.7.1 n-Type material

In this type of crystal the donor concentration is much greater than the acceptor concentration, or $N_d \gg N_a$. Also, since the electron density is much higher than its intrinsic value, the relationship $np = n_i^2$, which is always true, as we have seen, indicates that the hole density is much lower than the intrinsic density, n_i, or $p \ll n$. Thus, the first and last terms of Eqs (6.24) and (6.27) are negligibly small and

$$\frac{N_d}{1 + \exp[-(E_D - E_F)/kT]} = N_c \exp[-(E_g - E_F)/kT]$$

which can be rearranged to give

$$\exp[-(E_D + E_g)/kT]\exp(2E_F/kT) + \exp(-E_g/kT)\exp(E_F/kT) - N_d/N_c = 0$$

This equation is quadratic in $\exp(E_F/kT)$ and its solution is

$$\exp(E_F/kT) = \frac{-1 + \{1 + (4N_d/N_c)\exp[(E_g - E_D)/kT]\}^{1/2}}{2\exp(-E_D/kT)} \qquad (6.28)$$

The negative root has been omitted, since the exponential on the left-hand side of the equation can never be negative.

Equation (6.28) can usually be solved to obtain E_F at some particular temperature. Again, it is instructive to consider limiting cases to locate the Fermi level approximately in these conditions.

6.7.1.1 *n-Type at low temperature, or with high doping concentration*

In this case, the second term under the square root of Eq. (6.28) becomes much greater than unity and

$$\exp(E_F/kT) \simeq \frac{2(N_d/N_c)^{1/2}\exp[(E_g - E_D)/2kT]}{2\exp(-E_D/kT)}$$

$$= (N_d/N_c)^{1/2}\exp[(E_g + E_D)/2kT]$$

and

$$E_F = \tfrac{1}{2}(E_g + E_D) + \tfrac{1}{2}kT\log(N_d/N_c) \qquad (6.29)$$

Thus, at 0 K the Fermi level lies midway between the donor level and the bottom of the conduction band. If this Fermi energy is substituted in Eq. (6.13) we can obtain the number density of electrons in the semiconductor

$$n = N_c \exp(-E_g/kT)(N_d/N_c)^{1/2}\exp[(E_g + E_D)/2kT]$$

$$= (N_c N_d)^{1/2}\exp[(E_D - E_g)/2kT] \propto N_d^{1/2} \qquad (6.30)$$

6.7.1.2 n-Type at higher temperature, or low doping concentration

If the concentration of impurity atoms is low, such that $N_d \gg N_c$, or if the temperature is high, provided it is not so high that the material becomes near-intrinsic, which would invalidate our earlier assumptions, then $kT \gg E_g - E_D$ and the second term under the square root of Eq. (6.28) is much smaller than unity. Then, taking the first two terms of the binomial expression, this equation reduces to

$$\exp(E_F/kT) = \frac{-1 + 1 + (4N_d/2N_c)\exp[(E_g - E_D)/kT]}{2\exp(-E_D/kT)}$$

or

$$E_F = E_g - kT \log(N_c/N_d) \qquad (6.31)$$

Thus, the Fermi level falls as the temperature is increased. Again the density of majority carriers is given by Eqs (6.13) and (6.31), or

$$n = N_c \exp(-E_g/kT)(N_d/N_c)\exp(E_g/kT) = N_d \qquad (6.32)$$

and in this case the majority-carrier density is equal to the concentration of donor impurities.

We see from these extreme cases studied that for fairly high doping levels or low temperatures, the Fermi level remains in the vicinity of the donor level. However, if the donor concentration is small or the temperature high, the Fermi level falls until, when the number of thermally excited electrons is much greater than those donated by donors and the material is essentially intrinsic, it approaches the mid-gap position, E_{Fi}. This situation is illustrated diagrammatically in Fig. 6.19(a).

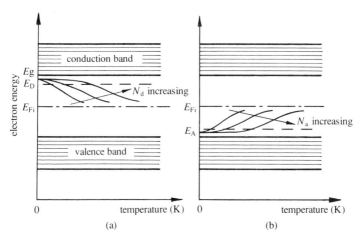

Fig. 6.19 Variation of Fermi level with impurity concentration and temperature for (a) n-type and (b) p-type semiconductors.

6.7.2 p-Type material

By arguments similar to those used for n-type material, it can be shown that for a p-type semiconductor the Fermi level lies midway between the top of the valence band and the acceptor level, E_A, at 0 K. Again, as the temperature is raised above this level, all the acceptors are ionized, $p = N_a$, and the Fermi level increases. At much higher temperatures the material becomes intrinsic and the Fermi level eventually approaches E_{Fi}, as shown in Fig. 6.19(b).

It should be emphasized that Eqs (6.28) to (6.31) are only approximate. It is always best in a particular case when the doping concentrations are known for a material to evaluate E_F directly using Eq. (6.37) and thus obtain the carrier concentration at a particular temperature.

6.8 Compensation doping

When both acceptor and donor impurities are added simultaneously to an intrinsic semiconductor, *compensation* takes place. Usually, all the added impurities are ionized at room temperature. Whether the material is n-type or p-type and also the concentration of the carriers depend on the degree of doping of each impurity type. Suppose, for example, that a silicon crystal contains 10^{22} donor impurities and 10^{21} acceptor impurities per cubic metre. It is clear that this crystal will be n-type since n-type impurities predominate. The free carriers supplied by the less concentrated dopant will recombine with an equal number of carriers of opposite type, supplied by the more concentrated dopant, leaving an excess of electrons or holes (in the above example, electrons). Some of the donor states have been effectively cancelled by the acceptor states, a process known as *compensation*.

If we assume that all the impurities are ionized, it is thus the quantity $(N_d - N_a)$, which may be positive or negative, that largely determines the carrier concentration. In this case, the condition for electrical neutrality is

$$n + N_a = p + N_d$$

Also, we make use of the relationship between n and p

$$np = n_i^2$$

Substituting this in the previous equation gives

$$n^2 - (N_d - N_a)n - n_i^2 = 0$$

which is a quadratic equation in n with solution

$$n = \frac{N_d - N_a}{2} + \frac{N_d - N_a}{2}\left[1 + \left(\frac{2n_i}{N_d - N_a}\right)^2\right]^{1/2} \qquad (6.33)$$

and likewise

$$p = -\frac{N_d - N_a}{2} + \frac{N_d - N_a}{2}\left[1 + \left(\frac{2n_i}{N_d - N_a}\right)^2\right]^{1/2} \qquad (6.34)$$

The negative roots have been omitted so that n and p remain positive.

Equations (6.33) and (6.34) can be simplified in two extreme cases, as follows.

6.8.1 Extrinsic compensated material

In this case the intrinsic carrier density is low compared with the difference between the densities of donors and acceptors, or $n_i \ll |N_d - N_a|$. The equations then reduce to

$$n \simeq N_d - N_a$$
$$p \simeq n_i^2/(N_d - N_a) \qquad \text{for } N_d > N_a$$

and

$$p \simeq N_a - N_d$$
$$n \simeq n_i^2/(N_a - N_d) \qquad \text{for } N_a > N_d$$

In such materials the majority-carrier concentration is large, approximately equal to $|N_d - N_a|$, and is almost temperature-independent, whereas the minority-carrier concentration is small and varies rapidly with temperature.

6.8.2 Intrinsic compensated material

In this case the difference in donor and acceptor concentrations is small compared to the density of intrinsic carriers, or $n_i \gg |N_d - N_a|$, and, in contrast to the previous case, the Fermi level lies near the middle of the forbidden gap. Equations (6.33) and (6.34) then have the approximate forms

$$n = n_i + (N_d - N_a)/2$$

and

$$p = n_i + (N_d - N_a)/2$$

Such materials are said to be *intrinsic* or *near-fully compensated*. This condition can also apply for the material in the previous case, at high temperatures, when the number of hole pairs generated in the intrinsic process can become large compared with $|N_d - N_a|$. The semiconductor is then said to be in the *intrinsic temperature range*.

6.9 Electrical conduction in semiconductors

Electrical conduction by electrons in the conduction band of a semiconductor has some similar characteristics to that for metals, which was described previously. The drift velocity of electrons in the presence of an applied field is again limited by scattering effects and is determined by an electron mobility, μ_e. The electron current density, by analogy with Eq. (4.34), is then

$$J_e = ne\mu_e \mathscr{E} \tag{6.35}$$

The expression for the mobility of free electrons, Eq. (4.33), must be modified to account for the motion of electrons in the lattice of a semiconductor by replacing the free mass m by the effective mass of electrons m_e^* and including the appropriate relaxation time τ_{re}.

In a semiconductor, there is a further contribution to the total conduction current from holes that occur in the valence band. The drift velocity of holes can again be specified by a hole mobility term, μ_h, which is given by Eq. (4.33) provided the effective mass of holes in the semiconductor, m_h^*, and the relevant relaxation time, τ_{rh}, are included. The drift current density due to hole conduction is then

$$J_h = pe\mu_h \mathscr{E} \tag{6.36}$$

Although an applied electric field causes electrons and holes to flow in opposite directions, the direction of conventional current flow due to the motion of each carrier is in the same direction as the field. The total current density in a semiconductor is thus

$$J = J_e + J_h = e(n\mu_e + p\mu_h)\mathscr{E} \tag{6.37}$$

and the electrical conductivity, σ, is given by

$$\sigma = e(n\mu_e + p\mu_h) \tag{6.38}$$

Since μ_e and μ_h are of the same order, typically about $0.1 \text{ m}^2 \text{ V}^{-1} \text{ s}^{-1}$ at room temperature, Eqs (6.37) and (6.38) must in general be used when n and p are comparable, for example when the semiconductor is near-intrinsic. For doped materials at room temperature, when one carrier predominates, it is usually only necessary to include the term pertaining to the majority carrier. For example, for an n-type semiconductor, $n \gg p$ and

$$\sigma_n \simeq en\mu_e \simeq eN_d\mu_e \tag{6.39}$$

Similarly, for a p-type semiconductor

$$\sigma_p \simeq ep\mu_h \simeq eN_a\mu_h \tag{6.40}$$

6.9.1 Diffusion of charge carriers in semiconductors

It is also possible for current to flow in a semiconductor, even in the absence of fields, owing to a concentration gradient of carriers in the crystal. Such a diffusion current can flow as a result of non-uniform densities of either electrons or holes.

Let us first of all consider diffusion in a neutral gas. We will assume an arbitrarily shaped concentration gradient of neutral particles, which varies only in the x direction, as shown in Fig. 6.20(a). When the density increases as x increases, as is assumed in the diagram, it is to be expected that diffusion of particles takes place in the negative x direction, from the high-pressure to the low-pressure region.

We consider an elemental volume, of thickness δx and area in the $y - z$ plane

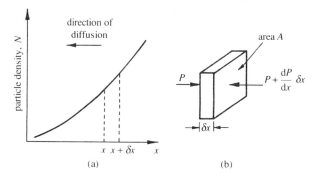

Fig. 6.20 Diffusion of particles when a density gradient exists.

A [Fig. 6.20(b)]. If $P(x)$ is the pressure at x, the overall force on the elemental volume is

$$\left[P(x) - \left(P + \frac{dP}{dx}\delta x \right) \right] A = -\frac{dP}{dx}\delta x A$$

Thus, if there are N molecules per unit volume at x, that is, $N\delta x A$ molecules in the volume considered, the force acting on each molecule is

$$F_{\mathrm{D}} = -\frac{1}{N}\frac{dP}{dx} \tag{6.41}$$

This diffusion force acts in a way entirely analogous to that in which the force due to an electric field acts, to accelerate particles in the opposite direction to the concentration gradient; the flow of particles is again limited by collisions or, in the case of charged-particle gradients in semiconductors, by scattering phenomena. The same arguments can be used to calculate the particle drift velocity as were used for calculating the mobility, but F_{D} replaces the force due to the electric field, $-e\mathscr{E}$. Thus, using Eqs (6.41) and (4.30), we see that the drift velocity of neutral particles is

$$v_{\mathrm{D}} = -\frac{\tau_{\mathrm{r}}}{M}\frac{1}{N}\frac{dP}{dx} \tag{6.42}$$

where τ_{r} is again the mean time between collisions and M is the mass of a particle. Now, the pressure is related to particle density and temperature T by the relationship

$$P = NkT$$

so the expression for drift velocity can be rewritten as

$$v_{\mathrm{D}} = -\frac{\tau_{\mathrm{r}}kT}{M}\frac{1}{N}\frac{dN}{dx} \tag{6.43}$$

This expression is a diffusion equation, which is often written

$$v_D = -D\frac{1}{N}\frac{dN}{dx} \qquad (6.44)$$

where D is a *diffusion coefficient*, which is in units of $m^2\,s^{-1}$ and is given by

$$D = \tau_r kT/M \qquad (6.45)$$

The minus sign in Eq. (6.44) confirms that particle diffusion occurs from the region of high concentration to the region of low concentration, as we presupposed.

Diffusion effects occur for any concentration gradient of particles; that the particles are charged does not affect the analysis. For example, for a density gradient of electrons in a semiconductor, the drift velocity is, from Eq. (6.44),

$$v_{De} = -D_e\frac{1}{n}\frac{dn}{dx}$$

which causes an electron diffusion current density, J_{De}, to flow, where

$$J_{De} = -nev_D = eD_e\,dn/dx \qquad (6.46)$$

Notice that this expression is positive since the electrons flowing in the direction opposite to the positive gradient of electron density correspond to a conventional current flow in the same direction as the positive gradient, as in Fig. 6.21(a). Similarly, for a gradient of hole concentration in the x direction in

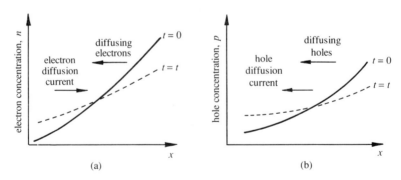

Fig. 6.21 Diffusion currents for (a) electrons and (b) holes, when a carrier concentration gradient exists.

a semiconductor, dp/dx, there is a resultant diffusion current density, J_{Dh}, given by

$$J_{Dh} = -eD_h\,dp/dx \qquad (6.47)$$

which flows in the opposite direction to the positive gradient, as in Fig. 6.21(b).

In Eqs (6.46) and (6.47), D_e and D_h are the diffusion coefficients for electrons and holes. Typical values, for germanium for example, are $D_e = 0.0093 \text{ m}^2 \text{ s}^{-1}$ and $D_h = 0.0044 \text{ m}^2 \text{ s}^{-1}$. The magnitude of the diffusion coefficient for either carrier is an indication of the ability of the carrier to move through the crystal; subsequently, it would not be surprising if it were related to the mobility of the carrier in some way. That there is a relationship between the two parameters can be seen by inserting the electron effective mass and relaxation time in Eq. (6.45) to obtain an expression for D_e and comparing this with Eq. (4.33). It follows that

$$D_e = (kT/e)\mu_e \tag{6.48}$$

Using similar arguments, it can be verified that

$$D_h = (kT/e)\mu_h \tag{6.49}$$

Thus, at any given temperature the diffusion coefficient and the mobility of carriers in a given material are not independent of each other, since

$$D_e/\mu_e = D_h/\mu_h = kT/e \tag{6.50}$$

This equation is known as *Einstein's relation*.

6.9.2 Total current flow in a semiconductor

In the two previous sections we have seen that current flow in a semiconductor is due to motion of the charged carriers under the influence of applied fields or concentration gradients. It is quite possible to have these two effects occurring simultaneously, and the net current flow is then the sum of drift and diffusion currents. Using Eqs (6.37), (6.46) and (6.47) we obtain the following general expressions for the total electron and hole currents

$$J_e = ne\mu_e \mathscr{E} + eD_e \nabla n \tag{6.51}$$

and

$$J_h = pe\mu_h \mathscr{E} - eD_h \nabla p \tag{6.52}$$

These flow relations have been generalized by allowing the electric field and the density gradients to be in any arbitrary direction.

An important effect that may be seen from these equations is that by suitable choice of $\mathscr{E}$ field the tendency for current to flow due to concentration gradients can be balanced out and the net current reduced to zero. For example, if we consider holes only, Eq. (6.52) shows that if

$$\mathscr{E} = D_h \nabla p / p\mu_h$$

then the hole current becomes zero. This compensation of diffusion and drift currents to produce zero net current is vital for understanding transistor and diode action, which will be discussed later.

6.10 The continuity equation

The continuity equation as applied to semiconductors describes how the carrier density in a given elemental volume of the crystal varies with time. The variation in density is attributable to two basic causes: the rate of generation and loss by recombination of carriers within the element, and drift of carriers into or out of the element. The problem can be simplified somewhat by first treating the two effects separately. We shall concern ourselves with the continuity equation for *minority* carriers since it is the density of these that changes most significantly in most practical cases of interest when carriers are injected into or created in a semiconductor.

6.10.1 No electron or hole current flow

To be more specific, let us consider a p-type semiconductor and derive the continuity equation for the minority electrons, first with no applied field and no concentration gradients, hence no current flow. In general terms, it is obvious that in this case, within an elemental volume of the semiconductor, the rate of change of electron density is equal to their rate of generation minus the rate at which they recombine with holes. Thus,

$$\mathrm{d}n/\mathrm{d}t = G - R \qquad (6.53)$$

where n is the minority electron concentration at any time t, and G and R are the generation and recombination rates for the minority carriers.

The generation rate, G, is a function of temperature only, since charge carriers are produced only by thermal excitation in the absence of current flow. Hence, at a constant temperature, $G(T)$ is constant. On the other hand, the recombination rate will be proportional to the product of the local densities of holes and electrons, both of which may vary with time, or

$$R = rnp \qquad (6.54)$$

where r, the constant of proportionality, is the *recombination coefficient*.

In equilibrium, $\mathrm{d}n/\mathrm{d}t = 0$ and the generation and recombination rates will be equal. Thus, we let n_0 and p_0 be the equilibrium densities of minority and majority carriers; Eqs (6.53) and (6.54) give

$$0 = G(T) - rn_0 p_0$$

or

$$G(T) = rn_0 p_0 = rn_\mathrm{i}^2 \qquad (6.55)$$

Now, let us assume that at $t = 0$, δn additional electrons are injected into the material so that the electron density becomes

$$n = n_0 + \delta n \qquad (6.56)$$

The hole concentration must change sympathetically so as to maintain charge neutrality; the density of holes is then

$$p = p_0 + \delta p \qquad (6.57)$$

and the concentration of electrons and holes injected into the volume must be equal, $\delta n = \delta p$. Under these circumstances, the time rate of change of minority-carrier concentration is given by Eq. (6.53), which with Eqs (6.54), (6.56) and (6.57) gives

$$(dn/dt)|_{J=0} = [d(\delta n)/dt]|_{J=0}$$
$$= G(T) - r(n_0 + \delta n)(p_0 + \delta p)$$

We can eliminate the generation term by assuming constant temperature throughout the process and using Eq. (6.55). Then

$$[d(\delta n)/dt]|_{J=0} = -rn_0\delta p - rp_0\delta n - r\delta n\delta p$$

Ignoring second-order quantities, and remembering that in a p-type material $p_0 \gg n_0$, this equation reduces to

$$[d(\delta n)/dt]|_{J=0} = -rp_0\delta n = -\delta n/\tau_{Le} \qquad (6.58)$$

where τ_{Le} is the *lifetime* of the electrons. The physical meaning of τ_{Le} can be made clearer by integrating Eq. (6.58) to give

$$\delta n(t) = \delta n|_{t=0} \exp(-t/\tau_{Le}) \qquad (6.59)$$

Thus, after a sudden change in electron density, $\delta n|_{t=0}$, the carrier density decays exponentially back to its equilibrium value by recombination, in a time comparable to τ_{Le}, as illustrated in Fig. 6.22(a). Thus, τ_{Le} is the average time

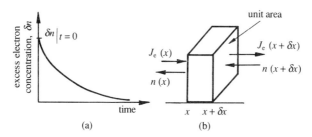

Fig. 6.22 (a) Decay of an excess electron concentration with time, due to recombination; (b) current and electron flow into and out of an elemental volume

that the electrons behave as free charge carriers before recombining or being lost by other mechanisms.

Equation (6.58) is the continuity equation for minority electrons in the absence of current flow. A similar equation can be derived in an exactly analogous manner for the change in excess hole density with time in an n-type semiconductor, viz.

$$d(\delta p)/dt = -\delta p/\tau_{Lh} \tag{6.60}$$

where τ_{Lh} is now the lifetime of minority holes.

6.10.2 Continuity equation with current flow

Let us now consider the situation when electrons are being continuously injected into the volume of p-type material under consideration, when an electron current flows. For the sake of simplicity, let us assume initially that the electron current, J_e, flows in the positive x direction and consider an elemental volume of semiconductor at some position x, of thickness δx and of unit area perpendicular to the direction of current flow, as shown in Fig. 6.22(b). Current flows in the direction shown as a result of electron flow in the negative x-direction.

The current density and electron concentration are related via Eq. (4.28). Hence, the number of electrons entering the volume at $(x + \delta x)$ is

$$n_{x+\delta x} = J_e(x + \delta x)/ev_D$$

and the number leaving at x is

$$n_x = J_e(x)/ev_D$$

Thus, the net increase in the number of electrons in the volume element is

$$n_{x+\delta x} - n_x = \delta n = [J_e(x + \delta x) - J_e(x)]/ev_D$$

$$= \frac{1}{ev_D}\left(J_e(x) + \frac{\partial J_e}{\partial x}\delta x - J_e(x)\right) = \frac{1}{ev_D}\frac{\partial J_e}{\partial x}\delta x$$

Now, $v_D = \partial x/\partial t$ and, in the limit, the equation reduces to

$$\left.\frac{\partial(\delta n)}{\partial t}\right|_{\substack{\text{current} \\ \text{flow}}} = \frac{1}{e}\frac{\partial J_e}{\partial x} \tag{6.61}$$

This is the continuity equation for minority electrons when current is flowing.

If we now allow recombination to take place inside the elemental volume, then the total rate of change of excess minority electrons, δn, due to both drift currents and recombination, is given by Eqs (6.58) and (6.61)

$$\left.\frac{\partial(\delta n)}{\partial t}\right|_{\text{total}} = -\frac{\delta n}{\tau_{Le}} + \frac{1}{e}\frac{\partial J_e}{\partial x} \tag{6.62}$$

There is, of course, a similar equation for the time dependence of the excess hole concentration in an n-type semiconductor. Now, the electron current density J_e is in general due to both drift in an externally applied field and diffusion. In the one-dimensional case, we have seen that

$$J_{ex} = ne\mu_e\mathscr{E}_x + eD_e\partial n/\partial x$$

We substitute this expression in Eq. (6.62), remembering that $n = n_0 + \delta n$ and that n_0 is not a function of x, to give

$$\frac{\partial(\delta n)}{\partial t} = -\frac{\delta n}{\tau_{Le}} + \mu_e \mathscr{E}_x \frac{\partial(\delta n)}{\partial x} + D_e \frac{\partial^2(\delta n)}{\partial x^2} \qquad (6.63)$$

This is the one-dimensional continuity equation for excess minority electrons. The continuity equation for excess minority holes is derived in exactly the same way to give

$$\frac{\partial(\delta p)}{\partial t} = -\frac{\delta p}{\tau_{Lh}} - \mu_h \mathscr{E}_x \frac{\partial(\delta p)}{\partial x} + D_h \frac{\partial^2(\delta p)}{\partial x^2} \qquad (6.64)$$

The continuity equations enable us to calculate the excess density of electrons or holes in time and space. These, together with eqs (6.51) and (6.52) giving the electron and hole current densities, are the basic equations for describing the behaviour of many semiconductor devices, as we shall see.

6.11 Semiconductor measurements

6.11.1 Hall effect measurements

If a current is passed through a semiconductor and a magnetic field, B, is applied at right angles to the direction of current flow, an electric field is induced in a direction mutually perpendicular to B and the direction of current flow. This phenomenon is known as the *Hall effect*. A Hall measurement experimentally confirms the validity of the concept that it is possible for two independent types of charge carrier, electrons and holes, to exist in a semiconductor. We shall also see that a measurement of the Hall coefficient determines not only the sign of the charge of the carriers but also their density. If the conductivity is measured simultaneously by, say, the method to be described in Sec. 6.11.2, this yields the mobility, and the two experiments can thus provide experimental information that is vital for evaluating the subsequent performance of devices made from the material.

Let us, for simplicity, consider a bar of p-type material, as show in Fig. 6.23, and assume the carriers to be positive holes, of charge e, only. We will assume that a current density J_x is produced in the bar by application of an electric field E_x and that a magnetic field of flux density B_z is applied in the z direction. Since the holes are flowing with some drift velocity v_{Dx} under the influence of the applied field, they experience a Lorentz force

$$F_L = ev \times B$$

of magnitude

$$|F_L| = ev_{Dx} B_z \qquad (6.65)$$

and directed in the negative y direction. This force tends to drive holes towards

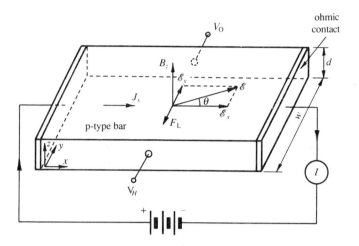

Fig. 6.23 **Hall effect in a p-type semiconductor.**

the front face of the block (see the figure); there is an excess of holes there and a deficiency of holes at the back face. Since there can be no net flow in the y direction, the movement of holes to front and back creates an electric field in the positive y direction, $\mathcal{E}_y$. This produces a force on the holes that exactly compensates for the Lorentz force field and prevents transverse current flowing. Hence, in equilibrium,

$$e\mathcal{E}_y = F_L = ev_{Dx}B_z \qquad (6.66)$$

Also, the current density J_x is given by

$$J_x \simeq pev_{Dx}$$

which can be substituted in (6.66) to give

$$\mathcal{E}_y = J_x B_z / pe \qquad (6.67)$$

Thus, if voltage probes are attached to front and back faces, assumed separated by a distance w, a voltage

$$V_H = \mathcal{E}_y w$$

can be measured.

The Hall coefficient, R_H, defined from Eq. (6.67), is

$$R_H = \mathcal{E}_y / J_x B_z = 1/pe$$

Alternatively, if the current through the bar of thickness d is I, then

$$R_H = \frac{V_H}{wIB_z} wd = \frac{V_H d}{IB_z} = \frac{1}{pe} \qquad (6.68)$$

It will be easily verified that for n-type material, where the current is carried by majority electrons, the polarity of the Hall voltage, V_H, is reversed and

$$R_{He} = -1/ne \qquad (6.69)$$

We see that the measurement of current and the magnitude and sign of the Hall voltage for any given magnetic flux density give the sign of the charge carriers, that is, it shows whether the doped semiconductor is n- or p-type, together with the density of the majority carriers.

It will be noticed that the net electric field in the semiconductor, which is the vector sum of $\mathscr{E}_x$ and $\mathscr{E}_y$, is not directed along the axis but is at some angle θ to it. From Fig. 6.23 it will be seen that θ, called the *Hall angle*, is given by

$$\tan \theta = \mathscr{E}_y/\mathscr{E}_x \qquad (6.70)$$

which, using Eq. (6.67) and remembering that $J_x = \sigma \mathscr{E}_x$, gives

$$\tan \theta = \frac{J_x B_z}{pe} \frac{\sigma}{J_x} = \mu_h B_z \qquad (6.71)$$

or alternatively

$$\mu_h = R_H \sigma \qquad (6.72)$$

Thus, provided the assumptions made in this simple model are correct, a simultaneous measurement of R_H and σ can lead to an experimental value for the carrier drift mobility.

We have tacitly assumed that all the carriers move with an average drift velocity, v_D, and hence have the same collision time. If the problem is treated on a more rigorous statistical basis, it can be shown that, if the principal scattering mechanism is thermal scattering, then an additional numerical factor is introduced into the expression for the Hall coefficient and

$$R_H = \frac{3\pi}{8} \frac{1}{pe} \qquad (6.73)$$

If other types of scattering predominate, the numerical factor has a different value. We see that, in this case, Eq. (6.70) becomes

$$\tan \theta = \frac{3\pi}{8} \mu_h B_z = \mu_H B_z \qquad (6.74)$$

where μ_H is known as the *Hall mobility*. It is related to the drift velocity via Eq. (6.74), i.e.

$$\mu_H = \frac{\tan \theta}{B_z} = \frac{3\pi}{8} \mu_h \qquad (6.75$$

and is equal to the product of Hall coefficient and conductivity, i.e.

$$\mu_H = R_H \sigma \qquad (6.76)$$

We have further assumed that the conduction process is by means of one type of carrier only. In some semiconductors, for example, in intrinsic or nearly fully compensated material, the assumption is not valid and the Hall coefficient must be modified to account for the presence of two types of charge carrier, as follows.

Both types of carrier will drift under the influence of the applied field $\mathscr{E}_x$ with drift velocities

$$v_{Dh} = \mu_H \mathscr{E}_x \qquad \text{in the positive } x \text{ direction}$$

and

$$v_{De} = \mu_H \mathscr{E}_x \qquad \text{in the negative } x \text{ direction}$$

using the same coordinate system as in Fig. 6.23. As a consequence, the holes and electrons each experience a Lorentz force given by

$$F_h = e(v_{Dh} \times B) = -ev_{Dh}B_z$$

and

$$F_e = e(v_{De} \times B) = -ev_{De}B_z$$

Both deflecting forces are in the same direction, deflecting electrons and holes to the front face, as before. The carriers recombine at the surface and the net charge there produces an electric field in the y direction, $\mathscr{E}_y$. In equilibrium, the current due to the deflected electrons and holes must be exactly cancelled out by current flowing in the opposite direction due to the Hall field. If the transverse velocities of the deflected carriers are v_{yh} and v_{ye}, the current in the transverse direction due to deflection of electrons and holes by the magnetic field is $(epv_{yh} - env_{ye})$ and, therefore, in equilibrium

$$\sigma \mathscr{E}_y = e(pv_{yh} - nv_{ye}) \tag{6.77}$$

Expressions for the transverse velocities can be obtained from the Hall angle equations, Eqs (6.70) and (6.71), which give

$$v_{yh} = \mu_h(\mathscr{E}_x \tan \theta) = \mu_h(\mathscr{E}_x \mu_h B_z)$$

and, similarly

$$v_{ye} = \mu_e^2 \mathscr{E}_x B_z$$

These are substituted in Eq. (6.77) to give

$$\sigma \mathscr{E}_y = e(p\mu_h^2 - n\mu_e^2)\mathscr{E}_x B_z \tag{6.78}$$

The generalized version of Ohm's law is used to eliminate $\mathscr{E}_x$ from this equation, which can then be rearranged to give the Hall coefficient

$$R_H = \frac{\mathscr{E}_y}{J_x B_z} = \frac{e(p\mu_h^2 - n\mu_e^2)}{\sigma^2}$$

Finally, we include the general expression for the conductivity, given in

Eq. (6.38), to give

$$R_H = \frac{p\mu_h^2 - n\mu_e^2}{e(p\mu_h + n\mu_e)^2} \tag{6.79}$$

Again, a more rigorous treatment, which does not assume that the time between scattering events is the same for all carriers, results in the introducing of an additional numerical factor in the Hall coefficient expression. For example, if lattice scattering is the principal loss mechanism

$$R_H = \frac{3\pi}{8} \frac{p\mu_h^2 - n\mu_e^2}{e(p\mu_h + n\mu_e)^2} \tag{6.80}$$

Comparison of this expression with Eqs (6.68) and (6.69) shows that the Hall coefficient, and hence the Hall voltage, are generally smaller for near-intrinsic materials than for the more highly doped extrinsic materials. Notice, too, that measurement of the Hall coefficient of the intrinsic material does not lead immediately to values for the carrier concentrations, as it does in the extrinsic material.

6.11.2 Four-point probe method for conductivity measurements

In principle, the conductivity of a semiconductor may be found by measuring the current drawn when a voltage is applied between two contacts formed at the ends of a bar of the material. In practice, such a measurement is not straightforward because of the difficulty of making uniform ohmic contacts to the sample. Further, it is not always convenient to produce a specimen of known dimensions for a conductivity measurement. The four-point probe method eliminates the difficulties referred to; it is used to measure, non-destructively and accurately, the conductivity of ingots or slices, both thick and thin, of semiconductor crystals.

The probe head basically employs four spring-loaded, equispaced needles, which make contact with a plane lapped surface on the specimen, as shown in Fig. 6.24. A stabilized current is passed through the outer pair of probes, A and

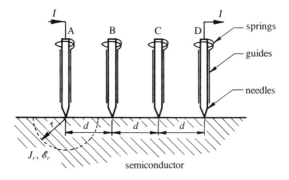

Fig. 6.24 Four-point conductivity measurement.

D, and the voltage, V, between the other two, B and C, is measured by, say, a digital voltmeter, which draws negligible current.

Let us first consider the sample dimensions to be large compared with the probe spacing, d. We can treat the source current at A and the sink at D independently, using the principle of superposition to find their combined effect later. Thus, the current density, J_r, at radius r from A, due to the current I entering at A, is

$$J_r = I/(2\pi r^2)$$

since current only flows in the bottom half-plane. We then use Ohm's law to find the corresponding electric field at r, $\mathscr{E}_r$:

$$\mathscr{E}_r = J/\sigma = I/(2\pi\sigma r^2)$$

where σ is the conductivity of the material.

By definition, the floating potential at some radius a is then

$$V_a = \int_{-\infty}^{a} \mathscr{E}_r \, dr = -\frac{I}{2\pi\sigma} \int_{-\infty}^{a} \frac{1}{r^2} \, dr = \frac{I}{2\pi\sigma a} \tag{6.81}$$

The potential difference between probes B and C due to the current source at A is then

$$V_{BC} = \frac{I}{2\pi\sigma d} - \frac{I}{2\pi\sigma(2d)} = \frac{I}{4\pi\sigma d} \tag{6.82}$$

If, now, we consider the current leaving at D separately, there exists a potential difference between B and C, which, because of symmetry, is the same as that due to the current source above, given in Eq. (6.82). By superposition, the total voltage measured between B and C is thus twice that for current source or sink alone, or

$$V = \frac{I}{(2\pi\sigma d)}$$

from which the conductivity is obtained as

$$\sigma = \frac{1}{2\pi d} \frac{I}{V} \tag{6.83}$$

This equation is valid only for materials whose dimensions are large compared with the probe spacing, d. If this is not the case, for example, when measurements are made on thin slices or small dice of material, then a correction factor, F, is introduced and the conductivity is given by

$$\sigma = \sigma_0 F$$

where σ_0 is the uncorrected value of σ, found using Eq. (6.83). Factor F is a function of slice thickness, breath and width, normalized to the probe spacing; the reader is referred to the more specialized literature for the derivation and tabulation of values of F for various slice geometries.*

6.11.3 Measurement of minority-carrier lifetime and mobility

It is possible to measure the drift mobility, μ, and the lifetime of minority carriers, τ_L, by a simple technique, the basic arrangement of which is as shown in Fig. 6.25. A bar of semiconducting crystal, length l, has ohmic contacts

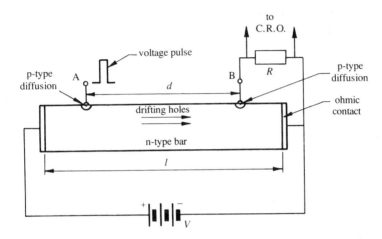

Fig. 6.25   Minority-carrier lifetime and mobility measurement.

formed at each end and a voltage V is applied between them. Two small pn junctions distance d apart are formed on one face of the bar as shown, by locally diffusing p-type impurity into the bar. We shall see later that if a voltage is applied at A such that the p-type material is biased positively with respect to the n-type material, minority holes can be injected into the bar at that point. These excess minority carriers drift to the other junction under the influence of the electric field due to the applied voltage, V. If the junction at B is biased in the reverse direction to that at A, any excess of holes at B will be collected, as explained below.

   Pulses of voltage are applied at A in a direction such that minority carriers are injected during the pulse. The voltage across resistor R connected to B is displayed on an oscilloscope, which thus monitors the collected current as a function of time. The time of flight of the minority carriers between injection at A and extraction at B, t_0, can be measured if both sets of pulses are displayed simultaneously on the oscilloscope. There will be some attenuation and spreading of the received pulse at B due to recombination and diffusion of the carriers as they traverse the bar.

*L. B. Valdes (1954) Resistivity measurements on Ge for transistors, *Proc. IRE*, **42**, 42.

The drift velocity of minority carriers is then found using

$$v_D = d/t_0$$

and hence the drift mobility can be found from

$$\mu E = \mu V/l = d/t_0$$

or

$$\mu = ld/t_0 V \tag{6.84}$$

The same experimental arrangement can be used to estimate the lifetime of the minority carriers, τ_L. If the excess hole density at A is δp_0 at $t=0$, when the voltage pulse is applied there, we know that the hole density will decay back to its equilibrium value exponentially with time, with a time constant τ_{Lh}, and from Eq. (6.59)

$$\delta p = \delta p_0 \exp(-t/\tau_{Lh})$$

Since the time of flight of holes is t_0, the excess minority-carrier density collected at B is

$$\delta p = \delta p_0 \exp(-t_0/\tau_{Lh}) \tag{6.85}$$

Now, if we assume that the height of the received pulse is proportional to the excess hole density at the collector, then, varying the transit time, t_0, by changing the voltage, V, and plotting the pulse height and transit time semi-logarithmically allows the lifetime τ_{Lh} to be estimated. It should be noted that the lifetime measured is influenced principally by recombination at the surface of the bar and may thus not be the same as the lifetime in bulk material.

It is also theoretically possible to estimate the diffusion constant for holes, D_h, by noticing the dispersion of arrival times of the excess carriers as the time of flight is varied, but in practice such a measurement does not yield accurate results.

Problems

1. Find the fraction of the electrons in the valence band of intrinsic germanium that can be thermally promoted across the forbidden energy gap of 0.72 eV into the conduction band at (a) 30 K, (b) 300 K and (c) the melting point, 937 °C.

Ans. $10^{-60}, 10^{-6}, 10^{-1.5}$

2. Estimate the fraction of electrons in the conduction band at room temperature of (a) germanium ($E_g = 0.72$ eV), (b) silicon ($E_g = 1.10$ eV) and (c) diamond ($E_g = 5.6$ eV). What is the significance of these results?

Ans. $10^{-6}, 10^{-9.3}, 10^{-47}$

3. The resistivity of intrinsic silicon at 270°C is 3000 Ω m. Calculate the intrinsic carrier density. Assume $\mu_e = 0.17$ and $\mu_h = 0.035\,\mathrm{m^2\,V^{-1}\,s^{-1}}$.

Ans. $1.1 \times 10^{16}\,\mathrm{m^{-3}}$

4. Compare the drift velocity of an electron moving in a field of $10\,\mathrm{kVm^{-1}}$ in intrinsic germanium with that of one that has moved through a distance of $10\,\mathrm{mm}$ in this field in a vacuum. Assume $\mu_e = 0.39\,\mathrm{m^2\,V^{-1}\,s^{-1}}$.

Ans. $3.9 \times 10^3\,\mathrm{m^2\,s^{-1}}, 5.93 \times 10^6\,\mathrm{m^2\,s^{-1}}$

5. Given that the mobilities in germanium vary with temperature as $\mu_e \propto T^{-1.6}$ and $\mu_h \propto T^{-2.3}$, show that the conductivity of intrinsic germanium as a function of temperature is of the form

$$\sigma = C_1 \exp\left(\frac{-3900}{T}\right)\left[C_2\left(\frac{290}{T}\right)^{0.1} + C_3\left(\frac{290}{T}\right)^{0.8}\right]$$

where the C's are constants to be found and $\mu_e = 0.38\,\mathrm{m^2\,V^{-1}\,s^{-1}}, \mu_h = 0.18\,\mathrm{m^2\,V^{-1}\,s^{-1}}$ and $E_g = 0.67\,\mathrm{eV}$ are assumed. Hence estimate the conductivity of germanium at its melting point, $958\,°\mathrm{C}$.

6. Measurements of the conductivity of germanium have shown that, in the intrinsic range, the conductivity varies with temperature as $\exp(-4350/T)$. Use this information to estimate the energy gap width in germanium.

Ans. $0.67\,\mathrm{eV}$

7. Show that a semiconductor has a minimum conductivity at a given temperature when

$$n = n_i(\mu_h/\mu_e)^{1/2} \qquad \text{and} \qquad p = n_i(\mu_e/\mu_h)^{1/2}$$

Find the intrinsic and minimum conductivities for germanium at a temperature such that $n_i = 2.5 \times 10^{19}\,\mathrm{m^{-3}}$, $\mu_e = 0.38\,\mathrm{m^2\,V^{-1}\,s^{-1}}$, $\mu_h = 0.19\,\mathrm{m^2\,V^{-1}\,s^{-1}}$. For what values of n and p (other than $n = p = n_i$) does this crystal have a conductivity equal to the intrinsic conductivity?

Ans. $2.28\,\mathrm{S\,m^{-1}}, 2.15\,\mathrm{S\,m^{-1}}, 1.25 \times 10^{19}, 5 \times 10^{19}\,\mathrm{m^{-3}}$

8. Show that the minimum conductivity attainable by a particular semiconductor at a given temperature is of the form

$$\sigma_{\min} = C(\mu_e\mu_h)^{1/2}$$

where μ_e and μ_h are the carrier mobilities and C is a constant to be determined. Why does the conductivity not possess its minimum value when the semiconductor is intrinsic? How might the minimum conductivity be realized in practice?

Ans. $C = 2en_i$

9. A current density of 10^3 A m^{-2} flows through an n-type germanium crystal that has resistivity $0.05\,\Omega$ m. Calculate the time taken for electrons in the material to travel $50\,\mu$m.

    Ans. $2.5\,\mu$s

10. Intrinsic silicon has a resistivity of $2000\,\Omega$ m at room temperature and the density of conduction electrons is 1.4×10^{16} m^{-3}. Calculate the resistivities of samples containing acceptor concentrations of 10^{21} and 10^{23} m^{-3}. Assume that μ_h remains as for intrinsic silicon and that $\mu_h = 0.25\mu_e$.

    Ans. $0.135\,\Omega$ m, $0.001\,35\,\Omega$ m

11. A rod of p-type germanium, 6 mm long, 1 mm wide and 0.5 mm thick, has an electrical resistance of $120\,\Omega$. What is the impurity concentration? Assume $\mu_h = 0.19$ m^2 V^{-1} s^{-1}, $\mu_e = 0.39$ m^2 V^{-1} s^{-1} and $n_i = 2.5 \times 10^{19}$ m^{-3}. What proportion of the conductivity is due to electrons in the conduction band?

    Ans. 3.29×10^{21} m^{-3}, 8.4×10^3

12. A 10 mm $\times$ 10 mm $\times$ 10 mm cube of silicon at room temperature has 10^{19} atom m^{-3} of gallium (p-type) impurities and 1.5×10^{19} atom m^{-3} of arsenic (n-type) impurities in the material. Determine the resistance of the bar between any two opposite faces. Assume $n_i = 1.5 \times 10^{16}$ m^{-3}, $\mu_e = 0.12$ m^2 V^{-1} s^{-1}, $\mu_h = 0.05$ m^2 V^{-1} s^{-1}.

    Ans. $1.04\,$kΩ

13. A resistor made from intrinsic silicon has a resistance of $2500\,\Omega$ at $20°$C, dropping to 1 per cent of this resistance at $100°$C. Estimate the energy gap for the material. Suggest a practical application for such a device.

    Ans. $1.08\,$eV

14. The resistance of a certain sample of intrinsic semiconductor is measured at 400 and 700 K and the ratio of the resistance is $100:1$. Assuming that the electron and hole mobilities vary with temperature as $T^{-3/2}$ over the measured range, determine the value of the forbidden gap energy, E_g.

    Ans. $0.88\,$eV

15. A sample of germanium is doped to the extent of 10^{20} donor atom m^{-3} and 7×10^{19} acceptor atom m^{-3}. At the temperature of the sample, the resistivity of intrinsic germanium is $0.6\,\Omega$ m. If the applied electric field is 200 V m^{-1}, find the total conduction current density. Assume $\mu_e = 0.38$ m^2 V^{-1} s^{-1} and $\mu_h = 0.18$ m^2 V^{-1} s^{-1}.

    Ans. 524 A m^{-2}

16. Show that for a two-carrier semiconductor the Hall coefficient is given by

$$R_H = \frac{1}{e} \frac{p\mu_h^2 - n\mu_e^2}{(p\mu_h + n\mu_e)^2}$$

What is the basic difficulty in applying this result to determine the properties of a two-carrier semiconductor?

17. Show that the ratio of the Hall coefficient in a p-type doped probe, R_{Hp}, to that for a probe made from the same intrinsic semiconductor, R_{Hi}, is given by

$$R_{Hp}/R_{Hi} = (n_i/p)f(\mu_e, \mu_h)$$

where $f(\mu_e, \mu_h)$ is a function of the mobilities to be determined. Comment on this result.

The Hall voltage obtained for a probe, made initially from $2000\,\Omega\,m$ intrinsic semiconductor, when placed in a magnetic field and with a certain current passing through it, is found to be $100\,mV$. After doping the probe material and subjecting it to the same conditions of field and current as before, the Hall voltage is found to be $10\,\mu V$ and is of changed sign. Find the carrier density and type in the doped material, assuming carrier mobilities of 0.13 and $0.05\,m^2\,V^{-1}\,s^{-1}$.

Ans. $3.9 \times 10^{20}\,m^{-3}$, p-type

18. A possible Hall effect multiplier might be constructed using an n-type semiconducting bar with suitable contacts, placed inside an air-cored solenoid. Briefly explain how this might operate.

If the coil has 1000 turns, is 100 mm long and has a current of 10 mA passing through it, and the Hall sample has a thickness of 0.1 mm, a conductivity of $100\,S\,m^{-1}$ and a current of 20 mA flowing along its length, find the resulting Hall voltage, assuming the sample is oriented so as to maximize this voltage. Assume also a majority carrier mobility of $7\,m^2\,V^{-1}\,s^{-1}$.

Ans. $1.76\,mV$

19. The resistivity of a doped silicon crystal is $9.27 \times 10^{-3}\,\Omega\,m$ and the Hall coefficient is $3.84 \times 10^{-4}\,m^3\,C^{-1}$. Assuming that conduction is by a single type of charge carrier, calculate the density and mobility of the carrier.

Ans. $1.6 \times 10^{22}\,m^{-3}$, $0.0414\,m^2\,V^{-1}\,s^{-1}$

20. A rectangular, n-type germanium bar has a thickness of 2 mm. A current of 10 mA passes along the bar and a field of 0.1 T is applied perpendicular to the current flow. The Hall voltage developed is 1.0 mV. Calculate the Hall constant and the electron density in the semiconductor. Find the Hall angle, assuming a mobility of $0.36\,m^2\,V^{-1}\,s^{-1}$ for the carriers.

Ans. $2 \times 10^{-3}\,m^3\,C^{-1}$, $3 \times 10^{21}\,m^{-3}$, $2.1°$

21. A bar of n-type germanium $10\,\text{mm} \times 1\,\text{mm} \times 1\,\text{mm}$ is mounted in a magnetic field of 0.2 T. The electron density in the bar is $7 \times 10^{21}\,\text{m}^{-3}$. If 1 mV is applied across the long ends of the bar, determine the current through the bar, the Hall coefficient and the voltage between Hall electrodes placed across the short dimensions of the bar. Assume $\mu_e = 0.39\,\text{m}^2\,\text{V}^{-1}\,\text{s}^{-1}$.

Ans. $43.6\,\mu\text{A}, 8.9 \times 10^{-4}\,\text{m}^3\,\text{C}^{-1}, 0.78\,\mu\text{V}$

22. A certain n-type semiconductor sample has a conductivity that is found to be $1.12 \times 10^4\,\text{S}\,\text{m}^{-1}$. When a transverse field of 0.1 T is applied, the magnitude of the Hall coefficient is found to be $6.25 \times 10^{-4}\,\text{C}^{-1}\,\text{m}^{-3}$. Estimate the Hall angle, θ, for the material and comment on the result.

Ans. $35°$

23. An initially intrinsic semiconductor is lightly doped with acceptors and donors to produce slightly p-type material of conductivity $3.72 \times 10^{-4}\,\text{S}\,\text{m}^{-1}$. Assuming an intrinsic carrier density of $1.47 \times 10^{16}\,\text{m}^{-3}$ and carrier mobilities of 0.12 and $0.05\,\text{m}^2\,\text{V}^{-1}\,\text{s}^{-1}$, find the difference between the densities of the acceptors and donors, commenting on the result.

Ans. $2 \times 10^{16}\,\text{m}^{-3}$

24. The resistance of a certain intrinsic silicon slice changes by a factor of 1000 when it is doped p-type. After compensation doping, the slice is found to be n-type and the resistance is then half that of the p-type slice. Assuming an intrinsic density of $1.38 \times 10^{16}\,\text{cm}^{-3}$ and carrier mobilities of 0.13 and 0.05 $\text{m}^2\,\text{V}^{-1}\,\text{s}^{-1}$, estimate the doping concentrations (a) after the p-type doping and (b) after the compensation doping.

Ans. (a) $5 \times 10^{19}\,\text{m}^{-3}$, (b) $8.8 \times 10^{19}\,\text{m}^{-3}$

25. A particular semiconductor, having mobilities of 0.13 and $0.05\,\text{m}^2\,\text{V}^{-1}\,\text{s}^{-1}$ and intrinsic carrier density of $1.38 \times 10^{16}\,\text{m}^{-3}$, is doped with $2.01 \times 10^{18}\,\text{m}^{-3}$ acceptors and $2.00 \times 10^{16}\,\text{m}^{-3}$ donors in an attempt to make its behaviour near-intrinsic. Show how successful or otherwise the attempt has been by estimating the conductivities of the intrinsic and compensation doped material.

Ans. $\sigma_p/\sigma_i = 0.90$

26. A germanium sample, with carrier mobilities of 0.38 and $0.18\,\text{m}^2\,\text{V}^{-1}\,\text{s}^{-1}$, is compensation doped with a donor density of $10^{21}\,\text{m}^{-3}$ and an acceptor density of $9 \times 10^{20}\,\text{m}^{-3}$. Estimate the conductivity of the semiconductor at (a) $20°\text{C}$ when the intrinsic density, $n_i = 2.5 \times 10^{19}\,\text{m}^{-3}$ and (b) $100°\text{C}$ when $n_i = 4.1 \times 10^{20}\,\text{m}^{-3}$.

Ans. (a) $6.6\,\text{S}\,\text{m}^{-1}$, (b) $38.6\,\text{S}\,\text{m}^{-1}$

27. When a particular intrinsic semiconductor is doped n-type, its resistance at 20°C drops to 1 per cent of its original value. Assuming a gap energy of 1.1 eV and carrier mobilities of 0.13 and $0.05 \, \text{m}^2 \, \text{V}^{-1} \, \text{s}^{-1}$, estimate the movement of the Fermi energy, as a proportion of the gap energy, when the intrinsic material is doped.

Ans. 11 per cent

28. Supposing that an intrinsic semiconductor is lightly doped n-type such that E_F moves from its intrinsic position by an energy aE_g, where a is a constant, derive an expression for the ratio of the electron density in the doped semiconductor to that in the original intrinsic semiconductor.

When a particular intrinsic material with $E_g = 1.1$ eV and carrier mobilities of 0.15 and $0.05 \, \text{m}^2 \, \text{V}^{-1} \, \text{s}^{-1}$ is lightly doped n-type, its conductivity at $T = 290$ K changes by a factor of 10^6. Estimate the corresponding change in the Fermi level.

Ans. $n/n_i = \exp(aE_g/kT)$, 0.35 eV

29. Describe in general terms the movement of the Fermi level when an initially intrinsic semiconductor is doped p-type.

Use a simplified energy band model to show that if, during such a doping operation, the Fermi level moves by aE_g, where a is a constant and E_g is the gap energy, the hole density, p, is given by

$$p \propto \exp[A(1-2a)]$$

where A is to be determined.

Samples of a particular semiconductor having a gap energy of 1.1 eV are p-type doped to varying degrees such that the constant a varies from 0.1 to 0.4. What is the ratio of the maximum to minimum conductivities of the samples at 290 K?

Ans. 5.34×10^5

7 Junction diodes

7.1 Introduction

A perfectly uniform semiconductor crystal has limited device possibilities. For instance, it may be incorporated in photocells, temperature-sensitive resistors and some bulk-effect high-frequency generators, which will be discussed later. Apart from these specialized applications, the most useful and interesting devices make use of semiconductors in which the type of impurity changes within the single-crystal material; for example, it may be p-type in one section and n-type elsewhere.

A pn *junction* may be formed from a single-crystal intrinsic semiconductor by doping part of it with acceptor impurities and the remainder with donors. We will establish that such junctions can form the basis of very efficient rectifiers. Further, and more important since many other solid-state devices contain several such junctions, if the mechanism of current flow in a simple single pn junction is understood, this will lead to a clearer conception of the operation of the more elaborate structures.

7.2 The pn junction in equilibrium with no applied voltage

A junction between p-type and n-type material may be fabricated in a variety of ways, as we shall see later. The precise distance over which the change from p- to n-type semiconductor occurs varies with fabrication technique, but an essential feature of all junctions is that the change in impurity concentration takes place in a very short length, typically much less than 1 μm. If the junction extends over a longer distance, then the characteristic behaviour to be described does not occur; the material then acts as an inhomogeneous conductor containing two carrier types, which vary in concentration from point to point.

For the sake of mathematical simplicity, an *abrupt* pn junction in which there occurs a sudden step change in impurity type at the junction plane will usually be discussed. Although this model is adequate for describing junctions fabricated in certain ways, for example, junctions in alloyed and some epitaxial diodes (see Sec. 7.13), it is not universally applicable to all junctions. For instance, a diffused junction may be *graded*, in which case the donor and

150

acceptor concentrations are functions of distance across the junction. Then the acceptor density, N_a, gradually decreases and the donor density, N_d, gradually increases as the junction is approached from the p-side until, when the junction is reached, $N_a = N_d$; thereafter, N_d becomes much larger and N_a decreases to zero. Such junctions are obviously more difficult to analyse than the abrupt junction, even if they are assumed to be *linear graded* and the net doping concentration $(N_a - N_d)$ is assumed to be a linear function of distance through the junction. Fortunately, however, many of the results obtainable for abrupt junctions also apply to graded junctions. Whenever this does not apply, for example when estimating junction capacitance, attention will be drawn to the fact.

It is not necessary for the abrupt junction to be symmetrical; indeed it is usually desirable, as we shall discover, that the doping concentrations at either side of the junction are dissimilar. Let us assume that the acceptor concentration, N_a, is somewhat greater than the donor concentration, N_d. Typical doping concentrations in germanium might be one acceptor per 10^7 germanium atoms, giving a conductivity in the p-region, σ_p, of order $10^4 \, \mathrm{S\,m^{-1}}$, the corresponding values in the n-region being, say, $1:10^5$ and $\sigma_n \approx 100 \, \mathrm{S\,m^{-1}}$. The majority- and minority-carrier densities in the extrinsic regions will then be as depicted in Fig. 7.1.

It will be noticed that a one-dimensional model of the pn junction has been adopted, in which current is only allowed to flow in the x direction, perpendicular to the junction. Whereas this model is not strictly correct, since in more realistic geometries the current flow is three-dimensional, as we shall

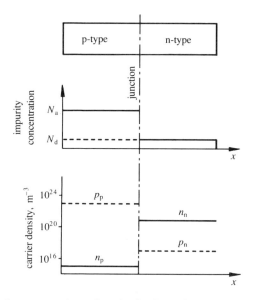

Fig. 7.1 Impurity concentration and carrier density variations in an abrupt pn junction.

see, it is not difficult to imagine a slice through a practically realizable junction that behaves in a very similar way to the model we have assumed.

The carrier density concentrations p_p, n_p, n_n and p_n, depicted in Fig. 7.1, apply only at relatively large distances from the junction. At the junction, these densities are modified since there is a tendency for holes to diffuse into the n-type material and for electrons to diffuse into the p-type material because of the severe concentration gradients appearing there. Thus, although the impurity concentration has been assumed to have an abrupt, step change, the carrier densities cannot change so abruptly but vary instead over a small distance either side of the junction, shown to an unrealistically large scale as d_1–d_2 in Fig. 7.2(a).

Even after the initial diffusion of majority carriers as described, large concentration gradients still exist but further diffusion is prevented in equilibrium by an electric field being set up in the junction region by the

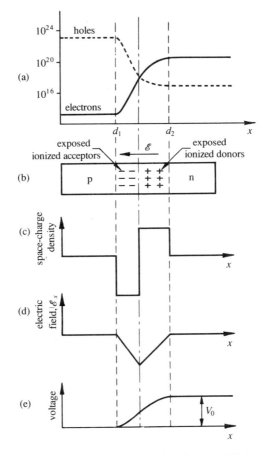

Fig. 7.2   Depletion layer in a pn junction in equilibrium.

following mechanism. The majority holes diffusing out of the p-region leave behind negatively charged acceptor atoms bound to the lattice, thus exposing negative space charge in a previously neutral region. Similarly, electrons diffusing from the n-region expose positively ionized donor atoms, and a double space-charge layer builds up at the junction, as shown in Figs 7.2(b) and (c). It will be noticed that the space-charge layers are of opposite sign to the majority carriers diffusing into them, which tends to reduce the diffusion rate; in other words the double space-charge layer causes an electric field to be set up across the junction directed from n- to p-regions, which is in such a direction to inhibit diffusion of majority electrons and holes, as shown in Figs 7.2(b) and (d). Thus, diffusion continues until the built-in electric field is great enough to inhibit the process; no current flows across the junction and the system is in equilibrium.

The double layer of charge, d_1-d_2, is known variously as the *space-charge layer*, for obvious reasons, the *depletion layer*, so-called since this region is depleted of its usual complement of majority carriers, or the *barrier layer*, since the electric field across it is an electrical barrier, which prevents diffusion currents flowing. The depletion layer is, typically, of order 1 μm thick.

As a consequence of the induced electric field across the depletion layer, an electrostatic potential difference is established between p- and n-regions, which is called the *contact* or *diffusion* potential, V_0, as shown in Fig. 7.2(e). The magnitude of the contact potential varies with doping levels and temperature, as we shall see, but it is typically of order 1 V. An expression for the contact potential can be obtained by finding the electric field at the junction from the continuity equation and integrating across the depletion layer, as follows.

Consider, for example, the continuity equation for holes in the depletion layer, Eq. (6.64). If we assume that the junction is in equilibrium, $\partial/\partial t = 0$, and if recombination in the depletion layer is ignored because of its extreme thinness, the first term on the right-hand side of the equation is zero and on integration Eq. (6.64) reduces to

$$\mathscr{E}_x = \frac{D_h}{\mu_h} \frac{1}{(\delta p)} \frac{d(\delta p)}{dx} \tag{7.1}$$

The diffusion forces are exactly balanced by electrostatic forces, as described. Using the Einstein relationship, Eq. (6.50), the equation for $\mathscr{E}_x$ becomes

$$\mathscr{E}_x = \frac{kT}{e} \frac{1}{(\delta p)} \frac{d(\delta p)}{dx} \tag{7.2}$$

Equation (7.2) is rearranged and integrated across the depletion layer to give

$$-e \int_{d_1}^{d_2} \mathscr{E}_x \, dx = -kT \int_{p_p}^{p_n} \frac{d(\delta p)}{\delta p}$$

or

$$eV_{d_1-d_2} = eV_0 = -kT \log_e (p_n/p_p)$$

which gives

$$p_p = p_n \exp(eV_0/kT) \qquad (7.3)$$

Thus, if the concentration of holes at either side of the junction is known, the contact potential, V_0, can be calculated at any given temperature T.

By considering the continuity equation for electrons in the depletion layer, we arrive at a similar expression for the electron concentrations in equilibrium, viz.

$$n_n = n_p \exp(eV_0/kT) \qquad (7.4)$$

Let us now consider how the pn junction can be represented using the band-structure concept of a semiconductor. First, it will be necessary to discuss what is meant by equilibrium in the context of a junction between two semiconductor types. In classical electrostatics, two bodies that are in equilibrium have the same electrostatic potential. However, in semiconductors or metals, considered on an atomic scale, electronic equilibrium occurs only between two bodies in contact when there is no net current flow between them. This does not necessarily imply that they must each have the same electrostatic potential; on the contrary the condition for equilibrium in such a case is that the Fermi level, E_F, in each material must be at the same energy level.

This statement can be verified by considering the simple case of two metals, A and B, in contact. The metals can be represented by the box model described in Chapter 4. The workfunctions of the metals, ϕ_A and ϕ_B, correspond to the minimum height of the potential energy barrier at the respective metal surface. The situations (a) before the metals come into contact, (b) immediately on contact and (c) a short while after contact are shown diagrammatically in Fig. 7.3(a)–(c). Immediately on contact, electrons spill over from the material with the highest Fermi level, A, to occupy empty levels in the other material. Thus, metal B becomes charged negatively and A positively. This charging process changes the potential of B relative to A and all electron energy levels in metal A, including E_{FA}, are lowered while those in B are raised. This process continues until the Fermi levels are aligned, as in Fig. 7.3(c), whereupon no further electron current flows. It is thus the Fermi level, rather than the electrostatic potential, that remains constant across the boundary between the two metals in contact.

The constancy of the Fermi level across a junction is quite general and applies equally well to semiconductors in equilibrium. Let us first of all consider two separate n- and p-type doped crystals of the same material, as depicted in Fig. 7.4(a), which are brought together to form a pn junction, as in Fig. 7.4(b). It should be realized that this process is not physically possible

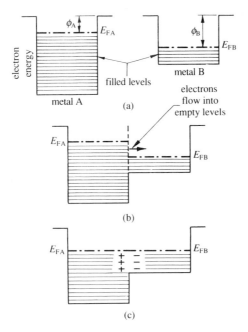

Fig. 7.3 **Energy level diagrams of two metals in contact: (a) before contact, (b) immediately contact is made and (c) in equilibrium.**

since the pn junction is usually formed in single-crystal material, but it will be instructive to develop the argument assuming for the moment that it is. Immediately contact is made, electron current flows as electrons transfer from the n-type material into empty levels in the conduction band of the p-type material, as in Fig. 7.4(b), and similarly holes transfer from the valence band of the p-type into that of the n-type material. The process stops only when the Fermi level is continuous through both materials, at which stage no net current flows across the junction and equilibrium is attained. Since the Fermi level in each material is fixed relative to its respective band structure, alignment of the Fermi levels is only achieved if all energy levels in one region move relative to those in the other; in other words, if each side of the junction takes up a different electrostatic potential. Thus, holes diffusing from the p-type region leave it negatively charged, raising all energy levels, and similarly electrons migrating in the opposite direction cause all levels in the n-type material to be lowered until, when the difference of potential existing between the two sections is equal to the contact potential, V_0, the Fermi level is continuous across the junction and equilibrium is established, as shown in Fig. 7.4(c). When these conditions prevail, a potential hill prevents electron flow from n- to p-regions and a similar barrier prevents hole flow in the opposite direction. (Note that the 'hill' in the latter case is still a barrier to holes

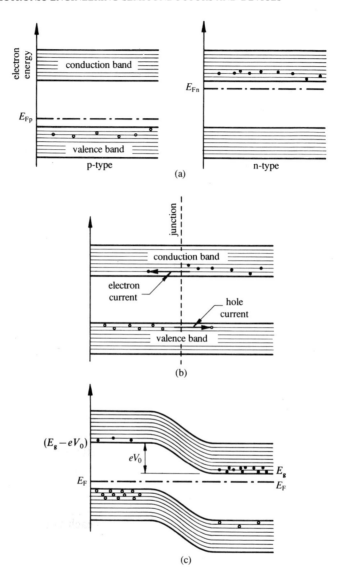

Fig. 7.4 Hypothetical contact between p- and n-type semiconductors. Band structure: (a) before contact, (b) immediately after contact and (c) in equilibrium.

even though sloping downwards, since the energy band diagrams are conventionally drawn for electron energies.)

Using the notation for the energy levels as shown in Fig. 7.4(c), the number of electrons in the conduction band of the n-region, n_n, is (from Eq. (6.13))

$$n_n = N_c \exp[-(E_g - E_F)/kT] \tag{7.5}$$

and the number of electrons in the p-region, n_p, is

$$n_p = N_c \exp\{-[(E_g + eV_0) - E_F]/kT\} \tag{7.6}$$

Dividing the two equations, we obtain

$$n_n/n_p = \exp(eV_0/kT) \tag{7.7}$$

which agrees with the expression obtained previously, Eq. (7.4). The contact potential can be obtained by rearrangement and, using Eq. (6.19), is

$$V_0 = \frac{kT}{e} \log_e\left(\frac{n_n}{n_p}\right) = \frac{kT}{e} \log_e\left(\frac{n_n p_p}{n_i^2}\right)$$

Also, if all impurities are assumed ionized, $n_n \approx N_d$ and $p_p \approx N_a$ or

$$V_0 \simeq \frac{kT}{e} \log_e\left(\frac{N_d N_a}{n_i^2}\right) \tag{7.8}$$

7.3 Current flow in a pn junction with forward bias

We have seen that no current flows across a pn junction in equilibrium because of the existence of a potential barrier, which prevents diffusion of majority carriers across the depletion layer. In order for current to flow, equilibrium must be disturbed in such a way as to reduce the height of the potential barrier. This can be accomplished by externally biasing the junction with a voltage, V, such that the positive terminal of the voltage source is connected to the p-region, as shown in Fig. 7.5. The junction is then said to be *forward-biased*.

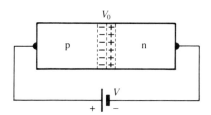

Fig. 7.5 A pn junction with forward bias voltage.

Now, the bulk materials at either side of the depletion layer have high conductivity because of their doping concentrations, whereas the depletion layer has a much lower conductivity, as a consequence of its lack of current carriers. As a result, the applied voltage is dropped almost entirely across the depletion layer, lowering the effective voltage across the depletion layer to $(V_0 - V)$. Thus, the potential barrier is lowered and current flows across the junction because drift and diffusion currents are no longer balanced, as for the junction in equilibrium.

How is this situation described in terms of the band picture of a pn junction? In order to answer this question we must first digress slightly to discuss the effect of an applied field on the band structure of a uniform semiconductor, as shown in Fig. 7.6(a). The complete band structure tilts, as shown in Fig. 7.6(b).

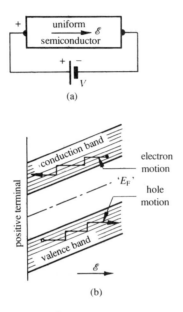

Fig. 7.6 Band structure of a uniform semiconductor with an applied electric field.

Electrons moving in the field acquire kinetic energy, moving horizontally in the diagram to higher energy levels in the conduction band, only to return to nearer the band edge after a scattering event. Similarly, holes, on acquiring energy from the field, move to lower levels in the valence band (since the band diagram is drawn for electron energies), again returning to nearer the top of the band after each collision.

When forward bias is applied to a pn junction, equilibrium is disturbed, and, at the instant V is applied, the energy bands tilt as described and as illustrated in Fig. 7.7(a). Majority carriers are now able to surmount the lowered potential barriers and diffuse across the junctions; large currents can flow. Carriers then flow in from the external circuit to restore near-equilibrium conditions in the material far from the junction; almost all the applied voltage appears across the junction and the steady-state situation depicted in Fig. 7.7(b) exists. Notice that, as soon as equilibrium is disturbed by the application of an external voltage, the Fermi level is no longer continuous across the junction.

The reduction of barrier height by the applied bias thus causes holes to be

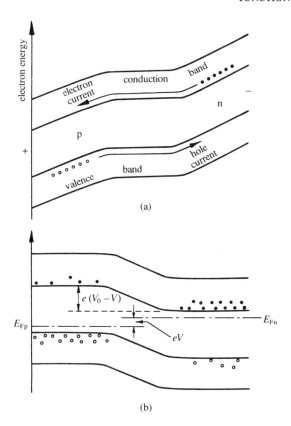

Fig. 7.7 **Band structure of a pn junction with forward bias voltage: (a) immediately on applying the bias and (b) in the steady state.**

injected from the p- to the n-region and electrons from the n- to the p-region, and current flows round the external circuit in the conventional direction. The majority carriers become minority carriers immediately they are injected and their densities at the edge of the depletion layer, n_{p0} and p_{n0}, fall rapidly with distance from the junction because of recombination until, eventually, the minority-carrier densities equal the equilibrium values n_p and p_n. This situation is illustrated in Figs 7.8(a) and (b). The majority-carrier densities are unaffected by the application of bias provided the forward bias voltage is not too large; each region acts as a reservoir of majority carriers, which is hardly perturbed by the injection of a relatively small fraction of carriers across the depletion layer.

The diffusion forces at the junction must still be almost exactly balanced by the electric field corresponding to the reduced potential barrier, since only a very slight imbalance of these forces causes large currents to flow. We shall therefore assume that the equilibrium state of the depletion layer is not

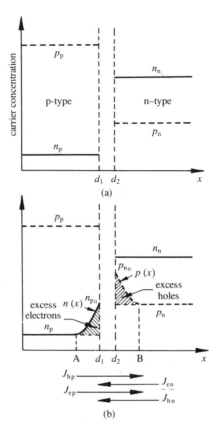

Fig. 7.8 Carrier concentrations and currents flowing in a pn junction: (a) with zero bias and (b) with forward bias applied.

seriously disturbed when forward current flows and that to a first approximation the electric field at the junction is again given by Eq. (7.2). Integrating across the depletion layer, as before, and remembering that the barrier voltage is $(V_0 - V)$ for the forward-biased junction, gives

$$e(V_0 - V) = -kT \log_e(p_{n0}/p_p)$$

or

$$p_{n0} = p_p \exp[e(V - V_0)/kT] \tag{7.9}$$

But, when $V = 0$, $p_{n0} = p_n$ and therefore Eq. (7.9) simplifies to

$$p_{n0} = p_n \exp(eV/kT) \tag{7.10}$$

Similar reasoning applied to electrons injected into the p-region from the n-region gives

$$n_{p0} = n_p \exp(eV/kT) \tag{7.11}$$

Thus, the minority-carrier densities just outside the depletion layer, p_{n0} and n_{p0}, are seen to increase exponentially with applied voltage as a result of the injection of carriers. This result can be substantiated using Boltzmann statistics; this implies that the number of electrons, for example, having sufficient energy to surmount the potential barrier at the junction is proportional to $\exp[-e(V_0-V)/kT]$, which equals constant $\times \exp(eV/kT)$, which agrees with our findings.

Let us now consider in a little more detail the various currents flowing across the junction when forward bias is applied, as shown in Fig. 7.8(b). Four components of the total current flow across the junction, namely

(a) a hole current density, J_{hp}, flowing from the p-region;
(b) a hole current density, J_{hn}, flowing from the n-region;
(c) an electron current density, J_{en}, flowing from the n-region; and
(d) an electron current density, J_{ep}, flowing from the p-region.

The relatively small currents J_{hn} and J_{ep} are due to thermally generated minority carriers in the n- and p-regions, respectively, which fall down the potential hills at the junctions. Since all such carriers will be swept across the junction in this way, the associated current densities are largely independent of the bias voltage and are limited only by the thermal generation rate.

On the other hand, the larger currents J_{hp} and J_{en} are a result of the diffusion of majority carriers across the reduced potential barrier at the junction. These currents will be proportional to the number of majority carriers injected, p_{n0} and n_{p0}, and hence to $\exp(eV/kT)$. The constant of proportionality can be found by remembering that in equilibrium, when $V=0$, the electron and hole currents at the junction must exactly balance so that there is no net junction current, or

$$J_{hp}=J_{hn} \qquad \text{and} \qquad J_{en}=J_{ep}$$

Hence, when bias is applied the electron and hole currents are related by the expression

$$J_{hp}=J_{hn}\exp(eV/kT)$$

and

$$J_{en}=J_{ep}\exp(eV/kT)$$

The net hole current across the junction is thus

$$J_h=J_{hp}-J_{hn}=J_{hn}[\exp(eV/kT)-1]$$

and, similarly the net electron current is

$$J_e=J_{en}-J_{ep}=J_{ep}[\exp(eV/kT)-1]$$

Thus, the total junction current, J, which is the sum of hole and electron currents, is

$$J=J_h+J_e=(J_{hn}+J_{ep})[\exp(eV/kT)-1]$$

or

$$J = J_0[\exp(eV/kT) - 1] \tag{7.12}$$

where J_0, the sum of the current densities carried by minority carriers across the junction, is called the *saturation current density*, for reasons that will become clearer later. We have, by these simple considerations, arrived at the so called *diode* or *rectifier equation*, which, as we shall discuss in more detail later, describes the electrical behaviour of a junction diode.

It is convenient to derive Eq. (7.12) more rigorously, first to provide a quantitative expression for the evaluation of the constant J_0 and secondly as an example of an analytical technique that we will be able to apply to more complicated structures later. We commence by considering how the minority-carrier densities just outside the depletion regions vary as a function of distance. Consider, for example, the continuity equation (6.64) for minority holes injected into the n-region at $x = d_2$. The term on the left-hand side of the continuity equation is zero in the steady state and the term containing $\mathscr{E}$ is also zero, since we assume that there is no voltage drop and hence no electric field outside the depletion region. With these assumptions

$$\frac{d^2(\delta p)}{dx^2} = \frac{\delta p}{\tau_{Lh} D_h} = \frac{\delta p}{L_h^2} \tag{7.13}$$

where we have defined

$$L_h = (\tau_{Lh} D_h)^{1/2} \tag{7.14}$$

which has the dimensions of length.

The general solution of Eq. (7.13) is of the form

$$\delta p = C_1 \exp(-x/L_h) + C_2 \exp(+x/L_h) \tag{7.15}$$

where C_1 and C_2 are constants. However, since δp, the excess minority hole concentration, falls to zero at large distances away from the junction because of recombination, constant C_2 must be zero. Now the excess concentration, δp, is equal to the difference between the local hole density, $p(x)$, and the equilibrium hole density, p_n; thus Eq. (7.15) becomes

$$p(x) = C_1 \exp(-x/L_h) + p_n \tag{7.16}$$

Finally, if for convenience we take the origin of x at d_2, when $x = 0$, $p(x) = p_{n0}$, which on substituting gives

$$p(x) = (p_{n0} - p_n) \exp(-x/L_h) + p_n \tag{7.17}$$

This is the equation for the minority hole density in region d_2–B in Fig. 7.8(b) as a function of x, the distance from the edge of the depletion layer, d_2. We see that $p(x)$ falls off exponentially to the equilibrium value p_n with distance, x, the distance d_2–B being of order L_h, the *diffusion length for minority holes*, which is typically of order 1 mm.

A similar expression can be derived for minority electrons injected into the p-region, the corresponding diffusion length for electrons, L_e, again being about 1 mm.

Now, we have already argued that there is negligible electric field outside the depletion region, so any hole current flowing in the region d_2–B can only be due to diffusion of holes because of their gradient of density. Thus the hole current is given by Eq. (6.47), which on substituting for dp/dx using Eq. (7.17) gives

$$J_h = -eD_h[-(1/L_h)(p_{n0}-p_n)\exp(-x/L_h)] \tag{7.18}$$

We see that the minority hole current decreases exponentially away from the junction because of recombination. However, electrons flow into the region from the end-contact of the n-type material to replace those lost by recombination and hence maintain charge neutrality. This electron current is itself x-dependent and just compensates for the decrease in hole current, thus ensuring that the total current at any point remains constant and independent of position, which is itself a consequence of Kirchoff's law. It follows that the net current flowing in the n-region due to injected holes is equal to the hole current at the edge of the depletion layer, at $x = 0$, which from Eq. (7.18) is

$$J_h|_{d_2} = (eD_h/L_h)(p_{n0}-p_n)$$

or, substituting for p_{n0} from Eq. (7.10)

$$J_h|_{d_2} = (eD_h/L_h)p_n[\exp(eV/kT)-1] \tag{7.19}$$

This type of analysis may be repeated for electrons injected into the p-region, to obtain an expression for the electron current at the boundary between depletion layer and p-region, d_1, given by

$$J_e|_{d_1} = (eD_e n_p/L_e)[\exp(eV/kT)-1] \tag{7.20}$$

Again, on moving further into the p-region, the total current density stays constant at the value given by Eq. (7.20), although an increasing proportion of current is carried by holes, which flow into the area to replace those lost by recombination with the injected electrons.

The total forward current density at the junction is the sum of hole and electron currents, or from Eqs (7.19) and (7.20)

$$J = J_h + J_e = e\left(\frac{D_h p_n}{L_h} + \frac{D_e n_p}{L_e}\right)[\exp(eV/kT)-1] \tag{7.21}$$

Comparing this equation with Eq. (7.12), we see that the saturation current density is given by

$$J_0 = e\left(\frac{D_h p_n}{L_h} + \frac{D_e n_p}{L_e}\right) \tag{7.22}$$

where the diffusion lengths are defined by Eq. (7.14). It is sometimes more convenient to rewrite the expression for J_0, remembering that

$$p_p n_p = n_i^2 = p_n n_n$$

or, for operation at such a temperature that all impurities are ionized

$$p_n = n_i^2/N_d \qquad \text{and} \qquad n_p = n_i^2/N_a$$

and then

$$J_0 = e n_i^2 \left(\frac{D_h}{L_h N_d} + \frac{D_e}{L_e N_a} \right) \tag{7.23}$$

Equation (7.21) confirms that the diode equation derived earlier by a less rigorous method is correct; it also provides expressions from which the saturation current density, J_0, can be calculated. Notice that for all voltages such that $V \gg kT/e$, which equals about 25 mV at room temperature, the exponential term in the diode equation dominates and the current for the forward-biased junction increases approximately exponentially with applied voltage

$$J \simeq J_0 \exp(eV/kT) \tag{7.24}$$

7.4 Current flow in a pn junction with reverse bias

We now consider the case when the junction is reverse-biased, as illustrated in Fig. 7.9(a). With the applied voltage of the polarity shown, it is evident that the effective junction voltage is now greater than the equilibrium value, V_0, by an amount equal to the applied voltage, V, and the height of the potential barrier preventing diffusion of majority carriers is correspondingly increased. This situation is illustrated in the band-structure diagram of Fig. 7.9(b). The potential barrier is now of such a height as to inhibit almost completely the diffusion of majority carriers across the junction; J_{hp} and J_{en} are almost zero. The currents flowing due to minority carriers crossing the junction, J_{hn} and J_{ep}, are unaffected by the increased junction voltage, since, as was explained earlier, any minority carriers in the vicinity of the junction can travel freely down potential hills, as shown in Fig. 7.9(b). Thus the current flow in the reverse-biased junction, assuming positive currents to flow from p- to n-regions as before, is

$$J \simeq -(J_{hn} + J_{ep}) = -J_0 \tag{7.25}$$

We see that the reverse current is almost constant and is equal to the saturation current density, J_0.

Let us now consider in a little more detail the mechanism of charge transport in the reverse-biased junction. The equations derived for the forward-biased junction in Sec. 7.3 will still be applicable provided the sign of the applied voltage, V, is made negative. For instance, the diode equation is

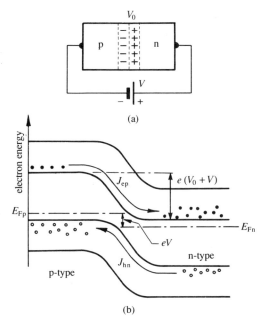

Fig. 7.9 A pn junction with reverse bias voltage.

still applicable and Eq. (7.12) becomes

$$J = J_0[\exp(-eV/kT) - 1] = -J_0[1 - \exp(-eV/kT)] \qquad (7.26)$$

Whenever the reverse bias voltage is greater than a few tens of millivolts, then the exponential term in this equation becomes negligible at room temperature and the current flowing corresponds to the saturation current density, J_0; this result is in agreement with that argued on more physical grounds in the previous section.

The minority electron and hole densities just outside the depletion layer, n_{p0} and p_{n0}, can be obtained directly from Eqs (7.11) and (7.10) by replacing V by $-V$, thus

$$n_{p0} = n_p \exp(-eV/kT) \qquad (7.27)$$

and

$$p_{n0} = p_n \exp(-eV/kT) \qquad (7.28)$$

The variation of minority-carrier density as a function of distance from the depletion layer can be obtained by a similar procedure; for example, for holes, Eq. (7.17) becomes

$$p(x) = p_n \{[\exp(-eV/kT) - 1] \exp(-x/L_h) + 1\} \qquad (7.29)$$

The various minority-carrier densities, described by Eqs (7.27)–(7.29), are

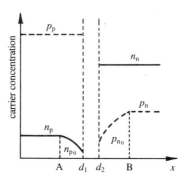

Fig. 7.10 **Carrier concentrations in a reverse-biased pn junction.**

shown graphically in Fig. 7.10. The decrease in minority-carrier densities below their equilibrium values near to the depletion layer is a result of minority electron injection from p- to n-region and hole injection in the opposite direction, as described.

Since there is a continuous flow of minority carriers across the junction, electrons, for example, must be continually created in region d_1–A by thermal electron–hole pair generation, excess holes generated by the process drifting to the negative contact. Hence, since the junction currents are due to minority carriers generated by the intrinsic process, J_{hn}, J_{ep} and hence the saturation current, J_0, are all relatively small and are temperature-dependent.

7.5 Current–voltage characteristics of a junction diode

The diode equation (7.12) is, as we have seen, applicable to a junction biased in either direction. If sufficient information is known about the materials forming the junction, the saturation current density, J_0, can be estimated using Eq. (7.23), say. Thus the I–V characteristic of a particular device can be predicted; it will be of the form shown in Fig. 7.11. The voltage appearing in the diode equation is the part of the applied voltage that appears across the junction. In most circumstances there is little voltage drop across the bulk p- and n-regions because of their relatively high conductivity, and the applied voltage appears almost entirely across the depletion layer. However, if the conductivity of the bulk materials is low, or for large forward currents, the voltage drops across these regions are no longer negligible but must be subtracted from the applied voltage, to give the effective voltage at the junction, which is inserted in the diode equation.

7.6 Electron and hole injection efficiencies

It is instructive to find the relative magnitudes of hole and electron currents at the junction. This information will also be most useful for later discussions concerning more complex junction devices.

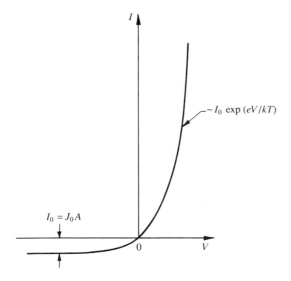

$$I_0 = J_0 A$$

$$\sim I_0 \exp{(eV/kT)}$$

Fig. 7.11 Current-voltage characteristics of a junction diode.

It can be seen from the diode equation (7.12) and Eq. (7.23) that the ratio of hole current to electron current at the junction for either forward or reverse bias is given by

$$\frac{J_h}{J_e} = \frac{J_{hn}}{J_{ep}} = \frac{D_h}{L_h N_d} \frac{L_e N_a}{D_e}$$

Now, for any particular temperature, Einstein's relation, Eq. (6.50), applies and so

$$\frac{J_h}{J_e} = \frac{L_e \mu_h N_a}{L_h \mu_e N_d}$$

Since the conductivities at the p- and n-regions are given by Eqs (6.39) and (6.40), and since the diffusion lengths are of the same order of magnitude, the equation reduces to

$$J_h/J_e \simeq \sigma_p/\sigma_n \tag{7.30}$$

Thus, the ratio of hole current to electron current is approximately equal to the ratio of the conductivities of the p-type and n-type materials constituting the junction.

It is sometimes convenient to define electron and hole *injection efficiencies*, η_e and η_h, as the fractions of the total junction current carried by electrons and holes, or

$$\eta_e = \frac{J_{ep}}{J_{hn} + J_{ep}} = \frac{1}{1 + (J_{hn}/J_{ep})} \simeq \frac{1}{1 + (\sigma_p/\sigma_n)} \tag{7.31}$$

and

$$\eta_h = \frac{J_{hn}}{J_{hn} + J_{ep}} = \frac{1}{1 + (J_{ep}/J_{hn})} \simeq \frac{1}{1 + (\sigma_n/\sigma_p)} \qquad (7.32)$$

It is evident from Eqs (7.30)–(7.32) that, for symmetrical junctions, when $\sigma_n = \sigma_p$, the electron and hole currents are equal but that, in asymmetrical devices, the current is mostly carried across the junction by carriers that are majority carriers in the more heavily doped material. For example, for the junction where it was assumed that $\sigma_p \approx 100\sigma_n$, the current is almost entirely carried by holes, only 1 per cent of the total current being carried by electrons.

7.7 The pn junction with finite dimensions

It has been assumed throughout this chapter that the p- and n-regions are of infinite extent, or at least that their lengths, l_p and l_n, respectively, are much longer than the diffusion lengths for minority carriers. Since this is not always the case in practice, a necessary refinement of the theory must be to consider the effects of finite dimensions on the saturation current and hence the diode equation.

We take as our starting point the expression for the minority hole density due to injected holes in the n-region of the forward-biased junction, Eq. (7.15), which gives

$$p(x) = C_1 \exp(-x/L_h) + C_2 \exp(x/L_h) + p_n$$

Constant C_2 can no longer be neglected since x can never become much greater than L_h, and l_n, the length of the n-region, might be less than L_h. If we assume the hole density at $x = l_n$ to be zero by virtue of there being a highly conducting metal end-contact there which acts as an effective recombination site for electrons and holes, the new boundary conditions on p become

$$p = p_{n0} = p_n \exp(eV/kT) \qquad \text{when } x = 0$$

and $\qquad p = 0 \qquad\qquad\qquad\qquad \text{when } x = l_n$

These conditions can be substituted in the equation to eliminate the constants C_1 and C_2 and obtain

$$p(x) = \left(\frac{p_n \exp(-l_n/L_h) + p_{n0} - p_n}{1 - \exp(-2l_n/L_h)} \right) \exp(-x/L_h)$$

$$- \left(\frac{p_n \exp(-l_n/L_h) + p_{n0} - p_n}{1 - \exp(-2l_n/L_h)} - (p_{n0} - p_n) \right) \exp(+x/L_h) + p_n \qquad (7.33)$$

As before, the hole current is given by

$$J_h = -eD_h \left(\frac{dp(x)}{dx} \right)\bigg|_{x=0} = \frac{eD_h}{L_h} p_n \tanh\left(\frac{l_n}{L_h} \right) [\exp(eV/kT) - 1] \qquad (7.34)$$

A similar expression can be derived for the electron current at the junction, so the saturation current, J_0, as defined in Eq. (7.12), becomes in this case

$$J_0 = e\left[\frac{D_h p_n}{L_h}\tanh\left(\frac{l_n}{L_h}\right) + \frac{D_e n_p}{L_e}\tanh\left(\frac{l_p}{L_e}\right)\right] \tag{7.35}$$

which reduces to the value given by Eq. (7.22) when $l_n \gg L_h$ and $l_p \gg L_e$, as was assumed originally. Since the diffusion lengths are typically of order 1 mm, Eq. (7.35) should always be used to calculate the saturation current whenever the thickness of p- and n-regions are less than this order, which is usually the case. Otherwise, the approximate expressions of Eq. (7.22) are sufficient.

7.8 The geometry of the depletion layer and depletion-layer capacitance

The existence of two space-charge layers of opposite sign at a pn junction, the amount of charge in each varying with bias voltage, gives rise to an effective junction capacitance, called the *depletion-layer capacitance*. It is with more than academic interest that we study the geometry of the space-charge layers and hence evaluate the junction capacitance since, for example, such capacitance is one of the factors that limit the high-frequency operation of junction devices. A further inducement is that, as we shall discover, the junction capacitance can be controlled to a large degree by the bias voltage; this voltage-dependent capacitance is deliberately exploited in the *varactor diode*, an extremely useful high-frequency generating device, which will be discussed later.

We again assume an abrupt pn junction, although this time the results obtained for the capacitance will have to be modified to suit graded junctions. In thermal equilibrium, as we have seen, holes move from the p- to the n-region, leaving uncompensated negatively charged acceptors and a net negative space-charge layer in the p-region. Similarly, electrons moving from the n-region leave fixed positively charged donors and a positive space-charge layer exists there. This variation of charge density across the depletion layer is indicated by the broken curves in Fig. 7.12(a). We have seen that it is justifiable to neglect mobile charge in the depletion layer so the charge density there is approximately equal to the product of the magnitude of the impurity concentration and charge in each region. Thus the charge density in the p-region is

$$\rho_p \simeq -N_a e \tag{7.36}$$

Similarly, the net charge density at the n-side of the junction is

$$\rho_n \simeq N_d e \tag{7.37}$$

We shall let the thickness of these layers be d_p and d_n and shall also assume that the space-charge density changes in the abrupt rectangular fashion, shown by the cross-hatching in Fig. 7.12(a), which can be done without too much loss in

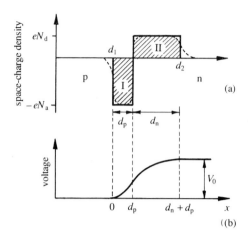

Fig. 7.12 Depletion layer in a pn junction in equilibrium with no bias voltage.

accuracy. It is convenient to take the origin of x at the junction between the depletion layer and p-region, as shown in the figure.

The potential at any position is related to the space charge via Poisson's equation

$$\partial^2 V/\partial x^2 = -\rho/\epsilon$$

Thus, in region I, for $0 \leqslant x \leqslant d_p$

$$\partial^2 V_1/\partial x^2 = eN_a/\epsilon \tag{7.38}$$

and in region II, for $d_p \leqslant x \leqslant d_p + d_n$

$$\partial^2 V_2/\partial x^2 = -eN_d/\epsilon \tag{7.39}$$

We can integrate Eq. (7.38) twice and if we arbitrarily choose a reference potential, $V = 0$ at $x = 0$, we get

$$V_1(x) = eN_a x^2/2\epsilon \tag{7.40}$$

Thus, the potential in the region rises parabolically with increasing x.

We next integrate Eq. (7.39) to give

$$\partial V_2/\partial x = -eN_d x/\epsilon + C_1 \tag{7.41}$$

Now the electric field must be continuous across the junction at $x = d_p$ or

$$(\partial V_1/\partial x)|_{d_p} = (\partial V_2/\partial x)|_{d_p}$$

Therefore we can substitute in the equation from Eqs (7.40) and (7.41) to evaluate the constant C_1:

$$eN_a d_p/\epsilon = -eN_d d_p/\epsilon + C_1$$

or

$$C_1 = ed_p(N_a + N_d)/\epsilon \tag{7.42}$$

Integrating (7.41) again we obtain

$$V_2 = -eN_d x^2/2\epsilon + C_1 x + C_2 \tag{7.43}$$

Again, for continuity of V at $x = d_p$, $V_1|_{d_p} = V_2|_{d_p}$ and Eqs (7.40), (7.42) and (7.43) give

$$\frac{eN_a d_p^2}{2\epsilon} = -\frac{eN_d d_p^2}{2\epsilon} + \frac{ed_p^2(N_a + N_d)}{\epsilon} + C_2$$

This value of C_2 can be substituted in Eq. (7.43) to give

$$V_2(x) = -\frac{eN_d x^2}{2\epsilon} + \left(\frac{ed_p(N_a + N_d)}{\epsilon}\right)x - \frac{ed_p^2(N_a + N_d)}{2\epsilon} \tag{7.44}$$

Now, at the further extremity of the depletion layer, $x = d_n + d_p$, the electric field strength, $\mathscr{E}$, is zero, and

$$(\partial V_2/\partial x)|_{d_n + d_p} = 0$$

or

$$0 = -\frac{eN_d}{\epsilon}(d_n + d_p) + \frac{ed_p(N_a + N_d)}{\epsilon}$$

which gives

$$d_p/d_n = N_d/N_a \tag{7.45}$$

It can be seen that the depletion-layer thicknesses are in inverse ratio to the doping concentrations, a result that can be justified on physical grounds, since we would expect there to be equality of net charge at either side of the junction in equilibrium, or $N_a d_p = N_d d_n$. We see that in an asymmetric junction the depletion layer and the potential rise occur mostly in the material that is more lightly doped; for example, in Fig. 7.12 it has been assumed that the p-type material is more heavily doped than the n-type and thus $d_p \ll d_n$.

Now, at $x = d_n + d_p$, the potential with respect to our assumed zero potential is V_0, the contact potential for the junction in equilibrium. Equation (7.44) thus gives

$$V_2|_{d_n + d_p} = V_0 = -\frac{eN_d}{2\epsilon}(d_n + d_p)^2 + \frac{ed_p(N_a + N_d)(d_p + d_p)}{\epsilon}$$
$$-\frac{ed_p^2(N_a + N_d)}{2\epsilon}$$

We now eliminate d_n from this equation, using Eq. (7.45), and rearrange to give the width of the layer in the p-region

$$d_p = \left(\frac{2\epsilon V_0 N_d}{eN_a(N_a + N_d)}\right)^{1/2} \tag{7.46}$$

Finally, this equation is used in conjunction with Eq. (7.45) to give the width of the depletion layer in the n-region

$$d_n = \left(\frac{2\epsilon V_0 N_a}{eN_d(N_a + N_d)}\right)^{1/2} \tag{7.47}$$

The total depletion-layer thickness is then

$$d_p + d_n = \left(\frac{2\epsilon V_0 (N_a + N_d)}{eN_a N_d}\right)^{1/2} \tag{7.48}$$

and the junction becomes narrower as the impurity concentration is increased.

The expressions we have derived, and in particular Eqs (7.46), (7.47) and (7.48), are still applicable to a biased junction provided the effective junction voltage is altered accordingly. For example, if a forward bias, V, is applied to the junction, V_0 in the equations is replaced by $(V_0 - V)$ and there is a decrease in the depletion-layer thickness. On the contrary, if a reverse bias of $- V$ volts is applied the junction voltage becomes $(V_0 + V)$ and the total depletion-layer thickness is increased to

$$d_p + d_n = \left(\frac{2\epsilon(V_0 + V)(N_a + N_d)}{eN_a N_d}\right)^{1/2} \tag{7.49}$$

The capacitance associated with the depletion layer can be obtained by the following argument. Suppose the junction is reverse-biased and the junction voltage, $(V_0 + V)$, equals V_j. The charge stored at either side of the junction is then represented by $\pm Q_j$ in Fig. 7.13. If the junction voltage is made less negative by a small amount, δV_j, there will be a reduction in the depletion-layer width and a corresponding reduction, $\pm \delta Q_j$, in the charge, as indicated by the cross-hatched regions in the figure. These changes are brought about by holes

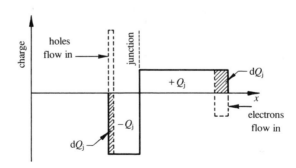

Fig. 7.13   Charge storage in a reverse-biased pn junction.

of total charge δQ_j flowing into the depletion layer from the p-region to neutralize some of the charge of the fixed exposed acceptors, and a similar negative charge flows from the n-type region as indicated by the broken lines in the figure. Thus, if the junction voltage is changed by δV_j, this results in a flow of mobile charge δQ_j, which, for small changes in δV_j, is proportional to δV_j. An incremental depletion-layer capacitance, C_j, is therefore defined as

$$C_j = dQ_j/dV_j \tag{7.50}$$

Now, the magnitude of the total charge stored at either side of the junction per unit junction area is given by

$$Q_j = eN_d d_n = eN_a d_p$$

or, using Eq. (7.46) or (7.47) in which V_0 is replaced by V_j

$$Q_j = \left(\frac{2\epsilon e V_j N_a N_d}{N_a + N_d} \right)^{1/2} \tag{7.51}$$

This is substituted in Eq. (7.50) to give the junction capacitance per unit area

$$C_j = \frac{d}{dV_j} \left(\frac{2\epsilon e N_a N_d}{N_a + N_d} \right)^{1/2} V_j^{1/2} = \left(\frac{\epsilon e N_a N_d}{2(N_a + N_d)} \right)^{1/2} \frac{1}{V_j^{1/2}} \, \text{F m}^{-2} \tag{7.52}$$

We see that, for an abrupt pn junction, the depletion capacitance varies as $V_j^{-1/2}$, where the junction voltage, V_j, is equal to $(V_0 \pm V)$, the sign depending on whether the bias is applied in the forward or reverse direction. Notice that, for large reverse bias voltages when $V \gg V_0$, C_j varies very nearly as $V^{-1/2}$.

The expression for C can be simplified further, using Eq. (7.48), which with Eq. (7.52) gives

$$C_j = \epsilon/(d_p + d_n) \, \text{F m}^{-2} \tag{7.53}$$

Thus, the abrupt junction behaves as a parallel-plate capacitor, filled with dielectric of permittivity ϵ and of thickness $(d_p + d_n)$. With forward bias, $(d_p + d_n)$ is small and C_j large, whereas when the bias voltage is in the reverse direction, $(d_p + d_n)$ becomes large and C_j is correspondingly smaller.

The capacitance of graded junctions varies with reverse voltage more slowly than the abrupt junction. For instance, a linearly graded junction in reverse bias has a junction capacitance that varies as $(V_j)^{-1/3}$, or $(V)^{-1/3}$ approximately. Therefore, an abrupt junction is superior to a graded junction in a varactor, because of its larger relative capacitance change for a given bias-voltage change. High impurity concentrations are also an advantage in varactor diodes since not only is the depletion-layer width reduced and hence the capacitance increased, but also the conductance of the bulk p- and n-regions in series with the capacitance is increased, which is desirable for high-frequency applications.

If the capacitance of a junction is measured as a function of the reverse bias voltage V, the information can yield experimental values for the doping levels of the p- and n-regions and also the contact potential. For example, if we assume the p-region to be the more heavily doped, such that $N_a \gg N_d$, we can rearrange Eq. (7.52) to give

$$V_j = V_0 + V \simeq \frac{1}{C_j^2}\left(\frac{\epsilon e N_d}{2}\right)^{1/2}$$

Thus, an experimental graph of V versus C_j^{-2} should be a straight line, with an intercept on the V axis giving V_0 and with a slope providing a value for N_d, the impurity level on the more lightly doped n-side. The acceptor concentration can then be found from N_d and V_0 using Eq. (7.8). A similar but slightly more complicated procedure can be used to find N_a and N_d when they have comparable values.

7.9 Diffusion capacitance and the small-signal equivalent circuit of a pn junction

When a junction is biased in the reverse direction its equivalent circuit will be of the form illustrated in Fig. 7.14(a). Here, r_{np} represents the resistance of the

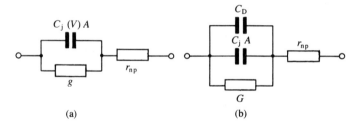

Fig. 7.14 Small-signal equivalent circuits of a pn junction: (a) reverse-biased and (b) forward-biased.

bulk n- and p-regions outside the depletion layer, C_j is the depletion-layer capacitance per unit area, which will be a function of the reverse bias, A is the junction area and g is the conductance of the junction due to the flow of reverse current across it. Usually r_{np} is very small and can be neglected and often g can be neglected when compared to the susceptance of $C_j A$. Thus, in many instances, the reverse-biased junction can be represented simply by a capacitance $C_j A$.

At first sight, it might be assumed that the same equivalent circuit would be satisfactory for the forward-biased junction, provided the appropriate values for C_j and g are introduced, but this is not the case. It is true that when the diode is forward-biased, its depletion capacitance increases, but a further capacitance, called the *storage* or *diffusion capacitance*, C_D, is introduced, which tends to swamp effects due to C_j. For the forward-biased junction, many minority carriers are injected into each side of the junction. It is the motion and charge storage of such carriers that give rise to the new capacitive effect. The small-signal equivalent circuit is then as shown in Fig. 7.14(b). The elements C_D and G in it can be estimated as follows.

We first assume that the steady-state forward bias voltage, V, has superimposed on it a small alternating voltage, $V_1 \exp(j\omega t)$. The injected hole density at the edge of the depletion layer, $x = 0$, is then given by Eq. (7.10) and is

$$p_{n0} = p_n \exp\left(\frac{e}{kT}[V + V_1 \exp(j\omega t)]\right) \tag{7.54}$$

Since we have assumed small-signal conditions, $V_1 \ll V_0$ and the exponential can be approximated to the first two terms of a power series, giving

$$p_{n0} = p_n \exp\left(\frac{eV}{kT}\right)\left(1 + \frac{eV_1}{kT}\exp(j\omega t)\right) \tag{7.55}$$

The alternating component of the forward bias thus causes an alternating hole current to be injected into the n-region. We therefore assume that the excess minority hole concentration, δp, is given by

$$\delta p = p(x, t) - p_n = p_0(x) + p_1(x)\exp(j\omega t) - p_n \tag{7.56}$$

where $p_0(x)$ represents the steady-state hole density and $p_1(x)$ the density of the alternating component, distance x from the edge of the depletion layer. If we substitute this in the time-dependent continuity equation for holes, Eq. (6.64), assuming $\mathscr{E} = 0$ as usual and retaining only a.c. components, we get

$$j\omega p_1 = \frac{-p_1}{\tau_{Lh}} + D_h\frac{\partial^2 p_1}{\partial x^2}$$

or

$$\frac{\partial^2 p_1}{\partial x^2} = \frac{1 + j\omega\tau_{Lh}}{L_h^2}p_1$$

By comparison with Eq. (7.15) this equation has a solution

$$p_1 = C\exp\left(-\frac{(1 + j\omega\tau_{Lh})^{1/2}x}{L_h}\right) \tag{7.57}$$

The constant C is obtained by comparing this equation, setting $x = 0$, with the alternating part of Eq. (7.55). Thus, the alternating injected hole density is

$$p_1 = p_n\left(\frac{eV_1}{kT}\right)\exp\left(\frac{eV}{kT}\right)\exp\left(-\frac{(1 + j\omega\tau_{Lh})^{1/2}x}{L_h}\right)\exp(j\omega t) \tag{7.58}$$

Proceeding as for the steady-state case, the alternating junction current due to injected holes is

$$J_{h1} = -eD_h\left(\frac{dp_1}{dx}\right)\bigg|_{x=0} = \frac{(1 + j\omega\tau_{Lh})^{1/2}}{L_h}\frac{p_nD_he^2V_1}{kT}\exp\left(\frac{eV}{kT}\right)\exp(j\omega t)$$

A similar expression can be derived for the alternating junction current due to electron injection into the p-region, and the total a.c. junction current is then

$$J_1 = \frac{e^2 V_1}{kT} \exp\left(\frac{eV}{kT}\right) \left(\frac{D_h p_n (1 + j\omega\tau_{Lh})^{1/2}}{L_h} + \frac{D_e n_p (1 + j\omega\tau_{Le})^{1/2}}{L_e}\right) \exp(j\omega t)$$

This current flows on application of an alternating forward bias voltage, $V_1 \exp(j\omega t)$, and the junction admittance, Y_1, is therefore

$$Y_1 = \frac{J_1}{V_1} = \frac{e^2}{kT} \exp\left(\frac{eV}{kT}\right) \left(\frac{D_h p_n (1 + j\omega\tau_{Lh})^{1/2}}{L_h} + \frac{D_e n_p (1 + j\omega\tau_{Le})^{1/2}}{L_e}\right) \tag{7.59}$$

If the frequency is not too high, $\omega\tau \ll 1$, and the admittance expression simplifies to

$$Y_1 \simeq \frac{e^2}{kT} \exp\left(\frac{eV}{kT}\right) \left(\frac{D_h p_n}{L_h} + \frac{D_e n_p}{L_e}\right) + \frac{j\omega e^2}{2kT} \left(\frac{D_h p_n \tau_{Lh}}{L_h} + \frac{D_e n_p \tau_{Le}}{L_e}\right)$$

Thus we see that the low-frequency admittance of the forward-biased junction can be represented by a conductance

$$G = \frac{e^2}{kT} \exp\left(\frac{eV}{kT}\right) \left(\frac{D_h p_n}{L_h} + \frac{D_e n_p}{L_e}\right) = \frac{eI_0}{kT} \exp\left(\frac{eV}{kT}\right) \tag{7.60}$$

where I_0 is the saturation current, in parallel with a capacitance

$$C_D = \frac{e^2}{2kT} \left(\frac{D_h p_n \tau_{Lh}}{L_h} + \frac{D_e n_p \tau_{Le}}{L_e}\right) \exp\left(\frac{eV}{kT}\right)$$

or

$$C_D = \frac{e^2}{2kT} (p_n L_h + n_p L_e) \exp\left(\frac{eV}{kT}\right) \tag{7.61}$$

Notice that this diffusion capacitance increases exponentially with forward bias or, alternatively, that it is directly proportional to forward junction current.

7.10 Switching characteristics

Since diodes are often used in a switching mode, a preliminary discussion of this form of operation will demonstrate which parameters must be optimized, and also forms the basis for the comparison of different diode configurations.

Consider, for example, the forward-biased n^+p junction diode shown in Fig. 7.15. The minority electron concentration in the p-region will be approximately linear, as shown, if the base is short compared to the diffusion length, L_e. If the junction is suddenly reverse-biased, at t_0, then, because of this stored electronic charge, the reverse current, I_R, is initially of the same magnitude as the forward current, I_F. However, as the stored electrons are

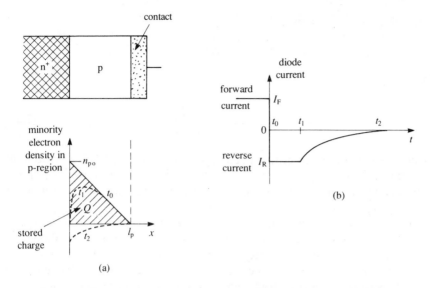

Fig. 7.15 Charge stored and current flow in a switched n^+p junction diode.

removed into the n^+ region and the contact, the available charge quickly
drops to an equilibrium level and a steady current eventually flows corresponding to the reverse bias voltage, as shown in Fig. 7.15.

To estimate the turn-off time, t_{off}, let us first calculate the minority charge
stored at the junction when a forward current, I_F, flows. If it is assumed for
simplicity that electron transport across the p-region is by diffusion only, then
the electron current density, J_{Fe}, is

$$J_{Fe} = ev(x)n(x) = eD_e \, dn/dx \qquad (7.62)$$

where D_e is the diffusion coefficient for minority electrons. Since a linear
gradient of electrons may be assumed, for simplicity

$$J_{Fe} = ev(x)n_{p0}(1 - x/l_p) = eD(-n_{p0}/l_p)$$

or

$$v(x) = D_e/(l_p - x) \qquad (7.63)$$

The charge density at x, $\rho(x)$, is then given by

$$\rho(x) = n(x)e = I_F/Av = I_F(l_p - x)/AD_e$$

and the total stored charge, Q, is then

$$Q = A \int_0^{l_p} \rho(x) \, dx = I_F l_p^2/2D_e \qquad (7.64)$$

When the voltage is reversed, the average reverse current flowing during the turn-off period, $\bar{I}_R$, is given by

$$\bar{I}_R = Q/t_{off}$$

or

$$t_{off} = \frac{Q}{\bar{I}_R} = \frac{I_F}{\bar{I}_R} \frac{l_p^2}{2D_e} \qquad (7.65)$$

Thus, the turn-off time, because of its dependence on I_F and $\bar{I}_R$ is a function of the external circuitry to some extent but it is also determined by diode parameters and can be reduced, for example, by shortening the length of the p-region, l_p.

The stored charge and consequently the switching time can also be greatly reduced by the introduction of gold impurities into the junction diode, by diffusion. The gold dopant, sometimes called a *lifetime killer*, provides a series of recombination centres, at which stored minority carriers are removed more quickly because of the increased recombination rate. This technique is used to produce diodes and other active devices for high-speed applications.

7.11 Metal–semiconductor junctions

It has been assumed that electrical contacts can be made successfully to the bulk regions of the semiconductor junction diode which do not affect the electrical performance of the junction. The contacts each constitute some kind of metal–semiconductor junction, which, ideally, should have a high conductance that is independent of the direction of current flow. Such a contact is said to be *ohmic*.

When a junction diode is provided with two ohmic contacts, contact potentials exist between the metal and semiconductor regions such that the algebraic sum of these and that at the pn junction is zero, and there is no net voltage across the device in equilibrium. Otherwise, a current could flow in an external load connected across the diode, which clearly would violate the second law of thermodynamics.

Not all metal–semiconductor junctions are ohmic; indeed many display unidirectional current-carrying properties and are thus rectifying. Whereas rectifying contacts are an obvious disadvantage when applied to a pn junction diode, rectifying metal–semiconductor junctions have important device applications in their own right, for example, in high-frequency diodes.

Whether a particular combination of metal and semiconductor forms a rectifying or ohmic contact is dependent to some extent on the relative workfunctions of each and to a larger degree on the surface condition of the semiconductor, which will now be discussed in more detail. To be specific, we shall first describe the junction formed by a metal and an n-type semiconductor. There are, of course, completely analogous arguments for the metal–p-type junction, which will be briefly mentioned subsequently.

7.11.1 The junction between a metal and an n-type semiconductor

Let us first consider the metal contact and semiconductor separately. Electrons in each are retained in the body of the material by a potential barrier at the surface, as discussed in Chapter 4. If we assume that the workfunction of the metal, ϕ_m, is greater than that of the semiconductor, ϕ_s, the energy band diagrams for each are shown in Fig. 7.16(a). When contact is made between the

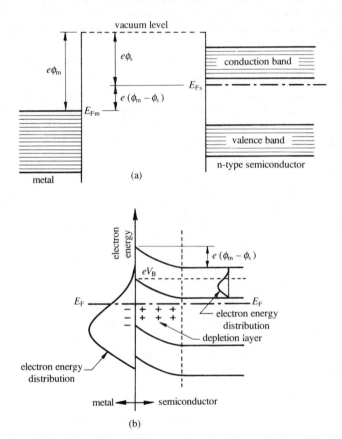

Fig. 7.16 Band structure of metal–n-type semiconductor junction (a) before and (b) after contact, when $\phi_m > \phi_s$.

two materials, charge is transferred from one to the other until, in equilibrium, the Fermi level is continuous across the junction, as illustrated in Fig. 7.16(b). This condition is brought about by electrons spilling over from the semiconductor into the metal, leaving a positively charged depletion layer in the semiconductor due to exposed donor atoms and producing a negative surface charge on the metal. The process continues until the electric field set up by the dipole layer is sufficiently strong to inhibit further electron diffusion. This is represented by a bending up of the band edges in the semiconductor, to form a potential energy barrier for electrons, of height $(\phi_m - \phi_s)e$, as shown. There will be some electron flow from metal to semiconductor due to thermal agitation; this is represented by the small tail of the electron distribution curve for electrons with energy greater than the barrier energy, eV_B. Similarly, a few electrons in the semiconductor will have sufficient energy to surmount the barrier and flow into the metal. In equilibrium, both electron currents will be equal and there will be zero net junction current.

When the semiconductor is made positive with respect to the metal by the application of an external bias voltage, V, all energy levels in the semiconductor are lowered by an amount eV, the effective barrier height is increased and the depletion layer becomes wider; the junction is therefore reverse-biased. Under these circumstances, electron current from semiconductor to metal is entirely prohibited by the high potential barrier, but the small electron current from metal to semiconductor is unaffected. The reverse current thus saturates at a low value independent of the reverse bias voltage.

Application of a forward bias voltage, V, by making the semiconductor negative lowers the effective junction height and reduces the width of the depletion layer. Again, the small electron current from metal to semiconductor remains constant but appreciable electron current is allowed to flow from the semiconductor into the metal contact. Minority hole injection from metal to semiconductor is now also possible since the top of the valence band is raised sufficiently for there to be appreciably more vacant energy levels in the metal than in the semiconductor at these levels.

We see that when $\phi_m > \phi_s$ the metal–semiconductor junction is rectifying, passing large currents in the forward direction and a small saturation current when biased in the reverse direction. A simple analysis of such a junction can be based on Boltzmann statistics. The electrons from metal to semiconductor have to surmount an energy barrier of height eV_B and at temperature T the electron current density from metal to semiconductor, J_{ms}, is

$$J_{ms} \propto \exp(-eV_B/kT) \tag{7.66}$$

On the other hand, the barrier to be surmounted by electrons in the semiconductor will be of height $e(V_B - V)$, which is greater or less than eV_B, depending on the direction of the applied bias and, hence, the sign of V. The electron current from semiconductor into the metal, J_{sm}, is thus

$$J_{sm} \propto \exp[-e(V_B - V)/kT] \tag{7.67}$$

Thus, if we ignore minority hole injection currents for the moment, the total junction current, J, will be the difference between the two electron components

or

$$J \propto \exp(-eV_B/kT)\,[\exp(eV/kT)-1] \qquad (7.68)$$

Notice that this is of the same form as the rectifier equation for a pn junction, Eq. (7.12), and hence the $I-V$ characteristics will be similar to that shown in Fig. 7.11. Hole injection will have the effect of reducing the forward resistance and increasing the slope of the $I-V$ characteristic in the forward-biased regime. A rectifying junction of this type is called a *Schottky* junction, after the original investigator.

Historically, metal–semiconductor junctions featured in one of the earliest electronic components, the metal 'cat's whisker' crystal detector, used in early radio receivers. Modern technology has now led to the production of much more reliable (and reproducible!) Schottky barrier diodes (see Sec. 10.5.6), which form a very important class of electronic device, particularly useful in high-speed integrated circuits.

Unlike the pn junction, which we have seen relies for its action on the behaviour of minority carriers, the current in a Schottky junction is transported by majority carriers. A limitation on the switching speed of a pn junction is due to minority-carrier storage, which prevents instantaneous current reversal when the bias voltage polarity is abruptly changed. This is because a finite time must elapse before the stored minority carriers are removed (see Sec. 7.10). This difficulty is considerably alleviated for a Schottky junction because there is no minority charge storage, which makes the storage time negligibly small, and the device is capable of switching relatively quickly, typically in around 0.1 ns. Hence Schottky diodes are preferred components for fast switching and high-frequency rectification applications.

A further point to note is that the majority electrons injected into the metal when the junction is biased in the forward direction will have energies much in excess of the Fermi energy in the metal; such electrons are called *hot electrons.* Subsequent lattice scattering reduces their energy to the average energy in the metal.

If now we consider the case when $\phi_m < \phi_s$, the band structures of metal and semiconductor before and after contact will be as shown in Figs 7.17(a) and (b). Before contact the Fermi level in the metal, E_{Fm}, is higher than that in the semiconductor, E_{Fs}. On contact there is an exchange of charge as electrons flow from metal to semiconductor, causing a surface-charge accumulation layer to build up as shown in the figure, until the Fermi level becomes continuous in equilibrium. In the energy level diagram, such behaviour is represented by a downward bending of the bands in the semiconductor in the vicinity of the junction. It is clear that in this case there is no potential barrier formed at the junction; electrons can move across the junction in either direction, depending on the direction of the bias voltage, without much hindrance. Such a contact is thus considered ohmic.

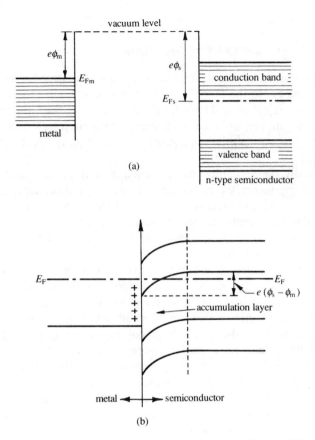

Fig. 7.17 Band structure of a metal–n-type semiconductor junction (a) before and (b) after contact, when $\phi_m < \phi_s$.

7.11.2 Metal–p-type semiconductor junction

It is clear from arguments similar to those set out in Sec. 7.11.1 that when $\phi_m < \phi_s$ and E_{Fs} is below E_{Fm}, a space-charge depletion layer is set up in the semiconductor and a barrier forms at the junction, which becomes rectifying. Conversely, when $\phi_s < \phi_m$ and E_{Fm} is below E_{Fs}, no barrier is formed, holes can flow freely across the junction in either direction, and the contact is ohmic.

7.11.3 Surface states

It would seem that the criterion for determining whether a metal–semiconductor contact is ohmic or rectifying are (a) whether the semiconductor is n- or p-type and (b) the relative workfunctions of metal and semiconductor. Unfortunately, these criteria are not always successful in practice for predicting the characteristics of a given contact. For example, it is found that the characteristics of microwave point-contact diodes are largely independent of the workfunction of the metal. The anomalous behaviour arises because of the existence at the surface of the semiconductor of a large number of localized energy levels situated in the forbidden gap. These *surface states* arise principally because of adsorbed impurity atoms or oxide layers at the surface. The surface states behave as either electron or hole traps, depending on the origin of the states.

Consider, for example, electron-trapping surface states in a moderately doped n-type semiconductor. The surface states deplete electrons from the bulk material, exposing donors and setting up a depletion layer and causing a natural potential barrier to occur independent of any metal contact being made, as shown in Fig. 7.18(a). It is also possible for electrons to be removed

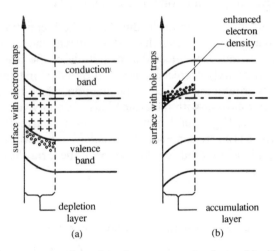

Fig. 7.18   Band structure near the surface of an n-type semiconductor: (a) with surface electron traps and (b) with surface hole traps.

from the valence band, if the electron-trap density at the surface is sufficiently high; the surface layer is then p-type in character and the Fermi level in the depletion layer is nearer the valence band, as is appropriate. Thus, a p-type inversion layer can be produced on the surface of an otherwise n-type crystal, causing a built-in pn junction. The importance of such layers will become apparent when the insulated-gate field-effect transistor is discussed later.

It is also possible for accumulation layers to be found at the surface of a semiconductor, for example, when an n-type semiconductor has surface states that trap holes, as in Fig. 7.18(b). The hole traps are normally occupied by electrons but these are given up into the conduction band to enhance the density in an accumulation layer, as shown.

Similar situations, i.e. depletion, accumulation and inversion layers, are of course possible with p-type material, depending on the number and type of surface states.

It is evident that the nature of the contact made by a metal to such layers is strongly dependent on the surface states, and effects due to relative work-function may be of secondary importance.

One method of practical importance for making successful ohmic contacts to a semiconductor is to alloy or interdiffuse a material that is itself of the same impurity type as that in the semiconductor to which the contact is being made. For example, the solder used to form a contact may be doped with n- or p-type impurity, whichever is appropriate; there is thus a gradual transition from metal to semiconductor and no rectifying effects will be observed.

7.12 Breakdown in pn junctions; the Zener diode

The simple theory for a pn junction leading to the diode equation predicts that, under reverse bias conditions, a small constant current, the saturation current, I_0, flows, which is independent of the magnitude of the bias voltage. This prediction is not entirely correct in practical diodes. First there are ohmic leakage currents around the surface of the junction, which result in a gradual increase of reverse current with increasing bias. Secondly, at some particular reverse bias voltage, V_{BD}, a sudden increase in reverse current can sometimes be observed, which is due to some sort of *breakdown*. The reverse characteristic of a practical pn junction diode could then be of the form shown in Fig. 7.19.

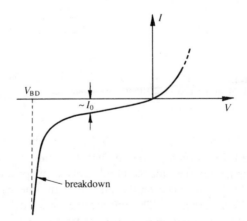

Fig. 7.19 *V-I* **characteristic of a pn junction with large reverse bias voltages.**

When breakdown occurs, the diode current is limited mostly by the resistance of the external circuit.

The presence of breakdown voltage sets a limit to the operating range of a practical pn diode rectifier, since, if it is exceeded in any part of the reverse cycle, the rectification properties are nullified by the large reverse currents that then flow. As is often the case, this deleterious limiting effect in rectifiers is usefully exploited in another class of device, the *voltage-stabilizer diode*, which relies for its action on the relatively constant voltage across a junction under breakdown conditions. Electrical breakdown in a junction therefore either limits the operating range of some junction devices or can determine the operating principles and characteristics of others; in either case it obviously merits further study.

In the reverse-biased pn junction, it has been seen that no current is transported by majority carriers and that the saturation current flowing is due to minority carriers, which are accelerated across the depletion layer by the favourably oriented electric field arising from the bias voltage. Minority electrons, for example, gain energy from the field, which is subsequently lost by interaction with the lattice of the semiconductor; roughly, the electron 'collides' with the lattice, or more precisely there is an exchange of energy between the electron wave and lattice vibrations, which results in Joule heating. As the reverse bias voltage approaches V_{BD}, an electron gains more energy from the field in a mean free path between collisions than it can give to the lattice, since there is a maximum possible energy that can be exchanged, which is limited by quantum-mechanical considerations. Hence an electron in this condition continuously gains energy and after several mean free paths may have acquired sufficient energy, i.e. has $E > E_g$, to ionize the lattice during a collision, so producing an additional electron-hole pair. The new electron and hole plus the original one, which by now has lost most of its energy, are then free to accelerate in the electric field in the depletion layer and so produce further carrier multiplication. Suppose, for a simple example, that the probability of a primary minority carrier producing an ionizing collision in the depletion layer is 50 per cent, i.e. one in two such carriers produce an electron-hole pair. The situation for four primary electrons is shown schematically in Fig. 7.20(a). Each ionizing collision is denoted by an asterisk. Note that the secondary carriers can also produce ionizing collisions; in this simple case, the two secondary holes also contribute one more electron after colliding with the lattice. The process of *avalanche multiplication* is perhaps seen more clearly in the energy band diagram, Fig. 7.20(b).

In the more general case, if the number of primary electrons is n_0 and the probability of ionizing collision is P_i, then, since each electron-hole pair created after each collision has the same probability, P_i, of producing a further pair, the total number of electrons emerging from the depletion layer, n_{out}, is given by

$$n_{out} = n_0 + P_i n_0 + P_i(P_i n_0) + \ldots = n_0 \sum_{m=0}^{\infty} P_i^m$$

The proportional increase in the number of carriers, in this case electrons, is

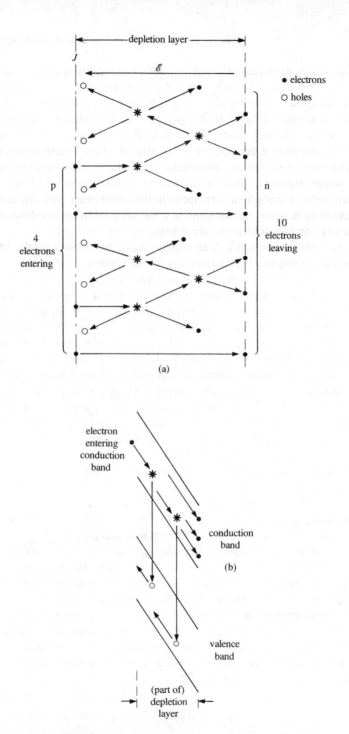

Fig. 7.20 Carrier multiplication by avalanche impact ionization: (a) four electrons entering depletion layer with 50 per cent probability of producing ionizing collision; (b) energy band representation of electron–hole pair production.

called the *multiplication factor*, designated M, so it follows that

$$M = n_{out}/n_0 = \sum_{m=0}^{\infty} P_i^m = 1/(1-P_i) \qquad (7.69)$$

Clearly, the ionization probability, P_i, is related to the accelerating field and hence the applied voltage, V. As V approaches the breakdown voltage, V_{BD}, then P_i becomes close to unity, and there is a rapid increase in carrier density, as shown by Eq. (7.69) and a corresponding increase in current. Because of the cumulative increase in carrier density after each collision, as described, the process is known as *avalanche breakdown*.

It has been found empirically that the bias voltage and ionization probability are related by a power law of the form

$$P_i = (V/V_{BD})^n$$

This relationship fits experimental data well and also satisfies the physical conditions that $P_i = 0$ when $V = 0$ and $P_i \to 1$ as $V \to V_{BD}$. The carrier multiplication factor, M, is then, substituting in Eq. (7.69), given by

$$M = [1 - (V/V_{BD})^n]^{-1} \qquad (7.70)$$

The empirical constant, n, in this equation, depends on the lattice material and the carrier type; for n-type silicon $n \simeq 4$, and for p-type $n \simeq 2$.

Now the electric field in the depletion layer of a pn junction can be obtained from Eqs (7.41) and (7.42) to give

$$\frac{\partial V_2}{\partial x} = \mathscr{E} = \frac{-eN_d x}{\epsilon} + \frac{ed_p(N_a + N_d)}{\epsilon}$$

Also, since the electric field has a maximum value when $x = d_p$, it follows that

$$\mathscr{E}_{max} = ed_p N_a/\epsilon = \left(\frac{2eV_j N_d N_a}{\epsilon(N_a + N_d)}\right)^{1/2} \qquad (7.71)$$

after substituting for d_p from Eq. (7.46), assuming a junction voltage, V_j. If we assume, as is usual, an asymmetrically doped junction, say n^+p, then $N_d \gg N_a$ and (7.71) becomes

$$\mathscr{E}_{max} = \left(\frac{2eV_j N_a}{\epsilon}\right)^{1/2} \propto (V_j N_a)^{1/2} \qquad (7.72)$$

In other words, the maximum field is a function of not only the junction voltage but also the doping at the *least* heavily doped side of the junction.

If it is assumed that avalanche breakdown occurs at some roughly constant maximum field, $\mathscr{E}_{max,B}$, which is of order 1 MV m^{-1} for silicon, then, from Eq. (7.72), it follows that

$$\mathscr{E}_{max,B} \propto (V_{BD} N_a)^{1/2} \qquad \text{and} \qquad V_{BD} \propto N_a^{-1} \qquad (7.73)$$

where the doping level concerned, in this case N_a, is again for the least heavily doped side of the junction. It is evident that the breakdown voltage for a particular diode can be controlled during manufacture by altering the doping levels in the junction.

Voltage reference diodes that utilize the almost constant voltage characteristic in the breakdown region are also called variously *avalanche diodes* or sometimes *Zener diodes*. We shall now see, however, that Zener breakdown, although resulting in *I–V* characteristics of similar general form, is quite a different type of behaviour from the avalanche process just described.

Whereas avalanche breakdown tends to dominate in wider junctions, for instance in graded junctions or when the doping concentration at one or both sides is only moderate, Zener breakdown occurs in relatively heavily doped pn junctions. Because of the high doping levels, the depletion layer is very thin and as a consequence large fields exist across it, often in excess of that required to initiate avalanche breakdown. However, since the depletion layer is so thin, the carriers have not enough distance in which to accelerate in the field to acquire sufficient energy to produce ionizing collisions and hence extra carriers. For example, suppose the field is 10^8 V m^{-1} and the depletion layer is 1 nm wide, then the energy gained by an electron is $10^8 \times 10^{-9} = 0.1$ eV, which is less than E_g, so there is no ionization of the lattice and no carrier pairs are produced.

However, it is possible for electron–hole pairs to be created as covalent bonds are broken down by the force due to the high electric field; large numbers of carriers are generated in this manner and the current increases rapidly, only to be limited by the resistance of the external circuitry.

The energy band explanation of Zener breakdown is illustrated in Fig. 7.21, which shows the band structure for a pn junction with a large reverse bias voltage applied. Notice that electrons at some point a in the valence band of the p-type side lie opposite equal-energy vacant levels in the conduction band of the n-type, at b, say. For normal doping levels, when the depletion layer is sufficiently thick to prevent a direct transition, an electron at a attempting to reach empty levels at b would be prohibited by the potential energy barrier that exists, i.e. it would have to follow the path indicated by the broken curve, gaining energy E_g to move into the conduction band and hence surmount the energy barrier before falling to b, which is not a probable occurrence. However, when the doping levels are increased, the depletion layer can become appreciably thinner and the energy barrier becomes narrow enough for there to exist a finite probability that an electron at a can *tunnel* through the barrier to b in the conduction band in the n-type, where it becomes available for conduction. Large currents can then flow as a result.

For a heavily doped junction, carrier tunnelling leading to Zener breakdown occurs at fields around 2×10^7 V m^{-1} and the breakdown voltage, V_{BD}, is usually less than 5 V. For more lightly doped junctions, the voltage to initiate Zener breakdown is much higher, so avalanche multiplication is

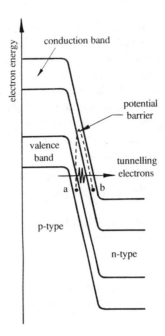

Fig. 7.21   Zener breakdown in a reverse-biased pn junction.

usually the predominant breakdown mechanism. Since Zener breakdown depends on gap energy and barrier thickness, it is largely temperature-independent. This contrasts with avalanche breakdown, for which, if the junction temperature is increased, the mean free path for phonon scattering is reduced and V_{BD} is correspondingly increased.

It is possible to design a pn junction such that the reverse breakdown characteristic is both nearly vertical and accurately reproducible. Such devices are called *Zener diodes* for historical reasons although it may well be that breakdown is due to the avalanche effect. They are used as voltage reference elements in voltage regulator circuits.

Another effect that can produce rapid increases in reverse current and characteristics closely resembling breakdown is known as *punch-through*. We have seen that the reverse current in a junction is due to a small number of thermally generated minority carriers. Now, as discussed in the previous section, the ohmic contacts to the bulk regions can generate large numbers of carriers. Consider, for example, the ohmic contact attached to the reverse-biased n-region. This will inject minority holes into the n-region, the holes being usually lost at small distances from the contact by recombination with majority electrons. However, if the n-region is relatively lightly doped, such that the depletion layer exists mostly in it, at some large reverse voltage, known as the punch-through voltage, V_{pt}, the n-region can become completely depleted of electrons and the depletion layer extends right to the end-contact. Under these conditions the large number of holes generated at the end-contact cannot recombine with electrons since there are none with which to recombine, but they add to the thermally generated minority holes and are swept into the p-region. When this happens, the diode is punched through and large reverse currents flow. We can estimate the punch-through voltage for an abrupt junction by using Eq. (7.47), replacing the contact voltage, V_0, by the junction voltage under reverse bias conditions, $(V_0 + V)$. Then the thickness of the depletion layer in the n-region is

$$d_n = \left(\frac{2\epsilon(V_0 + V)N_a}{eN_d(N_a + N_d)} \right)^{1/2}$$

If we assume that $N_a \gg N_d$, $V \gg V_0$, then the depletion-layer thickness equals the length of the n-region, l_n, when

$$V = V_{pt} = eN_d l_n^2 / 2\epsilon \tag{7.74}$$

Thus, the effect is critically dependent on device dimensions, which contrasts with the other breakdown mechanisms described earlier.

7.13 The tunnel diode

Operation of this device, alternatively called an Esaki diode after its inventor, is based on the possibility of quantum-mechanical tunnelling by electrons through a potential barrier as described in Chapter 1. If the p- and n-regions of a junction are extremely heavily doped, say with 10^{25} or more impurities per cubic metre, the Fermi level on the p-side of the junction lies *below* the top of the valence band and on the n-side it lies *above* the bottom of the conduction band; the materials are *degenerate*. Furthermore, the junction width becomes sufficiently narrow for tunnelling effects to become possible. Energy band diagrams for such a junction in equilibrium and with varying degrees of forward bias are shown in Fig. 7.22.

In equilibrium, Fig. 7.22(a), tunnelling currents across the junction in either direction are equal and no net junction current flows. For small forward bias voltages, Fig. 7.22(b), electrons from the conduction band in the n-type material can tunnel through the potential barrier at the junction, provided it is sufficiently thin, into vacant sites in the valence band at the p-side that have the same energy. For larger forward bias voltages, Fig. 7.22(c), there are no longer vacant sites in the valence band in the p-region at the same level as sites occupied by electrons in the conduction band in the n-region; tunnelling is no longer possible and the current reverts to the normal forward current. For bias voltages between these two extremes, the current decreases as the voltage increases, which results in an effective negative resistance and a forward $I–V$ characteristic of the form shown in Fig. 7.23. The positive peak in this characteristic corresponds to the condition in Fig. 7.22(b).

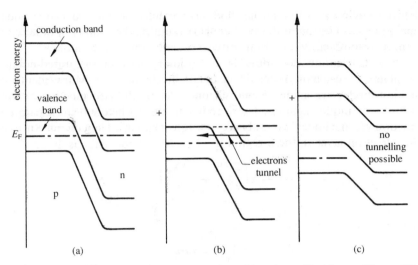

Fig. 7.22   Band structure of a tunnel diode: (a) with zero bias voltage, (b) with a small forward bias voltage and (c) with a large forward bias voltage.

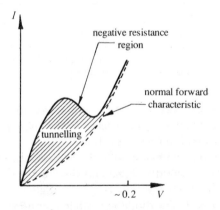

Fig. 7.23   I-V characteristic of a tunnel diode.

If the tunnel diode is incorporated in a suitable resonant circuit and biased appropriately, its negative resistance can be made to compensate for ohmic losses in the circuit and undamped oscillations can occur. Since tunnelling takes place almost instantaneously, very high frequency oscillators, up to say 100 GHz, can be built that utilize this effect in tunnel diodes.

7.14 Fabrication technology

Whereas it is beyond the scope of this book to discuss in detail all the manufacturing process involved in making devices, a brief description of the essential configuration of various diodes is presented here in order to provide a more realistic view of the devices but also to confirm the validity or otherwise of the simplified models used already to describe their performance.

Diode fabrication nowadays is a subsidiary part of integrated-circuit manufacture, which is discussed in detail in Chapter 10. Many identical junction diodes are made on one silicon wafer by diffused junction or ion injection techniques, again as described in Chapter 10, before being separated into discrete device chips ready for suitable external connections. In one possible arrangement, shown in Fig. 7.24, the diode chip is bonded to

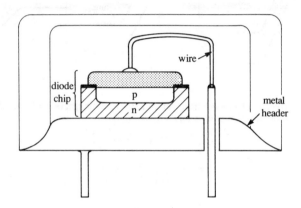

Fig. 7.24   A discrete junction diode.

a metallic *header*, which serves as one contact; the other is provided by an insulated lead-through, which is connected to the evaporated top contact by a bonded gold or aluminium wire, as shown.

Discrete diodes produced by the *epitaxial process*, described in Chapter 10, are now being manufactured commercially on a large scale. The process involves the growth of a very thin layer of single-crystal semiconductor, usually silicon, in which the diode (or other component) is formed, onto a supporting slice of parent material. Discrete diodes with a high breakdown voltage and a low capacitance can be produced in the high-resistivity epitaxial layer, while the low-resistance supporting substrate reduces the series resistance of the diode.

It is, of course, possible to use as a discrete diode any of the various pn junction configurations that exist in the standard silicon integrated circuit (IC) transistor technology described in Chapter 10. The choice of the most suitable depends on the circuit application of the diode, in particular the required switching speed and breakdown voltage.

The available diode configurations are shown in Fig. 7.25. A base–collector diode is shown in Fig. 7.25(a). The emitter diffusion is floating and can be omitted altogether. This diode has the advantage of a high breakdown voltage,

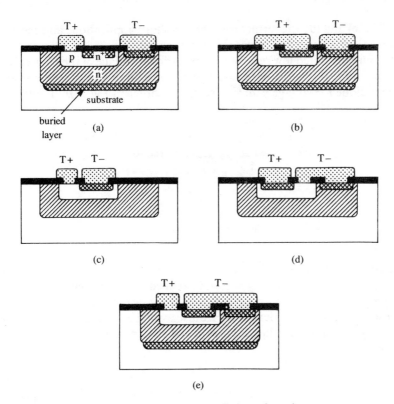

Fig. 7.25   Possible junction diode configurations.

V_{BD}, of around 5.0 V. However, it has a relatively long switching time, t_{off}, of about 100 ns, owing to the collector access resistance, R_{cc}. The switching time can be improved to about 70 ns by shorting the emitter and base to remove charge stored at that junction, while retaining the high breakdown voltage, as shown in Fig. 7.25(b).

The base–emitter junction diode with collector floating is illustrated in Fig. 7.25(c). The turn-off time, due mainly to charge stored in the base–collector junction, is about 80 ns and it has a low V_{BD}, associated with the highly doped emitter, of around 5 V. Again, the switching time can be reduced to as low as 20 ns, by shorting base and collector, to remove stored junction charge, as shown in Fig. 7.25(d), the low V_{BD} being unaffected.

Finally, it is possible to form a diode from the emitter–base and base–collector junctions in parallel, by shorting emitter and collector, as shown in Fig. 7.25(e). However, this arrangement is little used because of its high associated junction capacitance, which produces a low switching speed of around 150 ns, together with a poor breakdown voltage of about 5 V, associated with the base–emitter junction.

Obviously, the most widely used diode geometries are those illustrated in

Figs 7.25(b) and (d), the former for higher-voltage applications and the latter where switching speed is of paramount importance.

The already noted low reverse bias breakdown voltage of the emitter–base junction in the IC transistor technology can be used to advantage to produce a Zener diode for voltage control circuits, as shown in Fig. 7.26. Emitter and

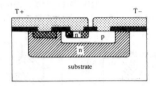

Fig. 7.26 Zener diode chip.

collector regions on the chip are shorted together, a reverse bias voltage exceeding V_{BD} and of polarity shown is applied between them, and the base and junction breakdown occurs as discussed in Sec. 7.12.

Discrete Schottky barrier diodes are also available via the integrated circuit technology route, but further discussion of their configuration will be deferred to Chapter 10.

Problems

1. The bulk of the n-type region of a particular germanium junction has a conductivity of $10^4\,\mathrm{S\,m^{-1}}$ at 300 K and that for the p-region is $10^2\,\mathrm{S\,m^{-1}}$. Find the voltage drop across the junction in equilibrium at 300 K, assuming $n_i = 2.5 \times 10^{19}\,\mathrm{m^{-3}}$, $\mu_e = 0.36$ and $\mu_h = 0.17\,\mathrm{m^2\,V^{-1}\,s^{-1}}$.

Ans. 0.36 V

2. An ideal abrupt pn junction has the following properties: doping concentration on the p-side, $10^{24}\,\mathrm{m^{-3}}$; doping concentration on the n-side, $10^{22}\,\mathrm{m^{-3}}$; area of cross section, $10^{-6}\,\mathrm{m^2}$; hole mobility, $0.2\,\mathrm{m^2\,V^{-1}\,s^{-1}}$; electron mobility, $0.4\,\mathrm{m^2\,V^{-1}\,s^{-1}}$; diffusion length of minority holes, $2 \times 10^{-4}\,\mathrm{m}$; diffusion length of minority electrons, $3 \times 10^{-4}\,\mathrm{m}$; bulk relative permittivity, 16; intrinsic carrier density, $10^{19}\,\mathrm{m^{-3}}$. Evaluate the following parameters of the diode at room temperature: (a) the majority and minority concentrations; (b) the conductivities of each region; (c) the contact potential; (d) the diffusion constants for each carrier type; (e) the diode saturation current; (f) the current flowing when 0.25 V is applied to the diode with the positive terminal connected to the p-type material; (g) the current flowing for large reverse bias voltages; (h) the battery polarity for the application of reverse bias; (i) the width of the depletion layer when a reverse bias voltage of 10 V is applied; (j) the incremental depletion-layer capacitance under the bias conditions as for (i); and (k) the approximate ratio of hole to electron current across the junction.

Ans. (a) 10^{24}, 10^{22}, 10^{16}, $10^{14} \mathrm{m}^{-3}$, (b) $3.2 \times 10^4 \mathrm{S\,m}^{-1}$, $6.4 \times 10^2 \mathrm{S\,m}^{-1}$, (c) 0.46 V, (d) 0.005, $0.01 \mathrm{m}^2 \mathrm{s}^{-1}$, (e) $0.04\,\mu\mathrm{A}$, (f) 0.88 mA, (g) $0.04\,\mu\mathrm{A}$, (h) negative terminal to n-side, (i) $1.4\,\mu\mathrm{m}$, (j) 100 pF, (k) 50:1

3. The current flowing in a certain pn junction at room temperature is 2×10^{-7} A when a large reverse bias voltage is applied. Calculate the current flowing when a forward bias of 0.1 V is applied.

Ans. $10\,\mu\mathrm{A}$

4. A pn junction has an observed saturation current of $1\,\mu\mathrm{A}$ at room temperature. Find the junction voltage corresponding to forward currents of 1 and 10 mA. The diode is constructed with p-type material of $2000 \mathrm{S\,m}^{-1}$ and n-type of $500 \mathrm{S\,m}^{-1}$. Each region is 1 mm long and 1×0.5 mm in cross section. Find the applied voltages corresponding to currents of 1 and 10 mA, including resistance drops in the n- and p-regions.

Ans. 0.175 V, 0.233 V, 0.180 V, 0.283 V

5. A planar pn junction in silicon has conductivities of $1000 \mathrm{S\,m}^{-1}$ and $20 \mathrm{S\,m}^{-1}$ and minority-carrier lifetimes of $5\,\mu\mathrm{s}$ and $1\,\mu\mathrm{s}$ for the p- and n-regions respectively. Calculate the ratio of hole current to electron current in the depletion layer, the saturation current density and the total current density flowing through the junction with a forward bias of 0.3 V. Assume room temperature, $n_i = 1.4 \times 10^{16} \mathrm{m}^{-3}$, $\mu_e = 0.12 \mathrm{m}^2 \mathrm{V}^{-1} \mathrm{s}^{-1}$ and $\mu_h = 0.05 \mathrm{m}^2 \mathrm{V}^{-1} \mathrm{s}^{-1}$.

Ans. 28.8:1, $0.5\,\mu\mathrm{A\,m}^{-2}$, $0.081 \mathrm{A\,m}^{-2}$

6. The reverse bias saturation current for a particular pn junction is $1\,\mu\mathrm{A}$ at 300 K. Determine its a.c. slope resistance at 150 mV forward bias.

Ans. $78\,\Omega$

7. An ideal silicon pn junction diode has a reverse saturation current of $30\,\mu\mathrm{A}$ at a temperature of 125°C. Find the dynamic resistance of the diode for a bias voltage of 0.2 V in the forward and the reverse direction.

Ans. $3.5\,\Omega$, $380 \mathrm{k}\Omega$

8. A germanium pn junction diode has a cross-sectional area of $10^{-6} \mathrm{m}^2$ and the distance from the metallurgical junction to each ohmic contact is 0.1 mm. The bulk resistivities and minority-carrier lifetimes are $4.2 \times 10^{-4}\,\Omega\,\mathrm{m}$ and $75\,\mu\mathrm{s}$ for the p-type and $2.08 \times 10^{-2}\,\Omega\,\mathrm{m}$ and $150\,\mu\mathrm{s}$ for the n-type. Assuming

$\mu_e = 0.30 \, \text{m}^2 \, \text{V}^{-1} \text{s}^{-1}$ and $\mu_h = 0.15 \, \text{m}^2 \, \text{V}^{-1} \text{s}^{-1}$ and that $n_i = 2.5 \times 10^{19} \, \text{m}^{-3}$, find the theoretical saturation current of the diode.

Ans. $3.82 \, \mu\text{A}$

9. The zero-voltage barrier height in a germanium pn junction is 0.2 V. The concentration of acceptor atoms on the p-side, N_a, is much smaller than that of donor atoms at the n-side and $N_a = 3 \times 10^{20}$ atom m^{-3}. The relative permittivity of germanium is 16. Calculate the width of the depletion layer for applied reverse voltages of 10 V and 0.1 V and a forward bias of 0.1 V. If the cross-sectional area of the diode is 1 mm^2, evaluate the junction capacitances corresponding to the various applied voltages. What significance have the results?

Ans. $7.6 \, \mu\text{m}$, $1.3 \, \mu\text{m}$, $0.77 \, \mu\text{m}$, 18.6 pF, 116 pF, 184 pF

10. In a particular junction diode, the n-side initially has a resistivity of $2 \times 10^{-4} \, \Omega \, \text{m}$ and the p-side $4 \times 10^{-3} \, \Omega \, \text{m}$. If the doping concentration on the n-side were to be increased by a factor of 10, what would be the corresponding percentage change in the contact potential? Assume electron and hole mobilities of 0.13 and 0.05 m^2 V^{-1}s^{-1} and an intrinsic carrier density of $1.5 \times 10^{16} \, \text{m}^{-3}$. Comment on the result.

Why does V_o not appear in the rectifier equation for a diode?

Ans. 7 per cent

11. Prove that the magnitude of the maximum electric field, E_m, in an abrupt pn junction with $N_a \gg N_d$ is given by

$$E_m = 2V_j/d_j$$

where V_j is the junction voltage and d_j is the depletion-layer width, which can be assumed to be given by the usual expression

$$d_j = (2\epsilon V_j/eN)^{1/2}$$

Zener breakdown occurs in germanium at a field intensity of $2 \times 10^7 \, \text{V m}^{-1}$. Find the resistivity of the n-type material in the above diode, if the Zener breakdown voltage of the diode is 10 V. Assume an electron mobility of 0.38 m^2 V^{-1}s^{-1} and a relative permittivity of 16.

Ans. $9.3 \times 10^{-4} \, \Omega \, \text{m}$

12. A pn junction diode is to be used as a voltage variable capacitor. How would it normally be biased? What would be the voltage dependence of the capacitance under these conditions?

The voltage-controlled capacitor is used in a tuned circuit where the resonant frequency, f_r, is proportional to (diode capacitance)$^{1/2}$. If it is required that f_r be tunable over a 2:1 frequency range, what ratio of applied bias voltages would achieve this result?

Ans. 16:1

13. Assuming that in an asymmetrically doped, abrupt p$^+$n junction the depletion-layer width is of the form

$$d \simeq (2\epsilon V/eN)^{1/2}$$

where the symbols have their usual meaning, derive an expression for the incremental depletion-layer capacitance of the junction, C.

For a particular silicon diode known to be of the above type, a graph of C^{-2} against reverse bias voltage is found to be a straight line of slope $5.88 \times 10^{23} \, \text{F}^{-2} \, \text{V}^{-1}$. Assuming the area of the diode to be $10^{-8} \, \text{m}^2$ and a relative permittivity of 12, find (a) the doping concentration in the n-region, and (b) the breakdown voltage of the diode. The breakdown field for silicon is $30 \, \text{MV} \, \text{m}^{-1}$.

Ans. (a) $2 \times 10^{21} \, \text{m}^{-3}$, (b) $150 \, \text{V}$

14. An abrupt silicon pn junction is made from material with $10^{24} \, \text{m}^{-3}$ acceptors in the p-type and $10^{21} \, \text{m}^{-3}$ donors in the n-type. Assuming a relative permittivity for silicon of 12, calculate the reverse bias required to generate a depletion width of $5 \times 10^{-6} \, \text{m}$. Will the depletion region be predominantly in the p-region or the n-region? What is the incremental depletion-layer capacitance under these conditions?

If the breakdown field of silicon is $30 \, \text{MV} \, \text{m}^{-1}$, will the diode have broken down at the above reverse bias? What is the depletion width and reverse bias at the point of breakdown?

Ans. $18.8 \, \text{V}$, n, $21 \, \mu\text{F} \, \text{m}^{-2}$, no, $20 \, \mu\text{m}$, $300 \, \text{V}$

15. An abrupt pn junction has an overall length of 2 mm, the junction being in the centre of its length, and a rectangular cross section 1 mm $\times$ 0.5 mm. On the p-side the doping concentration is $10^{22} \, \text{m}^{-3}$ and on the n-side it is $2 \times 10^{22} \, \text{m}^{-3}$.

When reverse biased by a few volts, at a temperature of 20°C, the diode passes a current of $1 \, \mu\text{A}$. Find the applied forward bias voltage across the diode terminals when diode currents of (a) 1 mA and (b) 10 mA flow, assuming carrier mobilities of 0.38 and $0.18 \, \text{m}^2 \, \text{V}^{-1} \, \text{s}^{-1}$.

What are the consequences of the results as far as a practical diode is concerned?

Ans. (a) $0.175 \, \text{V}$, (b) $1.15 \, \text{V}$

16. A particular silicon pn junction diode is doped with $10^{22} \, \text{m}^{-3}$ donors and $10^{21} \, \text{m}^{-3}$ acceptors. The diode has a cylindrical geometry, being $200 \, \mu\text{m}$

diameter and $20\,\mu\text{m}$ long; the junction is mid-way along the cylinder and normal to the axis and contacts are made at either end of the cylinder. Assuming carrier mobilities of 0.13 and $0.05\,\text{m}^2\,\text{V}^{-1}\,\text{s}^{-1}$, an operating temperature of $20°\text{C}$ and an intrinsic carrier density of $1.38 \times 10^{16}\,\text{m}^{-3}$, find: (a) the contact potential at the junction; (b) the junction voltage for a current of $100\,\text{mA}$ to flow assuming a saturation current for the diode of $1\,\text{nA}$; (c) the bulk internal resistance of the diode; and (d) the terminal voltage applied when the current conditions in (b) apply.

Ans. (a) 0.62 V, (b) 0.47 V, (c) 41 Ω, (d) 4.6 V

17. A certain abrupt pn junction diode has a cross-sectional area of $10^{-6}\,\text{m}^{-2}$ and the distance from the metallurgical junction to each ohmic contact is $1\,\text{mm}$. The resistivity of the p-side is $4 \times 10^{-4}\,\Omega\,\text{m}$ and the contact potential, V_0, is 0.35 V.

Assuming an intrinsic carrier density, n_i, of $2.5 \times 10^{19}\,\text{m}^{-3}$, carrier mobilities of 0.30 and $0.15\,\text{m}^2\,\text{V}^{-1}\,\text{s}^{-1}$ and an operating temperature of $20°\text{C}$, find (a) the carrier densities at either side of the junction, and (b) the ratio of hole to electron currents across the junction.

When a forward bias voltage of 0.6 V is applied across the ohmic contacts of the diode a current of $100\,\text{mA}$ flows. (c) What current would flow in the diode if the polarity of the applied voltage were reversed?

Ans. (a) $10^{23}\,\text{m}^{-3}$, $10^{17}\,\text{m}^{-3}$, $6.2 \times 10^{21}\,\text{m}^{-3}$, $6 \times 10^{15}\,\text{m}^{-3}$, (b) 8.4:1, (c) 15 μA

18. A certain asymmetrically doped n^+p diode supplies current in the forward direction from a 10 V d.c. source to a $100\,\Omega$ resistive load. If the supply is reversed, the turn-off switching time is found to be $0.1\,\mu\text{s}$. Estimate the average reverse current flowing during the turn-off period, assuming the effective length of the p-region to be $0.5\,\mu\text{m}$ and an electron diffusion coefficient of $0.0031\,\text{m}^2\,\text{s}^{-1}$, commenting on the result.

Ans. 40 μA

19. A certain n^+p, emitter–base diode has a turn-off time of $0.2\,\mu\text{s}$ when a current of $1\,\text{mA}$ is switched, the mean reverse current being $1\,\mu\text{A}$. Estimate the effective width of the p-type region under these conditions, assuming a diffusion coefficient for electrons of $0.003\,\text{m}^2\,\text{V}^{-1}\,\text{s}^{-1}$.

Ans. 1.1 μm

20. When a certain Schottky diode has an alternating voltage of peak instantaneous value 0.02 V applied at room temperature, the maximum instantaneous forward current flowing in the diode is $20\,\mu\text{A}$. Find (a) the saturation current for the device, and (b) the maximum instantaneous current flowing in the reverse direction

Ans. (a) 16 μA, (b) 9 μA

21. A particular Schottky diode is known to have an internal series resistance. When a 1 V d.c. voltage is applied to its terminals, at 20°C, currents of 10 mA and 25 pA flow, depending on the polarity of the voltage. Find the internal resistance of the device, explaining all reasoning.

Ans. 50 Ω

22. A particular Schottky diode, operating at 17°C, passes a current of 0.55 A when a forward bias voltage of 1 V is applied and 1.53 A when 2 V is applied. Assuming the metal section has negligible resistance, find (a) the resistance of the bulk semiconductor region, and (b) the diode saturation current.

Ans. (a) 1 Ω, (b) 7.4 nA

23. A particular Schottky diode has a cross-sectional area of 10^{-6} m^2 and the resistivity of the semiconductor side of the junction is 10^{-2} Ω m. When 0.2 V is applied to the diode so as to forward bias it, 1 mA flows, and when the voltage is increased to 0.25 V the current flowing is 5 mA. Find the thickness of the semiconductor side of the junction, assuming an operating temperature of 20°C.

Ans. 0.23 mm

24. Two ideal metal–n-type semiconductor Schottky diodes are operated in parallel. Under certain operating conditions one passes 10 times the current of the other and the dynamic resistance of the combination, at 20°C, is 2.7 Ω. Find the total current passed by the combination and the dynamic resistance of each rectifier separately.

Ans. 9.35 mA, 29.7 Ω, 2.97 Ω

25. A Schottky diode is sometimes used to prevent a transistor from saturating. A possible particular arrangement is shown in the diagram.

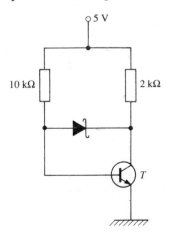

Assuming a drop of 0.4 V across the diode when it conducts, a base–emitter voltage of 0.7 V and $h_{fe} = 100$ (see Sec. 8.3), calculate the base and collector currents and also the current in each resistor.

Assuming an operating temperature of 300 K, determine the saturation current of the diode.

Ans. 0.028 mA, 2.75 mA, 0.43 mA, 2.35 mA, 78 pA

26. A hypothetical device consists of a p-type semiconductor, resistivity $0.1\,\Omega$ m, diameter 100 μm, thickness 2 μm and workfunction 1 eV, sandwiched between two different metal contacts, M_1 with workfunction 1.4 eV and M_2 with workfunction 0.6 eV. Explain in general terms how the structure would behave electrically.

When a voltage of 0.5 V is connected between M_1 and M_2 with M_2 positive, 10 nA flows. Estimate the current flowing with the polarity reversed, assuming operation at 290 K, explaining the reasoning behind the calculation. Neglect all surface-charge effects.

Ans. 6.5 mA

8 The bipolar junction transistor

8.1 Introduction

The physical processes that determine the electrical behaviour of bipolar junction transistors (BJT) will be discussed in this chapter. It will become evident that the operation of transistors in this class depends on the interaction of both majority and minority carriers, and because two carrier types are essential such devices are classified as *bipolar*. This term serves to differentiate the devices discussed here from another main class, unipolar transistors, in which the current is transported by majority carriers only; these will be discussed later.

8.2 Phenomenological description of current transport in the abrupt-junction bipolar transistor

The bipolar junction transistor consists essentially of a single-crystal semi-conductor, most often of silicon or germanium, which contains a narrow central region of opposite conductivity type to that of the rest of the material. For example, an npn transistor contains a narrow p-type layer sandwiched between two n-type layers with ohmic contacts made to each region, as shown diagrammatically in Fig. 8.1(a). Now, the physical construction of a modern

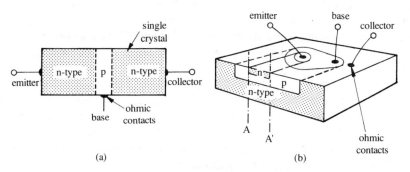

Fig. 8.1 (a) Model of an npn transistor; (b) a planar transistor structure.

transistor, which will be discussed in more detail later, may well be very different in appearance from the simple model shown; for example, a planar version of a discrete transistor structure may be as shown in Fig. 8.1(b). However, it will be seen that sections at A, A' encompass an npn sandwich not unlike the model and for the moment at least the simple model will be sufficient to help understand the basic transistor action.

The transistor thus consists of two back-to-back pn junctions closely coupled electrically by a narrow region of material common to both. In normal operation, one of the junctions is forward-biased and the other reverse-biased, as shown in Fig. 8.2. Briefly, what happens is that minority

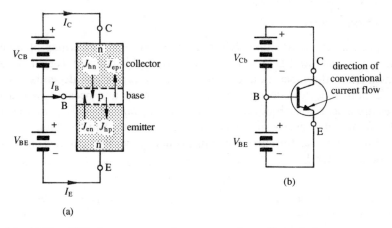

Fig. 8.2 (a) Normal biasing arrangement for an npn transistor; (b) circuit element representation.

carriers are injected from the *emitter* into the *base* region as a consequence of the forward bias appearing across the emitter–base junction and, because the base is deliberately made very thin, nearly all the injected carriers reach the reverse-biased base–collector junction, eventually determining the *collector* current. Although these physical processes have resulted in convenient labels for the various regions, not too much significance should be attached to the names, since in some applications a particular part of the transistor may not behave as its name suggests; for instance, the collector may be emitting rather than collecting electrons.

Of course, transistor action will also be possible if the conductivity type of emitter, base and collector regions is reversed so as to produce a pnp device. This will behave in a similar way to the npn transistor, provided the polarities of the bias voltages are reversed, so for simplicity we shall initially restrict our attention to the npn device, assuming that parallel arguments can easily be developed for the complementary device.

The npn transistor is usually preferred in Si since the mobility and diffusion coefficient for minority electrons in the base are higher than for the holes in the base of a pnp device, which, as we shall see, leads to a higher operating speed. Typically, pnp transistors have a turn-over frequency at which the gain falls to unity of around 1.5–2 GHz. The corresponding maximum operating frequency for an npn transistor approaches 10 GHz. There are also technological reasons for the preferential choice of the npn configuration. For example, the highly doped buried layer beneath collectors in planar epitaxial transistors used in ICs, which is necessary, as will become apparent, for low-resistance access to the collector, must be produced using a dopant that does not diffuse further during subsequent processing stages. Arsenic is an ideal material for such layers, creating an n^+ buried layer, which dictates that collectors must be n-type, in an overall npn device.

In most modern transistors, transport of carriers across the base is by diffusion and drift in the presence of density gradients of carriers and of an electric field. It will be convenient to consider these two processes independently so that initially the base region will be considered field-free, thus ensuring that transport of minority carriers across it is exclusively by diffusion. It will also be simpler at first to assume that the junctions are abrupt by virtue of a doping profile that changes rapidly with distance, as discussed in Chapter 7.

The band structure of an unbiased, symmetrically doped npn structure in equilibrium will be as shown in Fig. 8.3(a). The relative widths of base and depletion regions are much exaggerated for clarity. As a result of the necessary

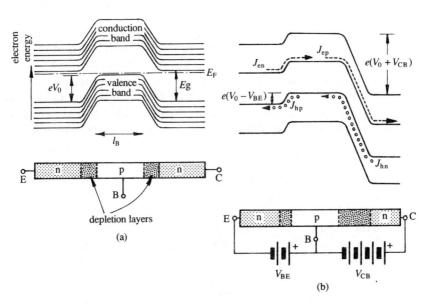

Fig. 8.3 Band structure of an npn transistor; (a) with no bias applied and (b) with external bias voltages.

continuity in the Fermi level through the transistor in equilibrium, contact potentials and potential barriers are established at the emitter–base and base–collector junctions which prevent diffusion of majority carriers across the junctions, as described earlier. When the correct d.c. bias voltages are applied the band structure is modified to that shown in Fig. 8.3(b). The potential barrier between emitter and base is reduced by virtue of the forward bias to $e(V_0 - V_{BE})$ and an electron current is injected from emitter in to base region, J_{en}, and a hole current in the reverse direction, J_{hp}. We shall see later that efficient transistor operation is achieved if most of the current across this junction is carried by carriers originating in the emitter, i.e. $J_{en} \gg J_{hp}$. This is achieved by doping the emitter to a higher degree than the base. The reverse bias at the collector–base junction causes an increase in barrier height to $e(V_0 + V_{CB})$; as a consequence, there is no majority-carrier diffusion and the only current flow across the junction is due to the motion of minority carriers. Thus a very small saturation current, I_{co}, called the collector leakage current, flows across the collector–base junction, which has contributions J_{ep} and J_{hn} from minority carriers in the base and collector regions.

Turning our attention now to the base region, let us first assume that its width l_B is longer than the diffusion length for minority carriers L_e and L_h. The density of minority electrons in the p-type base will be increased above the equilibrium value near the emitter–base junction because of the presence there of electrons injected from the base. This excess of electrons decays exponentially away from the junction because of recombination with majority holes, reaching the equilibrium value at a distance of order L_e from the junction. Near the reverse-biased base–collector junction the electron density is lower than the equilibrium value since minority electrons from the base are swept into the collector, but again the equilibrium density is restored at distances greater than about L_e from the junction. The electron density profile for the thick base transistor is then as shown in Fig. 8.4(a). This situation is artificial in that an essential property of the base region for efficient transistor operation is that it is very narrow and made of high-lifetime material, such that its length is much smaller than the diffusion length for minority electrons, or $l_B \ll L_e$. The

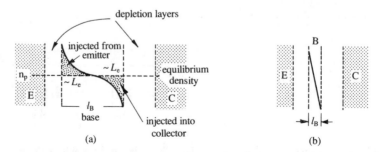

Fig. 8.4   Electron density profile in the base region for (a) a wide base and (b) a narrow base.

minority-carrier concentration in the base region is then as shown in Fig. 8.4(b). Under these circumstances there is little recombination in the base region and nearly all the electrons injected from the emitter diffuse across the base because of the steep electron density gradient existing in it and are eventually swept across the base–collector junction down the potential hill and into the collector. If, as is usually the case, the emitter is more heavily doped than the base, so that the majority of the current across the emitter–base junction is transported by electrons, then the collector current, I_C, is only slightly less than the emitter current, I_E.

It will be noticed that the electron current flowing into the collector is largely independent of V_{CB}, provided this is large enough to prevent majority-carrier diffusion across the junction; as a consequence, the collector circuit has a high impedance. This is in contrast to the emitter circuit, which has a very low impedance because small voltage changes at the emitter–base junction cause large changes in the current flowing across it. Since the collector current is nearly equal to the emitter current, as explained, the large difference in impedance level between collector and emitter circuits can result in potentially high power amplification. Incidentally, this description of the action of the device in terms of the transfer of current from a low- to a high-impedance circuit accounts for its original name, transfer resistor, which was subsequently contracted to transistor.

If we now consider the base current, I_B, it can be seen from Fig. 8.5 that it

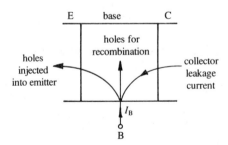

Fig. 8.5 Components of the base current in a transistor.

comprises three components: (a) the hole current flowing across the base–emitter junction, J_{hp}, which is made as small as possible by high emitter doping; (b) the collector leakage current, I_{co}; and (c) a hole current flowing into the base to replace holes that are lost by recombination with electrons flowing across it; this current can be made very small by reducing recombination to a minimum by ensuring that the base is thin and that the lifetime of minority electrons in it is high. Thus the base current is normally much smaller than either collector or emitter currents.

8.3 Gain parameters of the bipolar transistor

Let us assume initially that the external circuit of the transistor is as shown in Fig. 8.2; the arrangement is the *common-base* connection, since the base contact is common to both collector and emitter circuits. If the emitter current is changed by increment ΔI_E, there will be a corresponding incremental change in collector current, ΔI_C, and the relative change is described in terms of a gain parameter, which is defined by

$$\frac{\text{change in collector current}}{\text{change in emitter current}} = \left.\left(\frac{-\Delta I_C}{\Delta I_E}\right)\right|_{V_{CB}\text{const.}} = \alpha_B \qquad (8.1)$$

where α_B is the *common-base current gain*. Since by previous arguments a change in emitter current results in a change in collector current that is only marginally smaller, $\Delta I_C \simeq \Delta I_E$ and α_B is only slightly less than unity, typical values being in the range 0.900–0.999.

Although it has been convenient to discuss the common-base operation of a transistor first, it is far more usual for the device to be operated in the *common-emitter* circuit configuration shown in Fig. 8.6, in which the emitter

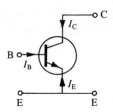

Fig. 8.6   Common-emitter configuration.

connection is common to input and output circuits. Arguing as before, a change in base current, ΔI_B, produces a corresponding change in collector current, ΔI_C, and a *common-emitter current gain*, α_E (or alternatively h_{fe} – see Sec. 8.5), is then defined by

$$\alpha_E = \left.\left(\frac{\Delta I_C}{\Delta I_B}\right)\right|_{V_{CE}\text{const.}} \qquad (8.2)$$

Since by Kirchhoff's law

$$I_B + I_C + I_E = 0$$

then

$$\Delta I_B = -(\Delta I_C + \Delta I_E)$$

Using Eq. (8.1) to eliminate ΔI_E gives

$$\Delta I_B = \Delta I_C[1 - (1/\alpha_B)]$$

and finally

$$\Delta I_C/\Delta I_B = \alpha_E = \alpha_B/(1-\alpha_B) \qquad (8.3)$$

Thus, we arrive at a relationship between the common-base and common-emitter current gains. Since α_B is usually very nearly unity, it is evident from Eq. (8.3) that α_E is much greater than unity, typically being in the range 10–1000.

It is possible to estimate the value of the gain parameter α_B, and hence α_E, through Eq. (8.3), in terms of the physical processes occurring in the transistor, its structure and its composition. It is evident that α_B will be dependent on (a) the number of electrons injected from the emitter into the base and (b) the proportion of these which diffuse across the base, without recombination, to the collector. It is therefore convenient to subdivide the gain parameter, α_B, into two components:

$$\alpha_B = \eta_E \beta \qquad (8.4)$$

where η_E, the *injection efficiency*, is the ratio of the electron current injected into the base from the emitter to the total emitter–base junction current, and β, the *base transport factor*, is the ratio of the electron current at the collector junction to that at the emitter junction. Current transport in an npn transistor in terms of these components of α_B is shown schematically in Fig. 8.7. Since the

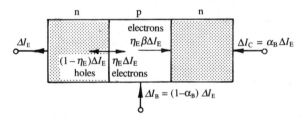

Fig. 8.7   Current transport in an npn transistor (note that I_{co} is omitted).

current flowing into the base to replace holes lost by recombination is $\eta_E \Delta I_E(1-\beta)$, and the base current flowing to provide the hole current across the emitter junction is $\Delta I_E(1-\eta_E)$, then, ignoring the collector leakage current, the total base current is

$$\Delta I_B = \eta_E \Delta I_E(1-\beta) + \Delta I_E(1-\eta_E) = (1-\alpha_B)\Delta I_E$$

which is the same as that required to satisfy Kirchhoff's law.

We now consider the components of α_B in more quantitative detail.

8.3.1 Emitter injection efficiency, η_E

From the definition given above and by referring to Fig. 8.2(a) it follows that the emitter injection efficiency can be written as

$$\eta_E = J_{en}/(J_{en} + J_{hp}) \tag{8.5}$$

It is tempting to use the arguments outlined is Sec. 7.6 and assume that the ratio of electron to hole currents at the junction is approximately equal to the ratio of emitter to base conductivities, but this would apply only if the base and emitter widths were substantially longer than the minority-carrier diffusion lengths, which we know is not usually the case in a transistor. A rigorous derivation of η_E would follow the analysis for the pn junction with finite dimensions, which is outlined in Sec. 7.7, but the following approximate approach will be sufficient for our purposes.

Consider first the density variation of minority carriers to the base as illustrated in Fig. 8.8. Electrons injected from the emitter raise the local

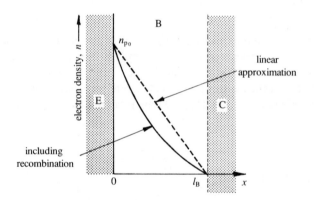

Fig. 8.8 Minority-carrier density in the base of an npn transistor.

electron density just inside the base from n_p, the equilibrium value, to n_{p0}, given by Eq. (7.11) as

$$n_{p0} = n_p \exp(eV_{BE}/kT) \tag{8.6}$$

At the collector end of the base region the electron density is depressed below the equilibrium value by virtue of the reverse bias there, to a value $n_p \exp(-eV_{BC}/kT)$, which can be assumed zero since usually $V_{BC} \gg kT$. Since the length of the base, $l_B \ll L_e$, the diffusion length for minority carriers, there will be little recombination in the base and the minority-carrier density falls off almost linearly with distance. Assuming such a constant density gradient, the electron diffusion current is

$$J_{en} = eD_e n_{p0}/l_B \tag{8.7}$$

On comparing this equation with (7.20) for the electron current flow in a wider n-type region of a pn junction, it will be noticed that the difference between the two expressions is that l_B in Eq. (8.7) replaces L_e in the earlier equation. It follows that Eq. (7.22) is applicable to the thin-base, thin-emitter junction provided that L_e and L_h are replaced by l_B and l_E, respectively. Using this modified equation and applying similar arguments to those used in Sec. 7.6 gives

$$\frac{J_{en}}{J_{hp}} \simeq \frac{D_e n_p}{l_B} \div \frac{D_h p_n}{l_E} \simeq \frac{\sigma_E}{\sigma_B} \frac{l_E}{l_B} \tag{8.8}$$

Substituting this expression in Eq. (8.5) gives the emitter injection efficiency as

$$\eta_E \simeq \left(1 + \frac{\sigma_B}{\sigma_E} \frac{l_B}{l_E}\right)^{-1} \simeq 1 - \frac{\sigma_B l_B}{\sigma_E l_E} \tag{8.9}$$

Thus, the emitter efficiency is largely controlled by the relative doping of base-emitter regions but it is also influenced by the ratio of the lengths of such regions. For example, if the conductivity of the emitter is made, say, 100 times that of the base, then provided the base length is less than the emitter length, which is usually the case, η_E is in excess of 99 per cent, and such values can be achieved in practice.

An alternative approach is used in the heterostructure bipolar transistor (HBT) described in Sec. 12.6. Semiconductors having different energy gaps are chosen to form the emitter and base such that the resulting heterojunction, as will be seen, contains band offsets that create a relatively small barrier to injected electrons from the emitter but a large barrier to minority holes flowing from base into emitter. As a consequence, a high emitter efficiency is ensured, which is independent of the relative doping of emitter and base.

8.3.2 Base transport factor, β

The base transport factor gives the efficiency with which minority carriers are transported across the base region, and for the particular case of an npn transistor

$$\beta = \frac{J_e|_{collector junction}}{J_e|_{emitter junction}} \tag{8.10}$$

Since in this expression the numerator is only slightly less than the denominator because of the usually slight recombination that takes place in the base, a more accurate method for determining J_e is required than that described in the previous section; the linear approximation to the electron density profile in the base region is no longer good enough to calculate the electron current, and the lower curve of Fig. 8.8 is now applicable. The shape of this curve can be found by applying the continuity equation for excess electrons in the base region, δn, which gives

$$d^2(\delta n)/dx^2 = \delta n/L_e^2$$

which, as we have seen, has the general solution

$$\delta n = C_1 \exp(-x/L_e) + C_2 \exp(+x/L_e) \tag{8.11}$$

The constants C_1 and C_2 can be determined from the boundary conditions, which are

$$\delta n = n_{p0} - n_p \qquad \text{at } x = 0$$

and

$$\delta n = -n_p \simeq 0 \qquad \text{at } x = l_B, \text{ since } n_p \ll n_{p0}$$

Applying these boundary conditions to Eq. (8.11) gives

$$\delta n = \frac{n_{p0} - n_p}{1 - \exp(2l_B/L_e)}[\exp(x/L_e) - \exp(2l_B/L_e)\exp(-x/L_e)] \tag{8.12}$$

Now, since the electron diffusion current is proportional to the gradient of the electron density given by Eq. (8.12), performing the differentiation and substituting in Eq. (8.10) gives

$$\beta = \frac{J_e|_{x=l_B}}{J_e|_{x=0}} = \frac{[\mathrm{d}(\delta n)/\mathrm{d}x]|_{x=l_B}}{[\mathrm{d}(\delta n)/\mathrm{d}x]|_{x=0}} = \frac{2\exp(l_B/L_e)}{1 + \exp(2l_B/L_e)}$$

Since $l_B \ll L_e$ usually, this expression simplifies to

$$\beta = [1 + \tfrac{1}{2}(l_B/L_e)^2]^{-1} \simeq 1 - \tfrac{1}{2}(l_B/L_e)^2 \tag{8.13}$$

Thus, for high base transport factors giving current gains as near to unity as possible, the lifetime of minority electrons in the base region must be high so as to make L_e large. An advantage of Si or Ge in this respect is that relatively high lifetimes in the range 1–100 μs are realizable; for base widths of 1–5 μm, which are readily fabricated using modern technology, transport factors in the range 0.95 to in excess of 0.99 can result.

8.4 Non-ideal transistor structures

Although the model used so far is satisfactory for explaining basic transistor mechanisms, it needs considerable modification before it is applicable to physically realizable transistors. Some of the additional effects that occur in real transistors and the manner in which these limit their electrical performance will now be discussed.

8.4.1 Avalanche breakdown and multiplication

An upper limit is set on the collector voltage V_{CB} by avalanche breakdown in the reverse-biased collector–base junction, as discussed previously in Sec. 7.12. Fields of order 10^6 V m^{-1} are required in the depletion layer for breakdown to occur, which usually limits V_{CB} to a maximum of several tens of volts.

For collector voltages lower than that necessary for the onset of avalanche breakdown, V_{BD}, there is a voltage range in which minority electrons in the base–collector depletion layer are accelerated sufficiently to cause electron-hole pair production by ionizing collisions with the lattice, but the holes produced by the process are not sufficiently energetic when accelerated in the field to produce the secondary ionization that is essential to maintain a self-sustained breakdown. In this voltage range, although no complete breakdown occurs, there is some electron multiplication and the number of electrons collected is greater than the number arriving at the base–collector layer edge. Under these conditions the current gain, α_B, is increased by a factor, η_c, called the *collector efficiency*, which has been found to be given empirically by

$$\eta_c = (1 - V_{CB}/V_{BD})^{-n} \qquad (8.14)$$

where n is usually in the range 2–4. It is thus possible for the effective current gain, which includes this additional factor due to avalanche multiplication, to exceed unity; the emitter circuit may then display negative-resistance effects, which can lead to undesirable instabilities.

8.4.2 Base-width modulation and punch-through

It has been tacitly assumed that the base width, l_B, is always constant, but in a real device the effective width of the base, $l_{B,eff}$, is dependent on V_{CB} and to a lesser extent on V_{BE}. This is because the boundaries that delineate the extent of the base are the edges of the depletion layers at the junction, which vary in position with changing bias voltages. For example, the assumed almost-zero minority-carrier concentration occurs at the edge of the collector–base depletion layer, which in turn is dependent on the collector voltage V_{CB}. If the base is relatively lightly doped, the collector–base depletion layer extends mostly into the base and the effective base width, using Eq. (7.49), is given approximately by

$$l_{B,eff} \simeq l_B - \left(\frac{2\epsilon V_{BC}}{eN_a}\right)^{1/2} \qquad (8.15)$$

where l_B is now the distance between the metallurgical junctions and N_a is the acceptor concentration in the base. Therefore, as V_{CB} is increased, the effective base width is reduced; there is a corresponding increase in the slope of the minority-carrier density profile as shown in Fig. 8.9, which leads to a higher collector current. It follows that the emitter efficiency, the base transport factor and current gains α_B and α_E are dependent in a non-linear way on the voltage V_{CB}.

The variation of collector current with changing V_{CB}, an effect known as *base-width modulation*, also affects the output characteristics of a transistor since the I_C–V_{BC} curves are no longer horizontal but take on a positive slope

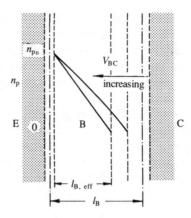

Fig. 8.9 Variation of effective base width with collector voltage V_{BC}.

indicating that the device has a finite output impedance that is voltage-dependent. The input characteristics are also affected; since the input circuit behaves as a forward-biased diode and the input current is given by

$$I \simeq I_0 \exp(eV_{BE}/kT)$$

where I_0 is dependent on the effective base width, the input impedance is then to some extent influenced by V_{CB}.

At high collector voltages or for low doping concentrations in the base, it is possible for the collector–base depletion layer to extend completely across the base region, thus effectively short-circuiting the collector to the emitter. The minimum voltage for this *punch-through* effect to occur can be obtained by letting $l_{B,eff}$ go to zero in Eq. (8.15), which gives

$$V_{BC}|_{max} = eN_a l_B^2 / 2\epsilon \tag{8.16}$$

Punch-through may thus set an upper limit to the permissible collector voltage but in many transistors the maximum value of V_{BC} is set by the onset of avalanche breakdown, which often occurs at lower voltages. It is of course possible to raise the punch-through voltage by increasing the doping concentration in the base, but since this automatically reduces the emitter efficiency, η_E, the particular choice of N_a is an engineering compromise.

8.4.3 Base resistance

It has been assumed in the simple transistor model that all externally applied voltages appear across the relatively high-resistance depletion layers and that voltages dropped across the bulk semiconductor regions are negligible. While it is usually permissible to ignore the resistance of collector and emitter regions in this way, the base region is relatively lightly doped in modern transistors and effects due to its finite resistance must normally be taken into account.

Consider, for example, the planar transistor shown in Fig. 8.1(b) and in section in Fig. 8.10. Base current, I_B, flowing to the active base region, passes through a region that can have a significantly high resistance, which is represented by a lumped resistance, r_B, in the diagram. As a consequence, the effective

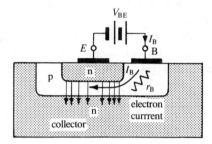

Fig. 8.10   Base resistance and emission crowding in a real transistor.

emitter–base voltage is given by

$$V_{BE,eff} = V_{BE} - I_B r_B \tag{8.17}$$

Unfortunately, the situation is more complicated than is suggested by Eq. (8.17), since I_B itself is dependent on $V_{BE,eff}$.

It should be noted that the arguments so far are also applicable to a.c. signals; indeed, whereas it is often possible to omit r_B when calculating biasing conditions, it is usually necessary to include an effective base resistance in the small-signal equivalent circuit.

A more serious consequence of a finite base resistance is that it causes *emission crowding* in the base. This arises because the base current, which is moving laterally in the active base region, i.e. perpendicular to the minority electron flow, causes a voltage drop across the face of the emitter that is in such a direction as to reduce the effective forward bias voltage in the centre of the emitter relative to that at the edges. Electron current from the emitter thus tends to concentrate towards the periphery of the emitter, as shown diagrammatically in Fig. 8.10. Under high-current conditions, irreversible damage can be caused by excessive current densities at emitter edges. At more modest current levels, emission crowding is not so serious and can be accounted for by inclusion of an additional resistance in the effective lumped base resistance.

8.4.4 Graded-base or drift transistors

The transistors discussed so far have been assumed to possess a uniformly doped base in which negligible electric fields exist; minority electron transport in this case is predominantly determined by diffusion effects. However, many modern transistors, particularly those made by the diffusion process, have an inherently non-uniform base doping profile. For many applications this is advantageous and in some devices, for example, high-frequency and switching transistors, such a doping profile is introduced deliberately. Devices with a non-uniform impurity concentration in the base are known as minority-carrier *graded-base* or *drift* transistors; the latter name arises since an electric field always exists in the base which causes the minority-carrier current in it to have a drift component.

The origins of the built-in electric field in the base of a drift transistor can be explained with reference to the typical doping profile and band structure of an npn device shown in Figs 8.11(a) and (b). It will be noticed that the net acceptor

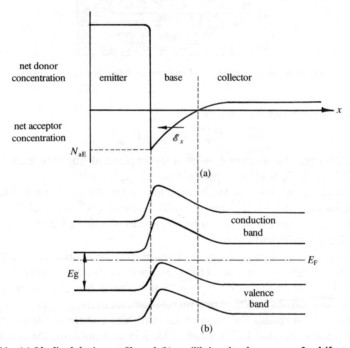

Fig. 8.11 (a) Idealized doping profile and (b) equilibrium band structure of a drift transistor.

level in the base has its largest value, N_{aE}, next to the emitter and falls to zero at the base–collector junction. The corresponding negative gradient in majority hole density causes holes to move towards the collector, thus exposing fixed ionized acceptors and creating negative space charge. The resulting electric field is in such a direction as to oppose further migration of holes and so maintains the net hole current due to drift and diffusion at zero. The base current due to majority holes will, in the absence of recombination, be given by Eq. (6.52):

$$J_h = e\mu_h p\mathscr{E}_x - eD_h \, \mathrm{d}p/\mathrm{d}x \qquad (8.18)$$

For low-level injection, the hole density at any point in the base, p, is approximately equal to the acceptor concentration there, so, for zero hole current, Eq. (8.18) yields

$$\mathscr{E}_x = \frac{D_h}{\mu_h N_a} \frac{dN_a(x)}{dx} = \frac{kT}{e} \frac{1}{N_a} \frac{dN_a}{dx} \qquad (8.19)$$

Since the impurity gradient is negative, $\mathscr{E}_x$ is directed towards the emitter as expected. Now, this built-in field, which prevents diffusion of majority carriers, is in such a direction as to cause electrons injected into the base to drift under the influence of the field towards the collector, and the net electron current in the base is given by

$$J_e = ne\mu_e \mathscr{E}_x + eD_e \, dn/dx \qquad (8.20)$$

Further, since carriers usually drift with faster velocities than they diffuse, the transit time of electrons across the base is very much shorter in a drift transistor than in a corresponding diffusion transistor in which $\mathscr{E}_x$ is assumed to be zero. This property has enabled drift transistors to be operated at frequencies in the gigahertz range.

It is desirable for some applications that the doping gradient be arranged such that the built-in field in the base is everywhere uniform. If the transistor is constructed so that the doping profile is exponential and of the form

$$N_a(x) = N_{aE} \exp(-Cx)$$

then Eq. (8.19) gives

$$\mathscr{E}_x = \frac{kT}{e} \frac{1}{N_a} (-C)N_a = -\frac{kTC}{e}$$

which satisfies the condition that $\mathscr{E}_x$ is constant and independent of position.

8.4.5 Geometrical effects

The flow of minority carriers through the base of a transistor has been assumed to be one-dimensional. Of course, in real transistors this is not so and the minority-carrier current flow from the emitter spreads out laterally to some extent in the base, before it arrives at the collector. The proportion of the current from the emitter that is collected thus depends in some part on the geometry of the transistor. For example, if the base region is thin and the area of the collector is made much greater than the area of the emitter, then the base transport factor approximates to the ideal value found for the one-dimensional model.

8.4.6 Transistors in the switching mode

So far, the bipolar transistor has been considered in its near-linear amplifying mode, but it is often used as an ON/OFF device in switching circuits, this being the dominant regime for the majority of integrated circuits.

To study switching behaviour, consider the drain characteristics of a bipolar transistor in the common-emitter configuration, shown in Fig. 8.12. The intersection of the load line corresponding to the load resistor, R_L, with a particular drain characteristic determines the collector voltage, V_{CE}, for a given base current, i_B. The amplifying region for a common-emitter amplifier

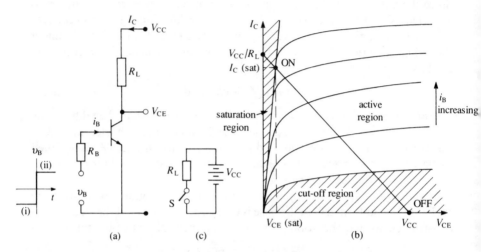

Fig. 8.12 (a) An npn transistor in common-emitter mode with a step voltage applied to the base; (b) collector characteristics; and (c) the ideal equivalent circuit for switching.

in which the output current responds almost linearly to changes in input current is shown as the active region (unhatched). However, if the base current is made zero or negative, by making the base voltage, v_B, zero or negative, region (i) in Fig. 8.12(a), thereby reverse biasing both transistor junctions, the operating point for the transistor enters the *cut-off* region (shown hatched). In such an OFF state, at the bottom end of the load line, the collector current becomes almost zero and the collector voltage almost equals V_{CC}, the collector supply voltage. The transistor is virtually open-circuit between collector and emitter, corresponding to switch, S, in Fig. 8.12(c) being open.

If the base current is subsequently driven large and positive, for example by a positive pulse change in v_B as shown in (ii) Fig. 8.1(a), the transistor switches into the *saturation region*, again shown hatched, via the active region, which is traversed at a rate that is dependent on factors such as gain and frequency response. In this ON condition, large collector currents flow and the collector voltage falls to a very low value, called $V_{CE}(sat)$, typically around 0.2 V for a silicon transistor, most of the supply voltage now being dropped across R_L. The transistor is virtually a short-circuit in this state, which almost duplicates the closed condition of the ideal switch, S, in Fig. 8.12(c).

In the ON state, the collector current, $I_C(sat)$, is dependent on the load resistance and is given by

$$I_C(sat) = [V_{CC} - V_{CE}(sat)]/R_L \qquad (8.21)$$

This corresponds to a base current required to drive the transistor into saturation of

$$i_B \geqslant I_C(\text{sat})/h_{fe} \tag{8.22}$$

As i_B is increased above this minimum value, to drive the transistor hard into saturation, the bias across the collector–base junction changes from zero at the outset to become *forward* biased when i_B is bigger. Under these conditions, both junctions are injecting electrons into the base, which accumulates an excess of stored electronic charge. The distribution of minority electrons in the base for the three operating regimes is shown in Fig. 8.13, the excess stored

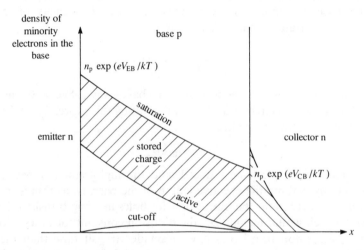

Fig. 8.13 **Minority electron distribution in the base for an npn transistor for cut-off, active and saturated operating conditions.**

electron charge in saturation being denoted by the hatched region. The electron density at the emitter–base junction, from Eq. (8.6), rises from around zero at cut-off to $n_p \exp(eV_{EB}/kT)$ for saturation, as shown, but the corresponding densities at the collector–base junction are $n_p\exp(-eV_{CB}/kT)$, which is again near zero, rising to $n_p\exp(+eV_{CB}/kT)$, corresponding to electron injection from the collector and the change in bias polarity. There will also be stored minority-carrier charge in the collector region of an ON transistor as shown, but this is much smaller and can often be ignored.

As for the pn diode discussed earlier, switching between ON and OFF states has to be accompanied by corresponding changes in the electronic charge stored in the base. For example, if the transistor is hard ON and is then switched into the OFF mode, all the stored charge shown in Fig. 8.13 has to be discharged before the current can adjust to the very small value corresponding to the cut-off condition. Such a process cannot occur instantaneously, so removal or insertion of stored charge out of or into the base leads to finite switching times. High-speed switching circuits are often designed to avoid this difficulty by arranging that transistors are not allowed to saturate, thus reducing switching times.

8.4.7 High-frequency performance of transistors

The minority carriers take a finite time to traverse the base region of a transistor and it is this *transit time*, τ_t, that is usually the major factor limiting high-frequency performance. As the operating frequency approaches τ_t^{-1}, the transistor becomes inoperative.

The transit time for electrons to cross the base region of length l_B in an npn transistor can be estimated as follows. Let x be the distance from the emitter–base junction, $v(x)$ the velocity and $n(x)$ the density of minority electrons. The transit time is then given by

$$\tau_t = \int_0^{l_B} [1/v(x)]\, dx \tag{8.23}$$

If the simple model for charge storage in the base given in Sec. 7.10 is assumed, the velocity $v(x)$ is given by Eq. (7.63). Substituting this expression in Eq. (8.21) and performing the integral gives

$$\tau_t \simeq l_B^2/2D_e \tag{8.24}$$

It is evident that τ_t can be reduced, and the high-frequency performance improved, by making the effective length of the base, l_B, as thin as possible. A similar expression for the transit time for holes in a pnp transistor could be derived, which would include D_h, the diffusion constant for holes. However, since $D_e > D_h$, npn transistors have a smaller transit time than their pnp equivalents and are to be preferred for high-frequency or high-speed operation. The base transit time can be reduced further if electrons are accelerated across the base by an electric field, rather than moving solely under the influence of diffusion. This property is exploited in drift transistors, which have a built-in drift field in the base region arising from a deliberately introduced doping profile there (see Section 8.4.4).

There is an additional transit time, τ_C, associated with carriers moving through the collector–base depletion layer. For collector voltages, V_{CB}, greater than a few volts, electrons in this region reach a saturated velocity, v_{sat}, of around 10^5 m s^{-1}. If the depletion-layer thickness, which itself is dependent on the collector voltage, resistivity and so on, is d_{CB}, then

$$\tau_c \simeq d_{cB}/v_{sat} \tag{8.25}$$

Finally there are two more time constants, associated with the capacitance of the pn junctions in a transistor. First, the time constant due to the emitter–base junction, τ_{EB}, is given by

$$\tau_{EB} \simeq r_B C_{EB} \tag{8.26}$$

where C_{EB} is the capacitance of the forward-biased junction and r_B its slope resistance. Substituting approximate values for these gives

$$\tau_{EB} \simeq A\left(\frac{kT}{eI_e}\right)\left(\frac{e\epsilon_r\epsilon_0 N_{OB}}{2V_j}\right)^{1/2} \tag{8.27}$$

where N_{OB} is the acceptor concentration in the base near to the emitter and V_j the net junction voltage. Hence in order to reduce τ_{EB}, the junction area, A, and N_{OB} must be kept as small as possible.

The time constant related to the base–collector junction, τ_{BC}, is given by

$$\tau_{BC} \simeq R_{cc}C_{CB} \tag{8.28}$$

which can be reduced by decreasing the collector access resistance, R_{cc}, or the doping concentration in the collector, or by increasing V_{CB}.

The sum of all four time constants provides a value for the total delay time, τ_T. The high-frequency operating limit corresponding to a particular value of τ_T, f_α, is then approximately

$$f_\alpha \simeq (\tau_T)^{-1} \tag{8.29}$$

Here, f_α is the alpha cut-off frequency, at which the gain, α_B, falls 3 dB below its low-frequency value.

8.4.8 High-frequency response

Gain parameters for a junction transistor have been determined in earlier sections assuming near-equilibrium conditions and are only applicable to low-frequency or slowly varying signals. The expressions obtained are modified when the transistor is operated at high frequencies. For example, the current gain in the grounded-base connection, α_B, is not constant but falls off with increasing frequency in the manner shown in Fig. 8.14. The fall-off in gain

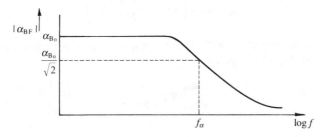

Fig. 8.14 **High-frequency common-base gain of a junction transistor.**

arises principally because of the finite time taken for minority carriers to diffuse across the base; when the transit time becomes comparable to the periodic time of an applied signal the minority carriers can no longer respond fast enough and the gain falls to zero. The *cut-off frequency*, f_α, is defined as that frequency at which the magnitude of the current gain, α_{BF}, falls by 3 dB to $1/\sqrt{2}$

of its low-frequency value, α_{B0}. The value of f_α can be estimated by considering the continuity equation for minority electrons in the base, assuming no electric field exists there, which is, from Eq. (6.63)

$$\frac{\partial(\delta n)}{\partial t} = -\frac{\delta n}{\tau_{Le}} + D_e \frac{\partial^2(\delta n)}{\partial x^2} \tag{8.30}$$

The term on the left-hand side of the equation, which was neglected in the time-independent solution, must be retained when discussing high-frequency effects. If small sinusoidal alternating signals of angular frequency ω are superimposed on the d.c. currents flowing in the base under equilibrium conditions, the solution to the continuity equation is expected to be of the form

$$\delta n(x, t) = \delta n(x) \exp[j(\omega t + \phi)] \tag{8.31}$$

Substituting this trial solution into Eq. (8.30) gives

$$0 = -\delta n \frac{(1 + j\omega\tau_{Le})}{\tau_{Le}} + D_e \frac{\partial^2(\delta n)}{\partial x^2}$$

or

$$\frac{\partial^2(\delta n)}{\partial x^2} = \frac{\delta n(1 + j\omega\tau_{Le})}{L_e^2} \tag{8.32}$$

Comparing this equation with that given in Sec. 8.3.2 it will be noticed that L_e^2 in the d.c. continuity equation has to be replaced by $L_e^2/(1 + j\omega\tau_{Le})$ to give the a.c. version of the equation. Equation (8.13) will therefore still be valid for the base transport factor in the a.c. case, provided L_e^2 is replaced by $L_e^2/(1 + j\omega\tau_{Le})$. Thus, the gain at high frequencies, α_{BF}, neglecting the frequency dependence of the emitter efficiency and assuming it to be near unity, becomes

$$\alpha_{BF} \simeq \beta_F = \left(1 + \frac{l_B^2(1 + j\omega\tau_{Le})}{2L_e^2}\right)^{-1} \tag{8.33}$$

This equation, together with the corresponding low-frequency equation derived from Eq. (8.13), gives

$$\frac{\alpha_{BF}}{\alpha_{B0}} = \frac{1}{1 + j\omega/\omega_c} \tag{8.34}$$

where

$$\omega_c = \frac{2L_e^2 + l_B^2}{\tau_{Le} l_B^2} \simeq \frac{2L_e^2}{\tau_{Le} l_B^2} = \frac{2D_e}{l_B^2} \tag{8.35}$$

It is evident from Eq. (8.34) that $|\alpha_{BF}|$ falls to $\alpha_{B0}/\sqrt{2}$ when $\omega = \omega_c$ and, hence, from Eq. (8.35),

$$f_\alpha = \omega_c/2\pi \simeq D_e/\pi l_B^2 \tag{8.36}$$

Although a transistor can be operated at higher frequencies, Eq. (8.36) gives an indication of the frequency at which the gain is falling off rapidly and at which phaseshift distortion becomes apparent. It will be noticed that, according to the approximate analysis presented, the α cut-off frequency is dependent only on the diffusion coefficient in and the width of the base. A thin base is again advantageous for good high-frequency performance and there is also some advantage in using npn transistor structures because the diffusion rate for minority electrons is higher than for holes. However, there are physical limitations to the reduction in base width and a more useful method of reducing the transit time, and hence increasing the high-frequency operating capabilities of a transistor, is to introduce a drift field in the base region by means of some degree of impurity grading, as discussed in Sec. 8.4.4.

8.5 Small-signal equivalent circuit

Although the bipolar transistor is inherently non-linear for large-amplitude signal variations, it may be considered to bahave in a linear manner over a limited range of its operating characteristics and a small-signal equivalent circuit can be derived to represent its electrical performance. Strictly, such a circuit will only therefore be applicable to small-amplitude a.c. or incremental d.c. signals. Many different equivalent circuits have been proposed; an equivalent circuit based on our discussions of the physical processes that take place in a transistor in the grounded-emitter configuration will be derived as an example. The parameters of the circuit will then be compared with those of a more usually encountered, more generally applicable circuit, that is based on a four-terminal network, 'black-box' approach.

Consider the base-emitter current of an npn bipolar transistor in the common-emitter connection (see, for example, Fig. 8.5). In the normal operating mode of the transistor, the base–emitter junction will be forward-biased and the small-signal input voltage, v_i, will be applied in series with the d.c. bias voltage V_{BE}. The input impedance presented to the signal includes the effective dynamic resistance of the forward-biased junction. Since

$$I_B \simeq I_0 \exp(eV/kT)$$

and

$$\partial I_B / \partial V = e I_B / kT$$

it follows that the dynamic resistance is given by

$$r = kT/eI_B \simeq 1/(40 I_B) \qquad \text{at room temperature} \qquad (8.37)$$

which is a low resistance at room temperature and normal bias voltage. For example, for typical standing currents, I_B, of order milliamps, r is a few tens of ohms. The input circuit also includes the bulk resistance of the base region and this is usually lumped together with the dynamic resistance to give an effective input resistance, r_{BE}.

Turning now to the output circuit, a large dynamic impedance will exist at the collector–base junction by virtue of the reverse d.c. bias across it and this results in a high value for the output resistance, r_{CE}, which is the resistance looking back into the collector–emitter terminals.

The signal current in the base, i_B, will be amplified and appear in the collector circuit as a current $\alpha_E i_B$, where α_E is the common-emitter current gain. Therfore, a simple, low-frequency, small-signal equivalent circuit based on the physical processes discussed so far might be of the form shown in Fig. 8.15(a), where a current generator $\alpha_E i_B$ is included in the collector circuit.

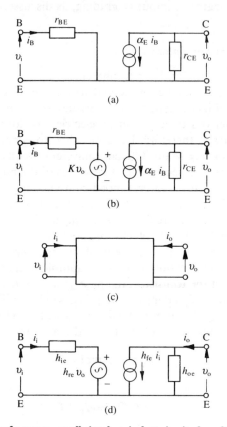

Fig. 8.15 Possible low-frequency, small-signal equivalent circuits for a bipolar transistor.

The simple equivalent circuit shown in Fig. 8.15(a) is highly idealized and can be refined somewhat so as to include base-width modulation effects by the addition of an extra component in the circuit to account for the built-in feedback between output and input circuits. The feedback arises because a change in output voltage changes the effective base width, resulting in changes in collector, emitter and hence base currents. The sign of the change is such that an increase in v_0 leads to a decrease in i_B. The effect is accounted for in the more comprehensive equivalent circuit shown in Fig. 8.15(b) by including a voltage generator proportional to the output voltage in the base circuit, which is of such a polarity as to cause a reduction in the base current when the output voltage is increasing.

It is sometimes more convenient to treat the transistor as a two-port active network, as shown in Fig. 8.15(c), and develop an equivalent circuit from measurements that can be made at the two ports, rather than devising a circuit based on the physical processes occurring in a particular device. Again, there are several different forms for the defining equations of the equivalent circuit, depending on which set of terminal characteristics is considered. As an example, one possible set of measured parameters might be

(a) the *input* impedance with output short-circuited,

$$h_{ie} = (v_i/i_i)|_{v_o = 0}$$

(b) the *reverse* open-circuit voltage amplification

$$h_{re} = (v_i/v_o)|_{i_i = 0}$$

(c) the *forward* current gain with output short-circuited

$$h_{fe} = (i_o/i_i)|_{v_o = 0}$$

and (d) the *output* conductance with input open-circuited

$$h_{oe} = (i_o/v_o)|_{i_i = 0}$$

These are the *hybrid parameters* of a transistor, so-called because they do not all have the same dimensions. The additional subscript 'e' is added to designate the circuit configuration, in this example common *emitter*. It follows that the small-signal voltages and currents at the input and output terminals of the equivalent circuit are then related by the equations

$$v_i = h_{ie}i_i + h_{re}v_o$$
$$i_o = h_{fe}i_i + h_{oe}v_o$$

(8.38)

The equivalent circuit shown in Fig. 8.15(c) used in conjunction with Eqs (8.38) can be used to define the small-signal performance of a transistor completely when it is included in a particular circuit.

The equivalence of the two circuit representations discussed can be seen by noting that Eqs (8.38) are also valid for the circuit shown in Fig. 8.15(d). This has obvious similarity to the equivalent circuit based on the internal physical processes occurring in a transistor, Fig. 8.15(b), and it follows by direct comparison of the two circuits that

$$h_{ie} \equiv r_{BE} \qquad h_{re} \equiv K \qquad h_{fe} \equiv \alpha_E \qquad h_{oe} \equiv 1/r_{CE}$$

8.6 Fabrication of junction transistors

Discrete transistors are usually made using one of the IC processes discussed more fully in Chapter 10, using various combinations of diffusion, epitaxy and ion-implantation techniques. Many transistors are usually fabricated per slice, the number being dictated by area-dependent considerations such as power rating and device complexity, transistor chips being separated subsequently and individually packaged.

Cross sections of a selection of possible transistor structures are shown in Fig. 8.16. The diffusion process for fabricating transistors of the type shown in

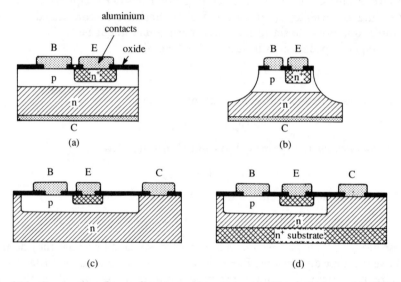

Fig. 8.16 Construction of some discrete transistor types: (a) diffused; (b) mesa; (c) planar; (d) epitaxial planar.

Fig. 8.16(a), which is described in Chapter 10, has many advantages. Until recently such technology was dominant but now newer techniques for junction formation, such as ion implantation and polysilicon emitters, are becoming progressively more important. In the *mesa* version, Fig. 8.16(b), the area of the collector junction is formed by an etching process, leaving the active portions isolated on a tapered plateau or mesa, which provides improved high -voltage performance. The planar discrete transistor, Fig. 8.16(c), is a direct derivative of the IC counterpart, all contacts being accessible from the top plane. Similarly for the epitaxial variant, Fig. 8.16(d), where the lightly doped epitaxial collector layer is grown on a highly doped supporting substrate, which provides a low-resistance path from collector terminal to the active collector region. Epitaxial diffused transistors have the additional advantage of reduced collector capacitance and higher breakdown voltages.

8.7 Silicon controlled rectifier

The silicon controlled rectifier, SCR, sometimes called a *thyristor*, is a four-layer, three-pn-junction, single-crystal silicon device, as shown in Fig. 8.17, which shows the principal features of a practical device, a possible model for the active silicon slice and the relative doping intensities in it. In an n^+pnp^+

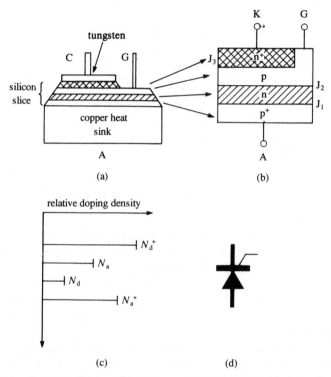

Fig. 8.17   The silicon controlled rectifier: (a) a cross section; (b) a model of the silicon slice; (c) relative doping in the layers; and (d) its symbolic representation.

device an *anode* connection, A, is made to the p^+-layer and contacts are also formed with the n^+ *cathode*, K, and the p-type *gate* region, G, as shown. Briefly, the SCR usually operates in a switching mode, when small injected gate currents control the point in the anode voltage cycle at which the device switches and large currents are allowed to flow.

SCRs are characterized by a very high-resistance OFF state in which the current flow between anode and cathode is very small, of order 1 mA, which on switching changes rapidly to a low-resistance ON state, via a negative-resistance region, in which large currents, typically kiloamps can flow. The general $I-V$ characteristics for such operations are therefore as indicated in Fig. 8.18. Three distinct operating regions are apparent, an OFF state, an ON

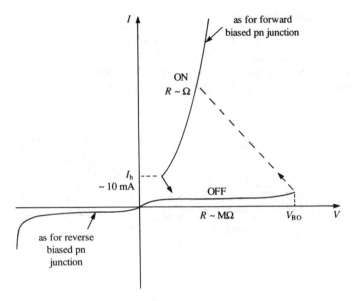

Fig. 8.18 *V–I* **characteristics of an SCR.**

state and, for negative applied anode voltages, a curve similar to that for a reverse-biased pn junction diode. Switching from the OFF state to the ON state can be accomplished either by applying an anode voltage, V_A, greater than the *breakover voltage*, V_{BO}, or by maintaining $V_A < V_{BO}$ and injecting an external triggering current, I_g, into the gate electrode. The device can be switched back into the OFF state by allowing the anode current to fall below a critical *holding current*, I_h, as shown. Currents in the range from milliamps to kiloamps can be switched, up to voltages in excess of kilovolts.

In operation, the SCR behaves in many respects like two bipolar transistors, n^+pn and p^+np, the properties of which are intimately linked by common regions, as shown in Fig. 8.19. Consider first the situation in which the gate current is zero, $I_g = 0$, for simplicity. In the forward bias condition

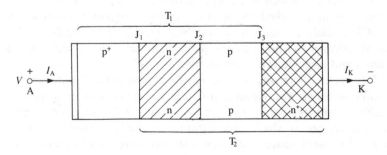

Fig. 8.19 **The four-layer diode,** $I_g = 0.$

assumed in the diagram when the anode voltage, V_A, is positive with respect to the cathode, junctions J_1 and J_3 are forward-biased but the current at low voltages is controlled by the reverse-biased junction, J_2. Under these conditions, only a very small saturation current flows across J_2, which consists of thermally generated minority electrons at either side of J_2 that are swept across the junction by the energetically favourable energy barrier that prevents majority carrier diffusion; see for example Fig. 8.20. However, since

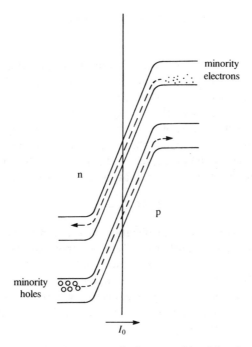

minority
electrons

n

p

minority
holes

I_0

Fig. 8.20 Carrier transport in the reverse-biased junction J_2.

J_2 also behaves as the base–collector junction of both overlapping transistors, there is an additional component of current at J_2 originating from the effective emitter sections. Consider, for example, the p^+n junction shown in Fig. 8.21. Since the acceptor concentration in the p^+-region, $N_a^+ \gg N_d$, the donor concentration in the n layer, most of the current crossing the forward-biased J_1 is transported by majority holes. The proportion of the junction current injected by holes is therefore $\eta_1 I_A$, where η_1 is the emitter injection efficiency of T_1, which will be almost unity. The injected holes become minority carriers in the n-region and the number of holes that survive recombination and reach the depletion layer of J_2 is $\beta_1(\eta_1 I_A)$, where β_1 is a 'base' transport factor. Contrary to the situation in a bipolar junction transistor (BJT), the length of the n-type 'base' region in an SCR, $l_n \gg L_h$, the diffusion length for minority holes, so

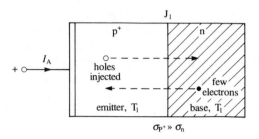

Fig. 8.21   Forward-biased junction J_1.

$\beta_1 \ll 1$ and most of the injected holes recombine in the base region and only a few reach J_2. Those that do are swept across the depletion layer at J_2 to add to the reverse bias saturation current, I_0.

At higher values of V_A, the voltage drop occurs mostly across the reverse-biased depletion layer of J_2, so the fields can become so high there that there exists a possibility of avalanche multiplication of the holes as they are accelerated across the depletion layers and collide with the lattice. So the total hole current reaching the p-region that originates from the p^+ 'emitter' is

$$\eta_1 \beta_1 M_1 I_A = \alpha_1 I_A \qquad (8.39)$$

where M_1 is an avalanche multiplication factor and α_1 is a common-base large-signal current gain. In other words the p^+np section behaves like a poor transistor because of the wide base region and, for low V_A, $\alpha_1 \ll 1$.

Turning to the n^+pn transistor section, by similar reasoning, an electron current $\alpha_2 I_K$ flows into its n-type collector, where $\alpha_2 = \eta_2 \beta_2 M_2$. Hence, provided $I_g = 0$, the total current crossing J_2, which must equal the external circuit current I_A, is given by

$$I_A = \alpha_1 I_A + \alpha_2 I_K + I_0 \qquad (8.40)$$

as illustrated in Fig. 8.22. It follows that, since $I_A = I_K$ if $I_g = 0$ as has been

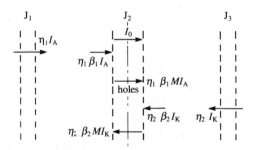

Fig. 8.22   Currents crossing junction J_2.

assumed, then the anode current becomes

$$I_A = \frac{I_0}{1-(\alpha_1+\alpha_2)} = \frac{I_0}{1-(\eta_1\beta_1 M_1 + \eta_2\beta_2 M_2)} \qquad (8.41)$$

At voltages below the breakover voltage, $V_A < V_{BO}$, the M and η factors are almost unity but the transport factors, β, are small so $I_A \simeq I_0$ and a very small reverse bias saturation current flows in this OFF condition.

At higher anode voltages, avalanche multiplication becomes more probable in the depletion layer of J_2, so that $M > 1$, until eventually, when $V_A = V_{BO}$ and

$$\eta_1\beta_1 M_1 + \eta_2\beta_2 M_2 \to 1 \qquad (8.42)$$

then I_A as indicated in Eq. (8.41) approaches infinity and the device switches into the ON state with current limited by circuit conditions.

The breakover voltage, V_{BO}, can be estimated, if it is assumed that $M_1 = M_2 = M$, which is given by Eq. (7.70) where $n \simeq 3$ for silicon and V is the applied junction voltage. Substituting this expression in Eq. (8.42) at breakover, when $V = V_{BO}$, gives

$$\eta_1\beta_1 + \eta_2\beta_2 = 1 - (V_{BO}/V_{BD})^n$$

or

$$V_{BO} = V_{BD}(1 - \eta_1\beta_1 - \eta_2\beta_2)^{1/n} \qquad (8.43)$$

It follows that the higher the gains, which are dependent on η and β, the lower the breakover voltage becomes.

Physically what is happening in the switching process is that the holes that are injected across J_1 are then swept across J_2 into the p-region where they become majority carriers again. Electrons are then injected across J_3 into the p-region to preserve charge neutrality. Similarly, electrons from the n^+-layer are eventually swept across J_2 into the n-region where neutrality is preserved by extra injected holes. Clearly, this is a regenerative process and it is possible for the total current to rise rapidly.

The holes swept into the p-region compensate some of the negative exposed acceptors fixed in the reverse-biased depletion layer of J_2, as shown in Fig. 8.23, thus reducing the effective depletion-layer charge. As the voltage is increased so that V_A approaches V_{BO}, the number of holes reaching the depletion layer increases, for the reasons explained, until sufficient ionized acceptors are compensated so that the *effective* depletion layer becomes similar to that for a forward-biased junction. In other words when the SCR is switched on, the potential barrier of J_2 is reduced until the junction is effectively forward-biased, as shown in Fig. 8.24. In this ON condition, large currents can flow but the anode voltage remains small.

The mode of operation discussed so far is known as *voltage triggering*. When gate currents are injected, so that $I_g > 0$, it is possible for an SCR to be *current triggered*, with $V_A < V_{BO}$. With gate current the situation is as depicted

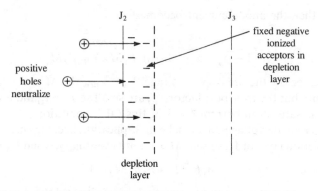

Fig. 8.23 Reduction in effective depletion-layer charge by injected carriers.

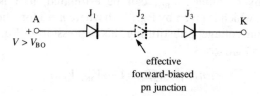

Fig. 8.24 Equivalent circuit of SCR for $V > V_{BO}$.

in Fig. 8.25. The current crossing J_2 is still given by Eq. (8.40) but this time Kirchhoff's law indicates that

$$I_K = I_A + I_g$$

so this can be substituted in the equation to give

$$I_A = \frac{\alpha_2 I_g + I_0}{1 - (\alpha_1 + \alpha_2)} \tag{8.44}$$

In the situation when $V_A < V_{BO}$ and I_g is increasing from zero, injection of I_g increases I_K and hence I_A. Now for gains $\ll 1$, the gain of a transistor is very

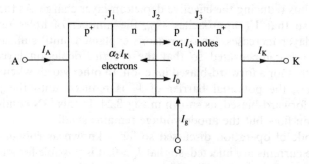

Fig. 8.25 SCR with gate triggering.

dependent on the emitter current, in this case I_A and I_K, so as I_g increases, so do I_A and I_K and hence α_1 and α_2. This continues until $(\alpha_1 + \alpha_2) = 1$ in Eq. (8.44), $I_A \rightarrow \infty$ and is only limited by the circuit resistance as the device switches ON. Since switching has occurred at $V_A < V_{BO}$, it follows that the gate current can be used to control the point at which switching occurs, and as I_g is increased, V_A for switching is reduced, as shown in Fig. 8.26. This is the most

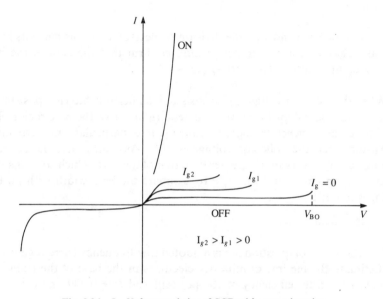

Fig. 8.26 *I – V* **characteristics of SCR with gate triggering.**

common triggering mode since it only requires low levels of power to control the switching of large currents. The ratio I_A/I_g at switching is of order 10^5, so only a very small gate current, typically of order 10–100 mA, is required to initiate switching.

Since the dissipation in the forward-biased gate junction is limited typically to a fraction of a watt, which it is possible to exceed with continuous gate currents of say 100 mA and corresponding gate voltages of a few volts, the trigger current is often pulsed, to prevent possible device failure. Provided the pulse of I_g is supplied for a sufficiently long time for the current multiplication process to occur and switching to take place, which is typically a few microseconds, then the duty cycle of the pulses can be made such that the average gate dissipation is never exceeded.

Dynamically the SCR is relatively slow. Its switching speed is typically of order 0.1 μs to turn ON and around 10 μs to turn OFF, which is much slower than a BJT. Its principal applications, therefore, are in the low-frequency, high-current field, for example for motor control in trains and drills or in light dimmers, and so on.

Problems

1. A certain pnp transistor has an effective base width of 20 μm under certain biasing conditions. The thickness of the emitter region is 5 μm and its resistivity is $50 \times 10^{-6} \, \Omega$ m. The effective base lifetime is 20 μs and $D_h = 0.0047 \, \text{m}^2 \, \text{s}^{-1}$. Estimate the common emitter current gain of the device.

   Ans. 81

2. Draw a diagram showing the division of electron and hole currents in the various regions of a pnp transistor and confirm that the sum of the base current components is $I_E(1-\alpha)-I_{co}$.

3. When the collector voltage of a transistor is sufficiently high it is possible for the collector–base depletion layer to extend right across the base region. This condition, called *punch-through*, is achieved in a particular npn germanium transistor when the collector voltage is 30 V. Assuming that the collector doping density is very much greater than that of the base, which contains 10^{21} acceptor atoms m^{-3} and that $\epsilon_r = 16$, estimate the base width with no bias voltages applied.

   Ans. 7.3 μm

4. The transistor of question 3 has a quoted low-frequency current gain α_E of 30. Estimate the lifetime of minority electrons in the base of the transistor. Assume an emitter efficiency of 100 per cent and $D_e = 0.0093 \, \text{m}^2 \, \text{s}^{-1}$.

   Ans. 0.09 μs

5. Show that the h_{fe} of a high-gain bipolar transistor is approximately of the form

$$h_{fe} \cong [C_1(\sigma_B/\sigma_E) + C_2/L_e^2]^{-1}$$

where constants C_1 and C_2 depend on the geometry.

   Under certain conditions of bias, the h_{fe} of a certain bipolar transistor is 450. Assuming the ratio of emitter to base conductivities is 100:1, a diffusion coefficient for carriers in the base of $3 \times 10^{-3} \, \text{m}^2 \, \text{s}^{-1}$ and emitter and base lengths of 10 and 20 μm, estimate the lifetime of carriers in the base.

   Ans. 3 μs

6. A certain npn transistor has a cross-sectional area of $10^{-8} \, \text{m}^{-2}$ and a doping density in the emitter of $10^{23} \, \text{m}^{-3}$. It is operated at 20°C with a base–emitter junction voltage of 0.65 V and a collector voltage of 10 V, when the collector current is 1.35 mA.

   Assuming electron and hole mobilities of 0.13 and 0.05 m^2 V^{-1} s^{-1} estimate:
(a) the base–emitter junction current, assuming a saturation current density of

$1\,\mu A\,m^{-2}$; (b) the density of majority carriers in the base, assuming that recombination in the base and collector leakage currents can be neglected; and (c) the ratio of electron to hole currents flowing across the base–emitter junction.

Ans. (a) $1.5\,mA$, (b) $2.5 \times 10^{22}\,m^{-3}$, (c) 10.4:1

7. If two bipolar transistors differ only in that one has a base width that is 90 per cent of the base width of the other, what will be the difference between the base–emitter voltages needed to maintain the same current flowing in the two transistors under reasonable forward bias and assuming an operating temperature of 290 K. It may be assumed that the number of injected majority carriers into the base is proportional to $\exp(eV/kT)$.

Ans. $1.1\,mV$

8. An npn bipolar transistor has a base region of thickness 0.025 mm. If the base region minority-carrier lifetime is $20\,\mu s$, what base transport factor might be expected?

The base region has a resistivity of $0.0005\,\Omega m$. Holes injected into the emitter region have a lifetime of $10\,\mu s$. What emitter efficiency might be expected for an emitter resistivity of $0.0001\,\Omega m$?

Estimate the common-base and common-emitter current gain of the device. Assume throughout that the hole and electron diffusion constants are 0.0044 and $0.0093\,m^2\,s^{-1}$ respectively.

Ans. 0.998, 0.975, 0.975, 39

9. A certain npn transistor has effective emitter and base lengths of $100\,\mu m$ and $20\,\mu m$ and the ratio of emitter to base conductivities is 50:1. Find the emitter efficiency of the transistor.

If the forward current gain of the transistor is 50 and the diffusion constant for carriers in the base is $3 \times 10^{-3}\,m^2\,s^{-1}$, estimate the lifetime of carriers in the base.

Ans. 0.996, $4\,\mu s$

10. A certain npn transistor has an effective base width of $20\,\mu m$ under given bias conditions. Estimate its forward current gain in the common-emitter configuration, h_{fe}, assuming a lifetime of minority carriers of $1\,\mu s$ and a diffusion constant of minority carriers of $0.01\,m^2\,s^{-1}$ in the base. Assume, for simplicity, an emitter efficiency of unity.

Ans. 49

11. In a particular bipolar transistor, the effective length of the base, l_B, under certain bias conditions, is $1\,\mu m$. When the bias is changed, so as to change l_B, the common-emitter forward current gain, h_{fe}, is found to decrease by a factor of 10. Estimate the effective length of the base under these new conditions, assuming an emitter efficiency of 100 per cent.

Ans. 3.2 µm

12. A particular bipolar transistor has a measured forward current gain, h_{fe}, of 200. Assuming an emitter injection efficiency of 99.8 per cent and a diffusion length for minority electrons of 13 µm, find the effective length of the base under these operating conditions.

Ans. 1 µm

9 Field-effect devices

9.1 General properties and types of field-effect transistor (FET)

This class of transistor may be distinguished from the solid-state devices discussed so far by several features that are common to all members of the class. First, the flow of carriers in a particular device is controlled by the application of an electric field, which permeates into the main conduction path in a semiconductor; this gives rise to the term *field effect*. Secondly, current flow along this main conduction path is almost entirely due to the motion of majority carriers. Injection of minority carriers, a mechanism that is essential for the operation of the bipolar transistor as described in Chapter 8, is not a necessary requirement in field-effect devices. The generic term *unipolar* is therefore used as an alternative to *field effect* to describe the devices, since they rely on only one type of carrier for current transport.

Before going on to describe the various types of field-effect transistor (FET), it will be useful to list some further properties that distinguish them as a class from bipolar transistors. First, they have extremely high input impedances, ranging from 10^{10} to $10^{15} \Omega$ depending on the type, often with lower noise levels than bipolar devices. The performance of bipolar transistors is degraded by neutron radiation because of the reduction in minority-carrier lifetime, whereas field-effect transistors can tolerate a much higher level of radiation since they do not rely on minority carriers for their operation. Field-effect devices are particularly attractive for use in integrated circuits because of their relatively small area and hence high packing density, their lack of need for additional isolation, and their ability to be directly coupled. Finally, their performance is relatively unaffected by ambient temperature changes.

The FET consists essentially of a semiconducting path the resistance of which is controlled by the application of a transverse electric field. There are two main categories of such devices, MOSFETs (metal–oxide–silicon field-effect transistors) or MOSTs (metal–oxide–silicon transistors) and JFETs (junction field-effect transistors). In the MOSFET there are no semiconductor junctions, the controlling electric field being applied to regulate the resistance of a main conducting path via an insulating layer. In contrast, in the JFET the resistance of the current path is modulated by the application of bias voltages to pn junctions adjacent to it.

There are two further subcategories into which a particular unipolar device can be classified. In a *depletion* device, the controlling electric field reduces the number of majority carriers available for conduction, whereas in the *enhancement* device, application of electric field causes an increase in the majority-carrier density in the conducting regions of the transistor.

Whereas, historically and for some specialist applications, JFETs are still significant, MOSTs have become much more important and are now the most widely used active electronic device, particularly because of their development for large-scale integrated (LSI) circuits. MOSTs have even threatened the well-entrenched BJT (bipolar junction transistor) in many integrated circuits, but, as we shall see, each transistor type has unique advantages that ensure its continued application.

The various field-effect transistors will now be discussed in more detail and some of the introductory points will be substantiated. In spite of the relatively minor role of the JFET in modern electronics, it will be useful to consider it as an initial example of a field-effect device that is perhaps more easily understood and can serve as a useful introduction to the now more technologically significant MOST.

9.2 The junction field-effect transistor (JFET)

The junction field-effect transistor exists in several practically realizable geometries. For example, Fig. 9.1(a) shows a cylindrical version sometimes

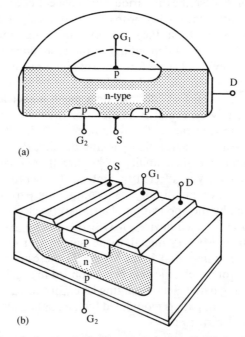

Fig 9.1 Construction of a junction field-effect transistor: (a) cylindrical and (b) planar geometry.

used for discrete transistor manufacture and Fig. 9.1(b) shows the planar form, which is most suitable for inclusion in integrated circuits. However, in order to investigate the basic general characteristics of such devices without added complications due to complex device geometry, we shall consider the most rudimentary configuration consisting of an n-type, single-crystal, semiconducting bar with ohmic contacts at either end and two p-type contacts on opposite sides, as shown in Fig. 9.2(a).

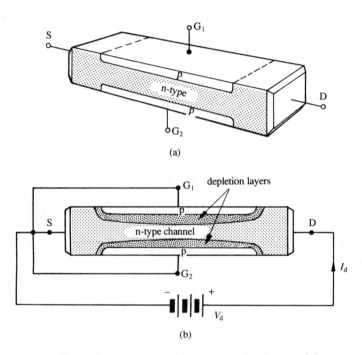

(a)

(b)

Fig 9.2 (a) Simplified model of a JFET and (b) its connection for grounded-gate operation.

If the two p-type regions called *gates*, G_1 and G_2, are shorted to the left-hand end-contact of the bar, called the *source*, S, and a small positive voltage, V_d, is applied to the right-hand contact, the *drain*, D, as in Fig. 9.2(b), a current I_d flows in the external circuit, as shown. Since the bulk of the material is n-type, the current is transported in the semiconductor by majority electrons flowing from source to drain; the left-hand contact is thus an electron source and the right-hand contact drains electrons out. Since the bar has ohmic resistance, the flow of current in it produces an IR drop and the potential at any point in the bar increases from the source to the drain end, becoming more positive towards the drain. Therefore, since the gates have been shorted to the source, the pn junctions between the p-type gates and the n-type bar are reverse-biased, becoming progressively more so towards the drain end. Now, we have seen that the thickness of the depletion layer of a reverse-biased pn junction varies approximately as the square root of the bias voltage, the layer becoming wider as the reverse voltage increases. Thus, the depletion layers between gates and bar become wider towards the drain end, as shown in Fig. 9.2(b). Further, the ratio of the thickness of the depletion layer in a p-type gate, d_p, to that in the n-type bar, d_n, at any point is given by Eq. (7.45) as

$$d_p/d_n = N_d/N_a$$

where N_a and N_d are the doping concentrations in the two regions. Therefore, if, as is usually the case, the p-type gates are much more heavily doped than the n-type conducting path, i.e. $N_a \gg N_d$ and $d_p \ll d_n$, most of the depletion-layer thickness occurs in the n-type region. As a consequence, current flow in the device is confined to the wedge-shaped region or *channel*, shown in Fig. 9.2(b), since there are no free carriers in the depletion layers.

As the voltage V_d is increased, the thickness of the conducting channel is reduced because of the widening of the gate depletion layers as the reverse bias is automatically increased. The source-to-drain resistance is increased correspondingly until at some particular drain voltage the space-charge regions from the gates meet; the channel is then said to be *pinched-off*. At drain voltages beyond pinch-off, the drain current becomes essentially saturated and remains at some value I_{d0}. The I_d–V_d or drain characteristic therefore has the form shown in Fig. 9.3(a), the particular curve applying to the bias conditions discussed being labelled $V_g = 0$.

If now an additional fixed reverse bias voltage, V_g, is applied to the gates G_1 and G_2 in parallel, as shown in Fig. 9.3(b), in the absence of drain current the space-charge layers would extend uniformly into the channel, region I in the diagram. When drain current flows due to the application of a drain voltage, V_d, the characteristic wedge-shaped depletion layers, region II, are superimposed on top of the uniform layers due to V_g alone, as shown. Under these conditions the IR drop and hence the value of V_d required to produce pinch-off is smaller and current saturation occurs at lower drain voltages. This results in a family of drain characteristics as shown in Fig. 9.3(a). Thus, application of a reverse voltage to the gates governs the effective width of the depletion layers, which in turn changes the effective channel dimensions, modulates the channel resistance and hence controls the drain current. When operating in this fashion the JFET behaves as a depletion-mode device, since the channel current is reduced as the gate voltage is increased. In the following section we will consider the operation of a JFET more quantitatively with a view to characterizing its d.c. and a.c. performance.

9.3 Drain and transfer characteristics of the JFET

9.3.1 Linear operation of the JFET

Let us first consider the situation when the drain voltage, V_d, is small enough for the conduction channel to be substantially uniform. This case is illustrated

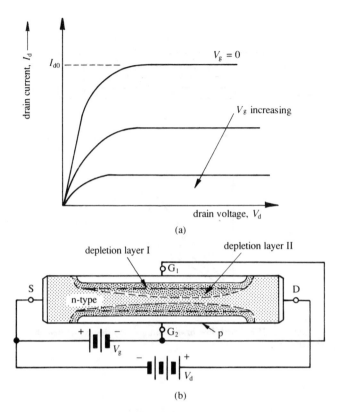

Fig. 9.3 (a) Form of the drain characteristics of a JFET; (b) the JFET with reverse bias voltage applied to its gates.

in Fig. 9.4(a), which includes relevant dimensions and coordinate system.

We shall assume that the channel is relatively lightly doped compared with the gates, i.e. $N_a \gg N_d$, so that nearly all the depletion-layer thickness occurs in the n-region. In these circumstances the thickness of the space-charge layer is given by Eq. (7.47) and

$$d_n = a - b \simeq \left(\frac{2\epsilon(V_0 + V_g)}{eN_d} \right)^{1/2} \tag{9.1}$$

where V_0 is the contact potential between n- and p-regions, as usual. This equation can be rearranged to give the channel half-thickness, b:

$$b = a - \left(\frac{2\epsilon(V_0 + V_g)}{eN_d} \right)^{1/2} \tag{9.2}$$

Notice that, as V_g increases, b decreases, as was argued qualitatively earlier.

Pinch-off occurs when the gate voltage is just sufficient to make the conduction channel width zero, or $b = 0$ when $V_g = V_p$, the pinch-off voltage.

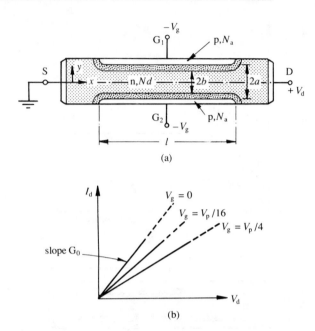

Fig. 9.4 Operation of a JFET with low drain voltages: (a) the model and (b) the linear drain characteristics.

Then

$$a^2 = \frac{2\epsilon(V_0 + V_p)}{eN_d}$$

or

$$V_p = \frac{eN_d a^2}{2\epsilon} - V_0 \tag{9.3}$$

The channel half-thickness can then be expressed in terms of the pinch-off voltage by substituting Eq. (9.3) in (9.2) to give

$$b = a\left[1 - \left(\frac{V_0 + V_g}{V_0 + V_p}\right)^{1/2}\right] \tag{9.4}$$

Now, the conducting channel has conductance G, where

$$G = \sigma_n 2bw/l = eN_d \mu_e 2bw/l \tag{9.5}$$

where w is the gate width and μ_e the electron mobility. Thus, the drain current, I_d, flowing when a small drain voltage, V_d, is applied, is given by

$$I_d = V_d G = eN_d \mu_e \frac{2bw}{l} V_d = \frac{eN_d \mu_e 2wa}{l}\left[1 - \left(\frac{V_0 + V_g}{V_0 + V_p}\right)^{1/2}\right]V_d \tag{9.6}$$

It is convenient to write the drain current in terms of the channel conductance, G_0, when zero gate-bias voltage is applied, where

$$G_0 = \sigma_n 2aw/l = eN_d \mu_e 2aw/l \tag{9.7}$$

which can be substituted in Eq. (9.6) to give

$$I_d = G_0 \left[1 - \left(\frac{V_0 + V_g}{V_0 + V_p} \right)^{1/2} \right] V_d \tag{9.8}$$

Further simplifications are possible since, at all but the lowest gate voltages, $V_g \gg V_0$, and in all cases except for very thin, high-resistivity channels, $V_p \gg V_0$. The drain current then becomes

$$I_d \simeq G_0 [1 - (V_g/V_p)^{1/2}] V_d \tag{9.9}$$

This relationship between drain current and voltage is only valid for $|V_g| < |V_p|$, that is, before pinch-off. The equation shows that for small drain voltages and fixed gate voltages, the drain current varies linearly with drain voltage. Hence, the device acts as a voltage-controlled resistor, whose resistance can be altered by altering the gate voltages as shown in the characteristics of Fig. 9.4(b). JFETs in this linear resistive mode have applications in voltage-controllable attenuators and variable-phaseshift networks.

9.3.2 Operation as far as pinch-off

We now turn our attention to the more usual case of larger drain voltages, which cause a not-insignificant axial electric field, $\mathscr{E}_x$, to exist. Since such a component of electric field is now present, there exists a corresponding change in potential along the channel. Thus, the net reverse bias voltage between the channel and the gate varies with distance along the channel, x, and the space-charge depletion layers take on their characteristic wedge shape, converging towards the drain end, as explained in Sec. 9.2.

This situation can be treated by assuming that the depletion layer converges in a gradual manner. A typical configuration might then be as shown in Fig. 9.5(a). If we now consider an element of thickness, δx, situated at x along the axis, as shown, and let the voltage at x be V_x, then, if the voltage at $x + \delta x$ is $V_{x+\delta x}$, the voltage drop across the element is given by

$$V_{x+\delta x} - V_x = \left(V_x + \frac{dV_x}{dx} \delta x \right) - V_x = \frac{dV_x}{dx} \delta x \tag{9.10}$$

We further assume that the gate junctions can be treated as equipotentials by virtue of the relatively high gate conductivity. Then, if the gate voltage is V_g, the total reverse bias at the element is $V_x + V_g$. Equation (9.6) will still be applicable to the element, provided l is replaced by δx, V_d by $(dV_x/dx) \delta x$, and V_g by

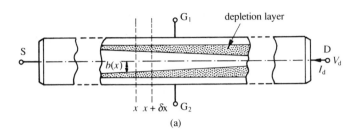

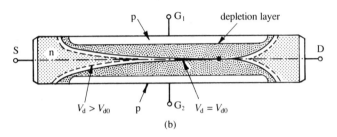

Fig. 9.5 Operation of the JFET with high drain voltages: (a) the model and (b) conditions after pinch-off.

$V_x + V_g$, and this gives

$$I_d = \frac{eN_d\mu_e 2wa}{\delta x}\left[1 - \left(\frac{V_x + V_g}{V_p}\right)^{1/2}\right]\frac{dV_x}{dx}\delta x \qquad (9.11)$$

Notice that we have again assumed that the contact voltage, V_0, is negligibly small. We next rearrange the equation and integrate over the length of the gate, l, as follows

$$\frac{I_d}{2eN_d\mu_e wa}\int_0^l dx = \int_0^{V_d}\left[1 - \left(\frac{V_x + V_g}{V_p}\right)^{1/2}\right]dV_x \qquad (9.12)$$

which, using Eq. (9.7), gives

$$I_d = G_0\left[V_d - \frac{2}{3}\left(\frac{V_d + V_g}{V_p}\right)^{3/2}V_p + \frac{2}{3}\left(\frac{V_g}{V_p}\right)^{3/2}V_p\right] \qquad (9.13)$$

This equation can be rewritten in terms of the drain-gate voltage, V_{dg}, which equals $V_d + V_g$, to give

$$I_d = G_0\left\{V_{dg}\left[1 - \frac{2}{3}\left(\frac{V_{dg}}{V_p}\right)^{1/2}\right] - V_g\left[1 - \frac{2}{3}\left(\frac{V_g}{V_p}\right)^{1/2}\right]\right\} \qquad (9.14)$$

It must be remembered that these equations are only valid for $V_{dg} = (V_d + V_g) < V_p$; otherwise, when $V_{dg} = V_p$, the channel has zero thickness and becomes pinched-off.

Detailed analysis of the behaviour of the device above pinch-off is complex but the following qualitative explanation might be useful. As V_{dg} is made bigger than V_p, most of the excess voltage appears across the pinched-off depletion layers at the drain end of the channel, causing a relatively strong axial electric field to exist there. Electrons arising from the source enter the pinched-off depletion layer, are immediately accelerated to their saturation velocity (around 10^5 m s^{-1} in silicon) by the high electric field across the region and are rapidly removed into the drain. Further increases in V_d above that required for the onset of pinch-off, given by $V_d = V_p - V_g$, increase the longitudinal field across the depleted region, but the electrons cannot move faster since they have already achieved a saturated velocity. Nor can there be an increase in the number of electrons flowing from the source end, since this is controlled by the channel voltage at the point at which electrons enter the depleted region, which stays constant, again determined by the pinch-off condition $V_d = V_p - V_g$. The drain current for increasing values of V_d above pinch-off therefore stays at a constant *saturated* value, I_{ds}; the excess drain voltage is then dropped across the depletion layer at the drain end, which lengthens slightly, to move the position where pinch-off occurs nearer to the source, as shown diagrammatically in Fig. 9.5(b).

The drain characteristics or I_d–V_d curves, derived from Eq. (9.13) or (9.14), and assuming that current saturation occurs after pinch-off, are then as shown in Fig. 9.6(a). The current in the saturation region for a particular gate voltage, I_{ds}, can be obtained from Eq. (9.14) by letting $V_{dg} = V_p$. Thus

$$I_{ds} = G_0 \left\{ \frac{V_p}{3} - V_g \left[1 - \frac{2}{3} \left(\frac{V_g}{V_p} \right)^{1/2} \right] \right\} \qquad (9.15)$$

and the saturated drain current for the particular case of zero gate voltage, I_{d0}, is from this equation

$$I_{d0} = G_0 V_p / 3 \qquad (9.16)$$

Finally, Eqs (9.15) and (9.16) can be rearranged to give the saturated drain current at any particular gate voltage in terms of the current at zero gate voltage, to yield

$$I_{ds} = I_{d0} \left[1 - \frac{3V_g}{V_p} + 2 \left(\frac{V_g}{V_p} \right)^{3/2} \right] \qquad (9.17)$$

The transfer characteristics of the device, or the variation of drain current with gate voltage in the pinch-off region, can be predicted using this equation and is of the form shown in Fig. 9.6(b). The incremental mutual conductance of the device beyond pinch-off, g_m, is also obtained from Eq. (9.17), as

$$g_m = \left(\frac{\partial I_{ds}}{\partial V_g} \right) \Bigg|_{V_d} = \frac{\partial}{\partial V_g} \left\{ I_{d0} \left[1 - \frac{3V_g}{V_g + V_d} + 2 \left(\frac{V_g}{V_g + V_d} \right)^{3/2} \right] \right\}$$

$$= -I_{d0} \frac{3V_d}{V_p^2} \left[1 - \left(\frac{V_g}{V_p} \right)^{1/2} \right] \qquad (9.18)$$

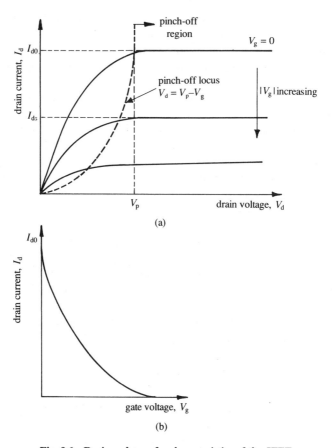

Fig. 9.6 Drain and transfer characteristics of the JFET.

The maximum transconductance, g_{m0}, occurs at $V_g = 0$ and is therefore

$$g_{m0} = -3I_{d0}/V_p = -G_0 \qquad (9.19)$$

We see that the maximum transconductance equals the conductance of the channel with zero gate voltage. The transconductance at any other gate voltage is then given approximately by

$$g_m \simeq g_{m0}[1 - (V_g/V_p)^{1/2}] \qquad (9.20)$$

9.4 The metal–oxide–silicon field-effect transistor (MOSFET)

9.4.1 Induced-channel MOSFET

In this type of transistor a metal gate electrode is completely insulated from a semiconductor by a thin insulating layer, but voltages on the gate can induce a conducting channel within the semiconductor and also modulate its conductivity. A more correct class name for such devices is the insulated-gate field-effect transistor (IGFET), which would naturally encompass FETs employing a variety of insulator and semiconductor types, but the generic title MOSFET or MOST is now almost universally used to describe all categories of FETs, even GaAs devices, for example.

Typical construction of a planar device is as shown schematically in Fig 9.7(a). Two heavily doped n^+-stripes are diffused into a p-type substrate by

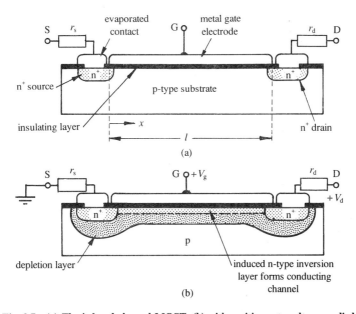

Fig. 9.7 (a) The induced-channel MOST; (b) with positive gate voltage applied.

the photo-masking techniques described in Sec. 10.4 and form the source and drain electrodes. An oxide layer of order 0.1 μm thick is thermally grown on the surface of the slice and etched away so that it just overlaps the n^+-regions, as shown. Finally, a metal gate electrode and metal connections to source and drain regions are evaporated on. Transistors fabricated in silicon will be seen to have a metal–(silicon) oxide–silicon structure, from which the MOST derives its name.

In order to understand the operation of this transistor, we will first assume for simplicity that no surface traps exist at the boundary between the substrate and the oxide layer. This approximation is often quite valid since it has been shown that thermally grown silicon dioxide on a silicon surface can so passivate the surface as to reduce substantially the number of deep traps to such an extent that for many devices they can be neglected. Under these conditions and with no gate voltage applied, no current can flow between source and drain contacts because of the existence of back-to-back pn junction diodes between them.

If, now, a small positive gate voltage, V_g, is applied, positive and negative charges are established at either side of the oxide layer, and the transverse electric field that is set up is terminated on negative charges induced in the p-type substrate. For small gate voltage, the necessary negative charges in the semiconductor are provided by holes being depleted from the channel, exposing negatively charged ionized acceptor impurities, which just balance the positive charge on the gate electrode; a depletion layer is formed in this manner and the band edges near the surface of the semiconductor begin to bend down as described in Sec. 7.11. As the gate voltage is further increased, conduction electrons are drawn to the surface of the semiconductor to provide the additional negative charge necessary to balance the extra positive charge placed on the gate electrode. Thus, as the gate voltage is increased, the channel changes from p-type to intrinsic and finally to an n-type induced layer, as shown in Fig. 9.7(b).

The energy band diagram after the induced channel has been formed is as shown in Fig. 9.8. Its evolution can be explained as follows. For small positive gate voltages, a depletion layer is formed in the semiconductor, as explained, and the bands begin to bend down near to the surface, as discussed in Chapter 7, the amount of bending corresponding to the potential drop across the layer. As the gate voltage increases, the depletion layer widens, the potential across the layer increases and the bands bend down further. Eventually, when the gate voltage is sufficiently high, the band edges bend down to such an extent that the Fermi level, which is necessarily constant throughout the semiconductor in equilibrium, approaches the bottom of the conduction band, as shown in Fig. 9.8. The band structure of the material near to the surface then corresponds to that for an n-type semiconductor and an inversion layer is formed.

Immediately the n-type layer is induced, ohmic conduction from source to drain becomes possible. Further increases in V_g cause more mobile electrons to be introduced into the induced channel, with a corresponding increase in the channel conductivity and drain current. The device therefore operates in an enhancement mode, since the number of mobile carriers and hence the drain current increases with increasing gate voltage; consequently, it is sometimes known as an induced-channel enhancement transistor.

The form of the characteristics of the enhancement-mode device can be derived as follows. At some distance x along the channel, the voltage is assumed to be $V(x)$, which is a function of distance because of the IR drop along the channel, and ranges from V_s at the source to V_d at the drain end. The voltage across the oxide layer at this point is then $V_g - V(x)$. If we assume the oxide layer to have thickness t_o, neglecting fringing fields and assuming that the oxide is relatively thick compared with the shallow conducting channel, the electric field in the oxide, $\mathscr{E}(x)$, is then

$$\mathscr{E}(x) = [V_g - V(x)]/t_o \tag{9.21}$$

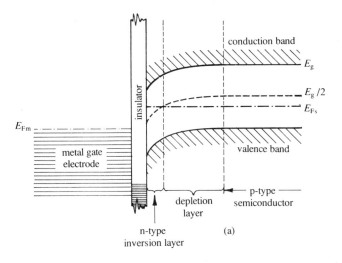

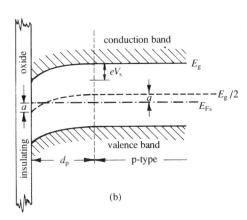

Fig. 9.8 **Band structure of the induced-channel MOSFET: (a) when the induced channel is fully formed and (b) just at the onset of inversion-layer formation, when $V_g = V_T$.**

The surface-charge density, ρ_i, induced in the channel at x, is related to the electric field by Gauss's law and is given by

$$\rho_i(x) = \epsilon \mathscr{E}(x) = \epsilon_0 \epsilon_{r0} [V_g - V(x)]/t_0 \ \text{C m}^{-2} \tag{9.22}$$

where ϵ_{r0} is the relative permittivity of the oxide layer.

Not all of this charge is available for conduction in the channel, since at low gate voltages a depletion layer but no inversion layer is formed. However, if a characteristic voltage, V_T, known as the *turn-on voltage*, is defined as that voltage across the oxide which just causes the mobile charge concentration in the channel to be greater than zero, then voltages in excess of V_T induce the inversion layer necessary for conduction. The excess mobile carrier density in the inverted channel, Δn, can then be related to an effective voltage across the oxide $[V_g - V(x)] - V_T$ and the mobile surface charge density is given by

$$e\Delta n(x) = \epsilon_0 \epsilon_{ro}[V_g - V(x) - V_T]/t_o \text{ C m}^{-2} \tag{9.23}$$

Now the gate electrode–insulator–semiconductor combination behaves as a capacitor of capacitance C_g, which is given approximately by the parallel-plate expression and is thus

$$C_g = \epsilon_0 \epsilon_{ro} lw/t_o \text{ F} \tag{9.24}$$

where w is the width of the gate electrode. Equation (9.23) can now be rewritten in terms of the gate capacitance to yield

$$e\Delta n(x) = (C_g/lw)[V_g - V(x) - V_T] \qquad \text{for } V_g - V(x) > V_T$$
$$e\Delta n(x) = 0 \qquad \text{for } V_g - V(x) < V_T \tag{9.25}$$

Now let us consider a portion of the assumed infinitesimally thin channel of length dx, and width w. This has conductance, $G(x)$, given by

$$G(x) = \sigma(x)w/dx \tag{9.26}$$

where $\sigma(x)$ is the surface conductance per square metre of channel. The conductance of the elemental channel length is then

$$G(x) = e\Delta n(x)\mu_e w/dx = (\mu_e C_g/l \, dx)[V_g - V(x) - V_T] \tag{9.27}$$

where μ_e is the mobility of electrons in the channel. The channel or drain current, I_d, is then given by Ohm's law as

$$I_d = G(x) \, dV$$

where dV is the voltage across the incremental length of channel, dx.
Substituting from Eq. (9.27) and taking to the limit then yields

$$I_d = (\mu_e C_g/l)[V_g - V(x) - V_T] \, dV/dx \tag{9.28}$$

This equation can be integrated over the length of the channel, l, to give

$$\int_0^l I_d \, dx = (\mu_e C_g/l) \int_0^{V_d} [V_g - V(x) - V_T] \, dV \tag{9.29}$$

Here, for simplicity, we have assumed the parasitic series resistances of source and drain, r_s and r_d in Fig. 9.7, to be negligibly small because of the relatively high doping level, but these could be included without too much additional complication. Carrying out the integration yields

$$I_d = (\mu_e C_g/l^2)(V_g - V_T - \tfrac{1}{2}V_d)V_d \tag{9.30}$$

Note that this equation is only valid when the net voltage across the oxide layer at any point exceeds the turn-on voltage, or $V_g - V(x) > V_T$. Thus, for any particular gate voltage, V_{g1} say, the equation is only valid for drain voltages in the range $0 \leqslant V_d \leqslant (V_{g1} - V_T)$. This upper limit for the drain voltage is physically explainable as that voltage which is just sufficient to prevent the formation of an inversion layer at the drain end. When the drain voltage is increased above this limit, the drain current saturates at some maximum value, I_{ds}, found by putting $V_d = V_g - V_T$ in Eq. (9.30) to give

$$I_{ds} = (\mu_e C_g / l^2)(V_g - V_T)^2 / 2 \tag{9.31}$$

The drain characteristics of a MOSFET operating in the enhancement mode are derivable from Eqs. (9.30) and (9.31) and are therefore typically of the form shown in Fig. 9.9. Notice that, except for the direction of increasing V_g, these

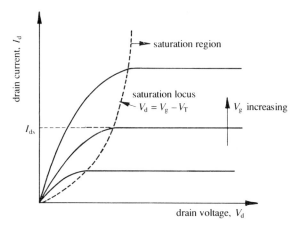

Fig. 9.9 Drain characteristics of the induced-channel MOSFET.

characteristics are similar in form to those shown for the JFET operated in the depletion mode, shown in Fig. 9.6(a).

The transconductance of the MOSFET when operated in the saturated region, g_m, is obtained from Eq. (9.30), which gives

$$g_m = \left(\frac{\partial I_d}{\partial V_g}\right)\bigg|_{V_d} = \frac{\mu_e C_g V_d}{l^2} = \frac{\mu_e C_g (V_g - V_T)}{l^2} \tag{9.32}$$

Measured transconductances are somewhat lower than would be predicted by this equation, partly because the parasitic resistances r_s and r_d have been neglected and partly because of the lowering of the mobility in the thin channel below that of the bulk material because of scattering at the oxide surface.

Since the turn-on voltage, V_T, appears in all the derived expressions, it will be useful to estimate its value. By definition, when the turn-on voltage is applied across the oxide layer, an inversion layer is about to be formed, but meanwhile the negative charge of the fixed ionized acceptors in an induced depletion layer just balance the positive charge supplied to the gate electrode. This onset of an inversion layer corresponds to the condition in which the electron concentration at the surface of the semiconductor is approximately equal to the acceptor concentration, $n = N_a$. For this to apply, it will be seen from Eqs (6.13) and (6.18) and the band structure shown in Fig. 9.8(b) that the Fermi level at the surface must be below the bottom of the conduction band by as much as it is above the top of the valence band in the bulk material, if the density of states in each band is assumed to be approximately equal, or $N_C \simeq N_V$. It follows that if the energy bands bend down a total amount eV_s, then, from Fig. 9.8(b)

$$eV_s = 2a = 2(E_g/2 - E_{Fs}) = E_g - 2E_{Fs} \tag{9.33}$$

The corresponding depletion-layer width can be obtained by integrating Poisson's equation across the depletion layer to give

$$d_p = (2\epsilon V_s/eN_a)^{1/2} \tag{9.34}$$

Now, assuming no contact voltage exists between oxide and semiconductor, the turn-on voltage is divided between the oxide and the depletion layer and

$$V_T = V_{ox} + V_s \tag{9.35}$$

where V_{ox} is the voltage across the oxide. Further, if the depletion-layer thickness is assumed to be much less than that of the oxide, so that the charge in the layer can be considered as a surface charge, then for the continuity of electric displacement across the interface

$$\epsilon_{ox} V_{ox}/t_o = eN_a d_p$$

or

$$V_{ox} = (wl/C_g)eN_a d_p \tag{9.36}$$

Finally, substitution of Eqs (9.33), (9.34) and (9.36) into (9.35) gives the required expression for estimating the turn-on voltage:

$$V_T = (wl/C_g)[2\epsilon_s N_a (E_g - 2E_{Fs})^{1/2}] + (E_g - 2E_{Fs})/e \tag{9.37}$$

9.4.2 Diffused-channel MOSFET

The diffused-channel MOST has similar constructional features to the induced-channel type, with the important difference that a thin channel of the same conductivity type as the source and drain is diffused into the surface of the semiconductor during manufacture, as shown in Fig. 9.10. The device can be operated in depletion or enhancement modes. If the gate has a negative bias voltage applied to it, the resulting transverse field must terminate on positive charges, which are induced in the n-type channel. In order to achieve this condition, majority electrons move away from the semiconductor surface to

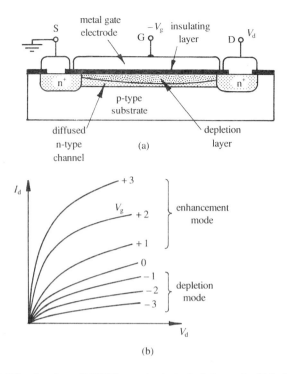

Fig. 9.10 (a) A diffused-n-channel MOST operating in the depletion mode; (b) drain characteristics of the diffused-channel MOST.

expose positively charged fixed ionized donors and so produce a depletion layer, as shown in Fig. 9.10. The channel conductance is reduced by the application of the bias voltage to the gate electrode, in much the same manner as for a JFET, and maximum drain current flows for zero gate voltages.

On the other hand, if the polarity of the gate voltage is reversed so as to apply forward bias, i.e. the gate is made positive, electrons are drawn into the channel to enhance the number of mobile carriers there, with a corresponding increase in the conductivity of the channel and hence the drain current. This enhancement mode of operation is possible because the complete isolation of the insulating layer results in negligibly small gate currents whatever the polarity of the gate voltage, but it is prohibited in a JFET, since forward biasing of the gate–channel junctions causes large gate currents to flow. A typical set of drain characteristics for the device is shown in Fig. 9.10(b).

It is not always possible to ignore the existence of surface states at the semiconductor–insulator interface, as has been assumed so far. If surface states exist, there will be curvature of the band edges in the semiconductor even when the applied gate voltage is zero in an induced-channel device (see Sec. 7.11.2). If the surface states act as donor impurities, there will be a tendency for an inversion layer to form, even when $V_g = 0$, and the turn-on voltage will be much lower than with no traps present. In the extreme case, when the surface trap density is very high, it is possible for an n-type inversion layer to be fully developed before any gate voltage is applied. The unit then operates in depletion or enhancement modes in a manner exactly similar to the operation of the depletion-type MOSFET with an n-type doped channel, and its drain characteristics are similar in form to those shown in Fig. 9.10(b).

A further point to note is that although n-channel devices have been discussed throughout this chapter, this does not preclude the possibility of devices being constructed with p-type channels. Our earlier discussions are readily extended to describe such components by interchanging p- and n-type and the polarity of the bias voltages throughout.

9.5 The small-signal equivalent circuit and frequency response of FETs

The normal symbolic representations of the JFET and the MOST, each with n-type channel, are shown in Figs 9.11(a) and (b). For a p-type channel the direction of the arrow on the gate connection is reversed.

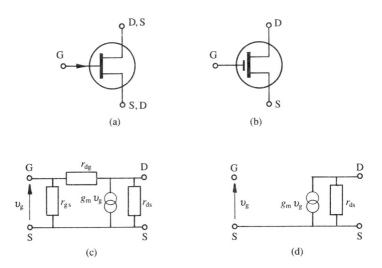

Fig. 9.11 Symbolic representations of (a) JFET and (b) MOSFET; (c) a low-frequency equivalent circuit of a field-effect transistor; (d) approximate low-frequency equivalent circuit.

Field-effect transistors have input characteristics that in no way resemble those of bipolar transistors. In the junction FET the gate–channel junctions are reverse-biased; the gate current flowing corresponds to the very small saturation current of a reverse-biased pn junction, and the input impedance in the common-source configuration is usually many megohms. The insulated-gate FET, by virtue of the electrical isolation of the gate electrode by the oxide layer, has even lower gate currents, typically a few nanoamperes, and its input impedance can be of order of many thousands of megohms. On the other hand, the bipolar transistor in common-emitter configuration has a low input impedance corresponding to the forward-biased base–emitter junction. In this respect, therefore, the FET is a voltage-controlled device, which contrasts with the bipolar transistor, which is current-controlled.

A convenient low-frequency equivalent circuit for a FET in common-source connection, based on its observed and calculated characteristics, is shown in Fig. 9.11(c). As we have just discussed, the input impedance is very high and the feedback resistance, r_{dg}, by similar reasoning, is of the same order, so both r_{gs} and r_{dg} can normally be omitted to give the simple low-frequency equivalent circuit shown in Fig. 9.11(d).

At high frequencies, interterminal capacitances must also be included in the equivalent circuit. For instance, associated with the reverse-biased junction gate of a JFET is an input capacitance of several tens of picofarads, typically, although this can be made considerably smaller for the MOST. Similarly, the gate–drain capacitance cannot be neglected at high frequencies. Both are at least partly depletion-layer capacitances and hence are voltage-dependent, but usually the capacitance change is only slight over the normal working range. A suitable equivalent circuit for a FET at high frequencies might then be as shown in Fig. 9.12(a). Alternatively, the two-port admittance or 'y' parameters are sometimes found convenient to characterize the FET for small-signal and high-frequency use, as shown in the equivalent circuit of Fig. 9.12(b).

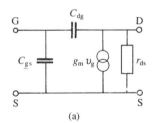

(a)

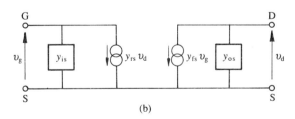

(b)

Fig. 9.12 (a) A high-frequency equivalent circuit of the FET and (b) an alternative high-frequency equivalent circuit.

As we have seen, current in a field-effect transistor is carried by majority carriers drifting under the influence of an electric field, whereas in the bipolar transistor, current is transported by means of diffusing minority carriers. Since drift velocities in semiconductors are usually very much higher than diffusion velocities, carrier transit times are much shorter in FETs than in bipolar transistors. For this reason, we might expect FETs to have a much more extended high-frequency range than bipolar devices.

A limitation to the high-frequency performance or the switching speed of a FET is the gate–channel capacitance, which must be charged via the channel resistance. The resulting time constant determines the upper limit of the frequency response. The gain × bandwidth product, which can be derived from the equivalent circuit and equals $g_m/2\pi C_g$, is normally taken as a figure of merit to indicate the high-frequency response of a particular device. Thus, ideally, for good high-frequency performance, the transconductance should be as large as possible and the total gate capacitance C_g should be as small as possible, both of which require the channel length to be short. Using this criterion it is possible for a MOSFET to have a higher g_m/C_g ratio and hence a higher high-frequency cut-off than a JFET because of the shorter gates that are more easily fabricated with insulated-gate technology.

The g_m/C_g ratio for a MOSFET can be found from Eq. (9.32), which gives

$$g_m/C_g = \mu_e(V_g - V_T)/l^2 \qquad (9.38)$$

Since the average field in the channel is $(V_g - V_T)/l$, and since the carrier velocity is the product of μ_e and this field, Eq. (9.38) shows that g_m/C_g is also approximately equal to the reciprocal of the transit time of majority carriers from source to drain.

9.6 Charge-coupled devices (CCDs)

The charge-coupled device is an example of a complete integrated circuit that performs a unique circuit function that cannot be exactly reproduced by any circuit employing discrete components. The circuit, or device, is shown in part in its most elemental form in Fig. 9.13. It will be seen that the structure consists

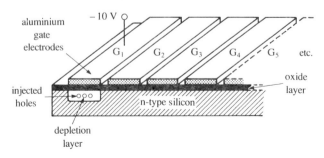

Fig. 9.13 The basic charge-coupled device structure.

essentially of a series of metal gate electrodes, separated from an n-type semiconducting substrate (for a p-channel device) by a thin oxide layer. Such structures are in principle fairly straightforward to make using conventional MOS technology as discussed in Chapter 10, but there are some additional difficulties that will be mentioned later.

If a negative-going pulse, of magnitude -10 V, say, is applied to the first gate, G_1, a depletion layer is initially formed, in typically less than 1 μs, as shown. Since the gate voltage is greater than the turn-on voltage, the additional charge required at the oxide–silicon interface would eventually be provided by minority holes, which move into the region to form the usual inversion layer. This process takes a relatively long time, typically of the order i s, and is usually inhibited. In times short compared with 1 s, say in the first few milliseconds after the application of the pulse, a non-equilibrium situation exists, in which the inversion layer has not had time to form, but a depletion layer exists, as shown in Fig. 9.13. This layer behaves as a potential well, and if minority holes are injected into it in some way, as shown, these will become trapped in the well. The trapped holes can then be transferred to a well under G_2, by a process to be described, and then to under G_3 and so on, sequentially down the whole structure. The device thus performs as a dynamic shift register. This process is only possible provided the charge is not stored in any well long enough for an inversion layer to form, in which case the charge would disappear and the corresponding information would be lost. The CCD structure thus behaves as a dynamic shift register, and charge has to be transferred in times short compared to 1 ms, say.

A three-phase clocked voltage pulse system supplied to the gates ensures that charge is transferred serially between gates and its direction is controlled, as follows. The first phase connects a -10 V pulse to G_1, to produce the potential well into which information, in the form of minority holes, is stored, as described. During the second phase, the adjacent gate, G_2, is biased to a greater negative voltage, say -15 V, to produce a deeper well under it, as shown in Fig. 9.14(a). The stored charge then transfers into the deeper potential well by diffusion down the potential gradient, which incidentally can be a relatively slow process. Note that, to ensure charge transfer, the potential wells must physically overlap. As depletion layers are typically only a few micrometres deep, the spacing between neighbouring gates has to be as small as possible, to ensure adequate overlap. The charge is then completely stored in the well under G_2, so the voltage on G_1 is reduced to a low value, say -5 V, and that on G_2 to a sustaining level of, say, -10 V, as shown in Fig. 9.14(b).

A third phase transfers a -15 V voltage pulse to the next gate, G_3, and the charge is transferred from G_2 to the well under it, as shown in Fig. 9.14(c). The voltages on G_2 and G_3 can then be relaxed as before as shown in Fig. 9.14(d), to complete one cycle of the clock frequency. The charge has been transferred from under G_1 to under G_3 in one cycle of the clocked three-phase pulse, which causes a series of voltages in the sequence of -15, -10, -5, -15, etc., to be applied to each gate electrode. The charge can be transferred in this

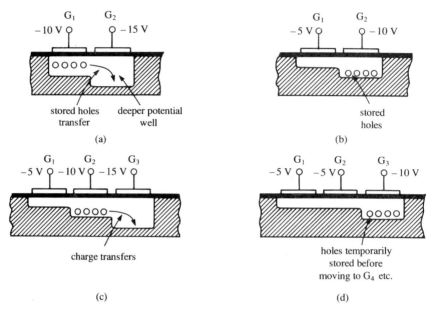

Fig. 9.14 The mechanism of charge transfer in a three-phase CCD.

manner down the structure, each storage cell of three adjacent electrodes accommodating the *bit* of information. As soon as charge is moved out of one set of three electrodes, say from G_3 to G_4, then the input gate is again put in a state to receive a further *bit* of information.

The practical realization of an eight-bit CCD shift register is shown diagrammatically in Fig. 9.15. The aluminium gate electrodes are typically

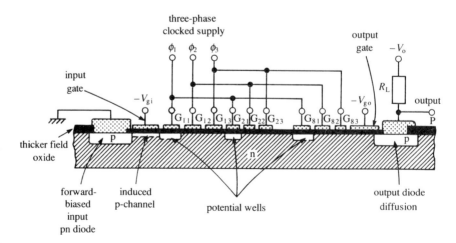

Fig 9.15 An eight-bit CCD dynamic shift register.

$250 \times 50\,\mu$m and the gaps between them $2\,\mu$m, which in itself is difficult to achieve technologically. Since there are three gate electrodes per storage cell, every third one is connected in parallel and connected to one phase of the three-phase supply, ϕ_1, ϕ_2, ϕ_3, as shown. Because of topological difficulties, one set of connections to a common phase is often by means of a diffused cross-under, but the remainder of the connections can be done by the usual metallizing process. The gate oxide is typically only 100 nm thick but the field oxide elsewhere is about 500 nm thick.

Input holes, generated by the forward-biased diffused input diode, are injected into the potential well under the first gate G_{11}, via an induced channel under the input gate, if this is biased sufficiently negative, as shown in Fig. 9.15. In other words, the potential well under G_{11} behaves as an equivalent drain in a MOST that is controlled by the input gate. The injected charge is then transferred down the structure by the clocked three-phase transport mechanism described and its presence at the output is detected by the reverse-biased diffused output diode shown. The minority holes that arrive under electrode G_{83} are switched to the output diode via the induced channel under the output gate. They are then swept across the reverse-biased pn junction to produce a corresponding current in the load resistor, R_L, and a signal output.

The input, output and information-shifting schemes described are only typical of several alternative arrangements. For example, it is possible to dispense with the input diode and to introduce minority carriers under G_{11}, say, by irradiation of the silicon by light, which produces holes there. The charge that is temporarily stored and is subsequently transferred down the structure produces an output that is proportional to the intensity of the illumination. A matrix version of a CCD, operated in this mode, has considerable potential in the imaging field and a solid-state colour TV camera, which uses the same principle, is now available.

Another possible application of CCDs, which is receiving considerable attention, is concerned with the processing of analogue signals. It is fairly straightforward to envisage the input of a simple linear CCD array sampling a slowly varying analogue signal, applied to its input gate, which can be reconstituted at the output of the CCD, after having experienced some delay. The delay can be varied over a wide range, electronically, by varying the clock frequency. It is also possible to process an analogue signal further, as it progresses along a CCD structure by, say, tailoring the geometry of the phase gates.

Problems

1. Show that if a JFET is operated at sufficiently low drain voltage it behaves as a resistance R given approximately by

$$R = R_0 [1 - (V_g/V_p)^{1/2}]^{-1}$$

where R_0 is the channel resistance for zero gate voltage.

2. A JFET has a circular geometry as shown.

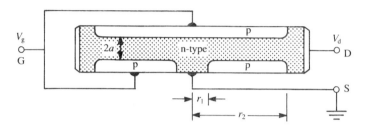

Assuming that the drain voltage, V_d, is always sufficiently low for the channel half-width, b, to be considered uniform, derive an expression for b in terms of V_g, the contact potential V_0 and a pinch-off voltage, V_p. Hence show that the channel resistance is given by

$$R = R_0 \left[1 - \left(\frac{V_0 + V_g}{V_0 + V_p} \right)^{1/2} \right]^{-1}$$

where R_0 is the channel resistance when $V_g = 0$.

3. A certain JFET operates in the linear region with a constant drain voltage of 1 V. When the gate voltage is 2 V a drain current of 10 mA flows, but when the gate voltage is changed to 1 V the drain current becomes 22.8 mA. Find (a) the pinch-off voltage of the device, and (b) the channel resistance for zero gate voltage.

Ans. 3 V, 18.6 Ω

4. A particular JFET has a channel of length 20 μm and cross section 200 μm × 12 μm. The device is operated in the linear region as a voltage variable resistor. When the gate is shorted to the drain, the source–drain resistance is 8 kΩ, and when the gate is shorted to the source, the resistance becomes 5 kΩ. Estimate (a) the doping concentration in the channel, (b) the pinch-off voltage of the device, and (c) the operating drain voltage.

Assume a majority-carrier mobility in the channel of 0.13 m² V⁻¹ s⁻¹ and a relative permittivity of 12.

Ans. (a) 8×10^{19} m⁻³, (b) 2.2 V, (c) 0.31 V

5. A particular JFET is to be used in the linear mode as a voltage variable attenuator. When the gate voltage, V_g, is 1 V, the channel resistance is 2 kΩ, and when $V_g = 2$ V, the resistance is 3 kΩ. The attenuator is required to have a resistance swing between a minimum possible value and 10 × this value. What variations of gate voltage would be necessary to achieve this design objective?

Why would such operation not be practicable at high drain voltage?

Ans. 0–4 V

6. A JFET operating at low drain voltages is to be used as a voltage variable resistor in an attenuator. The JFET is constructed using a semiconductor with a resistivity of $5 \times 10^{-3} \Omega$ m, which has a cross section of 1μm $\times 10 \mu$m and the effective gate length is 15μm. What is the minimum resistance achievable with the device?

What gate voltage, expressed as a fraction of the pinch-off voltage, would be required to double the minimum resistance?

Ans. 7.5 kΩ, $V_p/4$

7. A silicon n-channel JFET has a channel length 25μm and cross section 250μm $\times 2 \mu$m. If the pinch-off voltage has magnitude 7 V, estimate the drain–source resistance for low drain voltages and gate voltages of (a) 0 V and (b) half the pinch-off voltage.

Assume that the mobility of carriers in the channel is 0.13 m^2 V^{-1} s^{-1} and a relative permittivity for silicon of 12.

Ans. (a) 259 Ω, (b) 882 Ω

8. Show that the transconductance, g_m, the drain voltage V_d, and the saturated drain current, I_{ds}, of a MOST are related by

$$g_m = aI_{ds}/V_d$$

where a is a constant to be determined.

A particular MOST has a small-signal voltage gain of 50 when operated in the common-source mode with a load resistance of 10 kΩ from a supply rail of 125 V. Find the transconductance of the device and the drain current flowing under these conditions.

Ans. $a = 2$, 5 mS, 12 mA

9. A certain MOST, which has a gate capacitance of 0.5 pF and channel length of 12 μm, operates with a drain voltage of 3 V. Assuming a carrier mobility of 0.13 m^2 V^{-1} s^{-1}, find (a) the saturated drain current, (b) the transconductance, (c) the small-signal voltage gain when operated with a 15 kΩ load resistor, and (d) the supply rail voltage when operated as in (c).

Ans. (a) 2 mA, (b) 1.35 mS, (c) 20, (d) 33.5 V

10. A particular n-channel MOST has a transconductance of 2 mS at a drain current of 5 mA. Estimate the gate capacitance, assuming a channel length of 10 μm and an electron mobility of 0.13 m^2 V^{-1} s^{-1}.

Ans. 0.3 pF

11. A particular n-channel MOST has a turn-on voltage of 3 V and a g_m of 5 mS when the drain voltage is 5 V. It is desired to use the device as a voltage variable resistor in a small-signal automatic gain control circuit. How might this be achieved? Find an expression for the small-signal dynamic resistance between drain and source terminals and estimate its particular value when a gate voltage of 5 V is applied and the drain voltage is zero.

Ans. 0.5 kΩ

12. A particular MOST, with gate and drain strapped together, is used as a small-signal voltage variable resistor. A dynamic resistance of 1 kΩ is obtained between drain and source when the drain (and gate) voltage is 10 V, rising to 3.5 kΩ when the drain voltage is 5 V. What is the turn-on voltage of the MOST?

Ans. 3 V

13. A certain MOST when operated with a supply rail voltage of 50 V and a 12 kΩ load resistor draws a saturated drain current of 4 mA. Estimate the small-signal voltage gain of the arrangement, assuming a small-signal output impedance for the MOST of 100 kΩ.

Ans. 43

14. An n-channel MOST in silicon has a gate capacitance of 1 pF, a gate width of 1 mm and an oxide thickness of 100 nm. Find its g_m at a drain–source saturation voltage of 10 V, assuming an electron mobility of 0.02 m^2 V^{-1} s^{-1} and a relative permittivity of SiO$_2$ of 3.7.

Ans. 21 mS

10 Integrated circuits

10.1 Introduction

Although a complete discussion of the intricate and complex new technology associated with integrated circuits is not possible here, because of their extreme importance at present and their potential for the future, a brief description of their types and the processes used for their manufacture will now be given.

An integrated circuit (IC) is a complete electronic circuit, containing both active and passive components, that is fabricated in one small chip of semiconductor, usually silicon, complete with metallic interconnections. Consider, for example, the circuit shown schematically in Fig. 10.1(a).

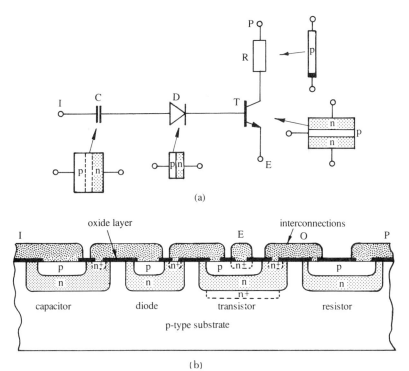

Fig. 10.1 (a) Hypothetical electronic circuit; (b) possible integrated-circuit realization.

Although the circuit is artificial in that it has been chosen so that the interconnections are more easily represented, a discussion of its realization in integrated form will be useful. It is evident that each of the circuit elements shown can be made using semiconductor material, with or without junctions, provided the component values are within prescribed limits. As a first step, the active devices might take the form of the discrete versions mentioned in the relevant earlier chapters: resistors could be rods of suitably doped semiconductor of a particular geometry to suit the resistance required; and capacitors may comprise, for example, the junction capacitance of a reverse-biased junction of suitable geometry, as shown in Fig. 10.1. It will be noticed that each component type can be fabricated from, say, an npn sandwich of semiconductor, although not all would utilize all three layers. It seems a reasonable concept, therefore, that each circuit element might be fabricated on the same semiconductor slice, using diffusion techniques, say, with metal circuit connections deposited on one face of the slice. Such an integrated-circuit unit, as shown in section in Fig. 10.1(b), will have obvious advantages of high packing density, reliability, robustness, reproducibility, and so on.

There are several difficulties to be overcome before a particular electronic circuit can be realized in integrated form as envisaged. First, every circuit component must be so designed that all electrical connections can be made at one surface of the semiconductor from which they are formed. Secondly, since each circuit element is to be fabricated in close proximity to its neighbours in the same slice of semiconductor, which is itself a conducting medium, some means of electrical isolation between elements must be introduced. This ensures that there is no unintentional coupling between components and that the only conducting paths joining them together are metallic interconnections.

A description of the elements of various possible isolation techniques follows.

10.2 Methods of isolation

Electrical isolation of integrated-circuit components has been achieved in several different ways. By far the most common technique is that of diode isolation. In this process, each component is formed in an island of, say, n-type semiconductor surrounded by bulk p-type, as shown for example in Fig. 10.1(b). In operation, the pn junction so formed is reverse-biased by making the p-type bulk material more negative than any other part of the integrated circuit; the reverse-biased junction provides a very high resistance between components and effectively isolates them electrically. An obvious disadvantage of the method is that additional parasitic capacitances are introduced as a consequence of the reverse-biased isolating junctions.

Integrated circuits with a higher degree of isolation can be fabricated using the *beam-lead method*. After the components have been formed in a semiconductor slice, to provide a circuit similar to that shown in Fig. 10.1(b), an especially thick metallizing layer is deposited to produce mechanically strong inter-connections. The bulk p-type semiconductor is then completely removed by applying a suitable etch to the back face of the slice, to leave each component completely isolated from all others and only supported by relatively massive interconnections, as shown in Fig. 10.2(a). The leads act as cantilever beams;

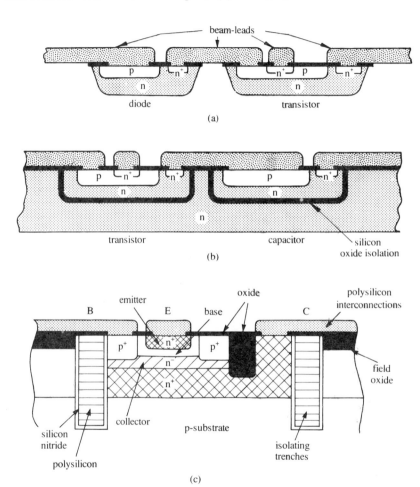

Fig. 10.2 Possible isolation techniques: (a) beam-lead method; (b) dielectric isolation; (c) trench isolation.

hence their name. It is usual to add additional mechanical rigidity by potting the complete circuit in a thermosetting plastic.

A further possible isolation process is that of *dielectric isolation*, in which a thick insulating layer of dielectric, often silicon oxide, completely surrounds each component and effectively isolates it electrically from its neighbours, as shown in Fig. 10.2(b). The process has the advantages of much-reduced parasitic capacitance, provided the insulating layer is thick enough. It also affords an opportunity for fabricating npn and pnp complementary transistor pairs on a single slice, which can be advantageous in certain circuit configurations.

Another more recently introduced possibility is to use *trench isolation* with dielectric infill, shown in Fig. 10.2(c), which provides electrical and physical isolation between adjacent transistors. The isolating trenches, 20 μm deep and 1 μm wide, are formed by plasma etching and are lined with a dielectric coating, usually silicon nitride. Polysilicon is then used to fill in the grooves to produce a planar surface. This modern isolation technique provides a significant improvement in packing density and performance because of reduced parasitic capacitance. An example of a trench-isolated integrated circuit is shown in Fig. 10.3.

Fig. 10.3 An 8 × 8 asynchronous time division switch. This 406 000-transistor chip is an eight-input switch for telecommunications applications capable of switching speech, data and video channels. The 6.5 × 5.5 mm chip is fabricated using a trench-isolated complementary metal–oxide–silicon (CMOS) process. (Courtesy of Plessey Research (Caswell) Ltd.)

The transistor shown in Fig. 10.2(c) also features a polysilicon emitter and interconnections, which will be discussed later. Also of note are the heavily doped n^+- and p^+-regions, which provide lower-resistance paths between the active collector (n^-) and base (p) and their respective surface electrodes.

The choice of a particular isolation technique is, of course, governed by economic considerations and depends on the degree of isolation required.

10.3 Monolithic integrated circuits

A *monolithic* (Greek: single-stone) integrated circuit is so-called because all passive and active circuit components are formed in one small chip of single-crystal semiconductor, usually slicon, using planar diffusion techniques, the metallic interconnecting matrix being incorporated on one face of the chip. Various other types of integrated circuit are possible, however. For example, thin-film integrated circuits have a pattern of passive components and interconnections deposited onto a ceramic insulating substrate by evaporation techniques; the active semiconductor devices are made by a separate process and are subsequently bonded and connected to the thin-film circuit. Thick-film circuits use a variant of this technique in that the passive components and interconnecting pattern are produced on the substrate by a silk-screen printing process. Although film microcircuits have been used, monolithic circuits have many advantages, not least of which is a considerable potential for large-scale integration, and they have largely superseded film circuits for most applications. For this reason, our attention will mostly be devoted to monolithic integrated circuits in silicon.

Monolithic circuits are made using essentially the same processes of masking, diffusion and evaporation that are employed to make discrete planar devices, as mentioned for example with reference to diodes in Chapter 7. The majority of integrated circuits and components are made from the semi-conductor, silicon. Certainly, some specialized circuits use the compound semiconductor, gallium arsenide, but these are in a minority for the present. Other semiconductors have applications in device technologies, such as GaAsP for light-emitting diode (LED) arrays, but again these do not constitute the major activity of the IC industry. Silicon technology has now become so refined and predominant that silicon seems unlikely to be ousted as a preferred material for most integrated circuit fabrication. For this reason, component and circuit manufacture starting with the basic material, silicon, will be described mostly, with only passing reference to alternative semi-conductors.

Another trend these days is for most modern integrated circuits to be composed entirely of active devices such as BJTs or MOSTs. Although integrated passive components such as resistors and capacitors are feasible, as we shall see, they are usually relatively extravagant in terms of area occupied, so are avoided where possible. We will therefore concentrate on the production of active devices and circuits in this brief summary of the technology.

Of course, complete details of the fabrication of semiconductor components and integrated circuits is a vast subject, of which only a brief outline of the principal manufacturing operations can be included here. We shall start with a brief résumé of the basic steps followed to make an integrated circuit, so as to obtain an overview of the subject, and then return to discuss some of the production stages in more detail, concentrating on those processes which are most likely to remain an integral part of semiconductor technology well into the foreseeable future.

It may be wondered what is the concern of an electronics engineer with semiconductor technology, other than a general interest, unless of course he or she is actually designing integrated circuits. As will become apparent, severe constraints are imposed on component specification and circuit configuration by the technology used, which will always have to be considered by engineers, whether they are designing circuits on the chip or applying the complete integrated circuit. Systems engineers will also have to be aware of the principal characteristics of particular integrated circuits, for example whether bipolar or MOS circuitry is to be preferred for a particular application, so it is necessary that they are aware of the advantages and constraints afforded by the different circuits available, which are to a large extent dictated by the semiconductor technology employed. There is also an increasing amount of customer participation in integrated-circuit design, particularly at the systems level, so the engineer can no longer afford to stay removed but must be aware, at least in broad outline, of what is possible and what cannot be achieved in a particular processing technology.

10.4 A brief outline of silicon planar integrated-circuit fabrication

As mentioned, the basic material used for most integrated circuits is single-crystal silicon. The initial raw material is obtained by the reduction of silicon oxides, such as sand, to yield materials that are typically 99 per cent pure (Figs 10.4(a) and (b)). Much more refined material is required for semiconductor components, where deliberate doping levels of one dopant atom per 10^8 silicon atoms are typical. A *zone-refining* technique is used to obtain the electronically pure grade of silicon required (see Fig. 10.4(c)) and a silicon bar in an inert atmosphere is melted locally using radio-frequency (r.f.) heating. The impurities are swept out by successive passes of the r.f. coil, which moves the molten region from one end to the other; the silicon then contains typically less than one impurity atom per 10^9 silicon atoms. Single crystals are grown from this material, most often from the melt, using a seed crystal that touches the surface and is then slowly withdrawn at a prescribed rate, again in an inert atmosphere (see Fig. 10.4(d)). Single crystals around 300 mm long and with diameters up to 100 mm are currently being grown by

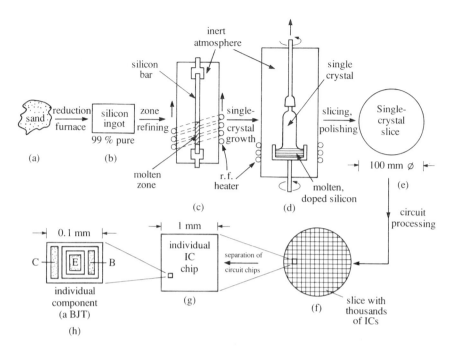

Fig. 10.4 Basic processing stages in the production of an IC chip: (a) silicon oxides; (b) reduction to 99 per cent pure Si; (c) zone refining; (d) crystal growth; (e) substrate slice; (f) slice with circuits; (g) individual circuit; and (h) individual component.

this method. Dopant is usually introduced in the melt during crystal growth so as to produce n- or p-type crystals. The single crystals are then cut using diamond saws, into thin discs or *slices*, each up to 100 mm diameter and of order 0.1 mm thick (Fig. 10.4(e)). Each slice when lapped and polished can be used to produce hundreds of integrated-circuit *chips*, as shown diagrammatically in Figs 10.4(f)–(h). The dimensions in the figure are variable, of course, but typical values are included to give some idea of scale. It will be noted that thousands of complete integrated circuits each containing possibly hundreds of thousands of components are produced on one slice of silicon. Since all components in each circuit are designed to be constructed by compatible planar diffusion processes, they are all formed at the same time by a sequence of masking and diffusion steps. That every such circuit and component on the slice is manufactured simultaneously by a few processing steps demonstrates the enormous power of the technology for producing relatively inexpensive, reliable, reproducible and often complex circuits on a very small area of silicon.

The stages of processing required to produce integrated circuits on the single-crystal silicon slice are dependent on the type of circuits required, for example whether bipolar or MOS, so the most prevalent bipolar silicon epitaxial technology will be described first, and its variants will be discussed later. The starting point is usually a p-type doped silicon slice, called a *substrate*, shown in the flow diagram (Fig. 10.5). Each particular stage in the fabrication of a component is carried out simultaneously on all similar components in the circuit and on the slice, using photographic techniques for location, as will be described. The first series of processes produce a highly conducting n^+ *buried layer* under all the active devices on the slice, the cross section of one such layer being shown in Fig. 10.5(a). The function of such a layer is to provide a low-resistance path from the planar collector contact to the collector region of a BJT. The buried layers are formed by first oxidizing the entire slice by heating it; many slices are processed together in an oxidizing atmosphere of oxygen and steam at 1150°C, the steam serving to accelerate oxidation, to produce an oxide layer around 1 μm thick. The next and subsequent stages involve cutting *windows* at appropriate places in the oxide through which the diffusion of dopants can take place. The oxide acts as an effective shield to prevent incursion of the diffusant, so the n^+ buried layers, for example, can only be deposited through windows opened in the oxide and nowhere else. Layer diffusion is often from the gas phase. In this instance, many slices are heated together to around 1200°C and a supply of arsene gas (AsH_3) is passed over them; n-type arsenic atoms in the gas that appear at windows in the oxide diffuse into the originally p-type substrate to dope the silicon substitutionally and change it locally by compensation to n^+-type (see Chapter 6). Arsenic is usually chosen as the donor for this operation because it has a lower rate of diffusion in silicon than do phosphorus or boron, which are used in subsequent diffusion. This ensures that the buried layers remain in position during succeeding processing stages. A doping concentration in excess of 10^{25} m^{-3} provides the low resistance required for the buried layer.

Next, any remaining oxide is removed and an n-type *epitaxial* layer (or *epilayer*) is grown over the entire p-type slice (see Fig. 10.5(b)). The layer, which is of order 10 μm thick, will eventually contain all the circuit components, so it must be single-crystal material and the crystal structure must be continuous across the interface to the substrate. The function of the substrate is to give mechanical support for the circuits developed in the epilayer, and also to provide electrical isolation between components, as will be explained later. The epitaxial layer is grown by heating the slices in an induction furnace at around 1200°C and exposing them to an atmosphere of silicon tetrachloride ($SiCl_4$), hydrogen and phosphine (PH_3). The silicon tetrachloride vapour is reduced by the hydrogen present and single-crystal silicon is grown on the slices. If the temperature is reduced, the epilayer becomes defective and eventually non-crystalline. The phosphorus atoms in the dopant gas ensure that the epilayer is doped n-type with a typical donor concentration around 10^{22} m^{-3} and a corresponding resistivity typically around 0.005 Ω m.

The next stage is a p-type *isolation diffusion* (see Fig. 10.5(c)), which is

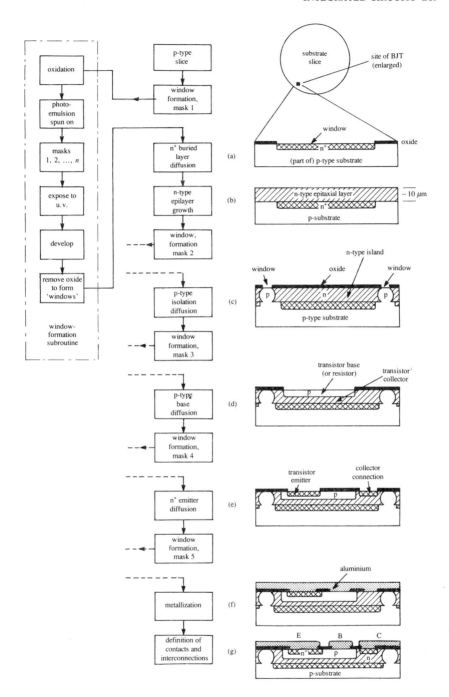

Fig. 10.5 Flow diagram of the fabrication of a bipolar IC, and cross sections of a planar BJT (bipolar junction transistor) in the circuit at each stage.

arranged to penetrate the epilayer completely through to the substrate and so provide islands of n-type in which one or more components can be formed. These are, as a consequence, then electrically isolated from others located in neighbouring islands, as has been discussed. The isolation diffusions are again located by etching windows in an oxide layer formed on the slice by heating, with others, in steam at around 1000°C. The mechanics of such *window formation*, which is outlined in the subroutine shown in the flowchart (Fig. 10.5), is common to this and all other diffusion stages in the process, so it merits a more detailed study later. The p-type compensation diffusion is usually carried out in two stages. Boron atoms are first deposited on the surface at high temperature using solid (boron trioxide) phases; the slice is then oxidized and the boron dopant is subjected to a *drive-in* diffusion at around 1000°C until the compensated p-type regions reach the substrate, as shown in Fig. 10.5(c).

There follows a further window-opening series of operations that locate windows in an oxide layer through which p-type *base diffusions* can occur in each n-type island on the slice, as shown in Fig. 10.5(d). Boron is again the usual dopant, but this time at a lower concentration. The resulting compensated p-type regions, typically around 2 μm deep, constitute the bases of all BJTs on the slice, together with any monolithic resistors and the anodes of diodes that may be required in the IC, as will be seen. The remaining n-type material forms the collectors of BJTs as well as isolation for monolithic resistors and possibly diode cathodes.

Next comes another now familiar set of window-opening routines to locate n^+-type emitter diffusions which follow. Phosphorus is usually the preferred dopant, often in the gaseous phase as phosphine (PH_3), but in a much higher surface concentration, typically around 10^{27} P atoms m^{-3}, which is close to the solid solubility limit of phosphorus in silicon. High-conductivity emitters, typically of order 1 μm deep, are produced in this manner. The same process simultaneously produces n^+ diffusions under collector contacts, which ensure a good ohmic contact to the relatively lightly doped n-type collector regions by the aluminium collector connection, which might otherwise locally dope p-type and provide an undesirable rectifying pn junction (see Fig. 10.5(e)).

All the various components of the bipolar integrated circuits are now developed, and it remains only to make contact to the various parts of each component and to provide circuit interconnections. Both these objectives are achieved in one series of operations. First, the slices are oxidized preparatory to a window-opening sequence that locates contact areas, and the entire slice along with others is covered in a thin aluminium layer by vacuum evaporation, as shown in Fig. 10.5(f). The contacts and interconnections are then defined in the aluminium, by selectively etching away superfluous areas of metal, using a series of processes very similar to the window-formation routine but omitting the oxidation stage (see Fig. 10.5(g)). The circuit interconnection metallization also includes terminal pads for external connection to the IC.

Each circuit is usually tested electrically while it is still on the complete slice, along with many similar circuits. A probe head containing many needle probes is lowered onto the aluminium terminal pads of each circuit on the slice in turn and a measurement is made of its major electrical characteristics. Any that are not within the specification are marked and eventually discarded.

The next stage is the separation of each IC chip on the slice, by diamond sawing, or scribing and cleaving, as shown in Fig. 10.5(g). Then each chip is bonded to a *header* and fine gold or aluminium wires are bonded between pads on the circuit and lead-throughs on the header to provide external connections for the circuit. The final product is encapsulated, often by encasing the entire header and IC chip in thermosetting plastic, as shown in Fig. 10.6.

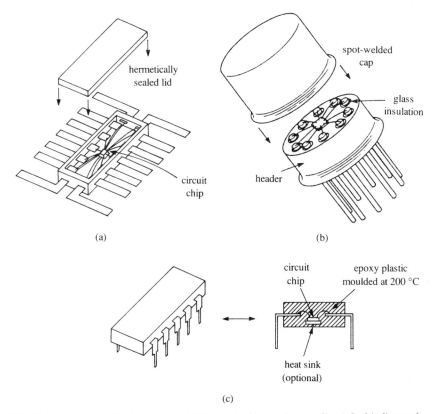

(a)

(b)

(c)

Fig. 10.6 Integrated-circuit packages: (a) flat pack; (b) transistor pack; (c) dual in-line pack.

This, then, is a brief summary of how integrated circuits are made, but it will be instructive to return now to discuss some of the more important stages in their production in a little more detail.

Consider, first, the window-opening subroutine, used to locate subsequent

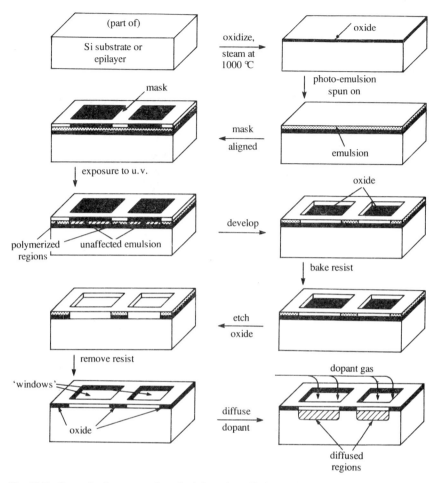

Fig. 10.7 **Stages in the preparation of windows in oxide layers for the location of diffusions in silicon.**

diffusions, which is outlined in Fig. 10.5 and illustrated in Fig. 10.7. The slice, along with many others, is first oxidized by heating in steam at a temperature of 1000°C. The oxidation time is chosen so that the resulting oxide layer is thick enough to be impenetrable by subsequent diffusants. Windows are formed in the oxide, through which diffusion of dopants can take place, by a photographic technique. First the slice is coated with a special photographic emulsion or *resist*, which is formulated so as to be insoluble to oxide etchant, by spinning it at high speed, say 10 000 rpm, while at the same time depositing a drop of the resist on the surface of the oxide. The photographic resist is hardened at this stage by baking at 150°C, which also removes solvents from the film. Meanwhile a photographic plate or *mask* has been prepared, by methods that will be discussed later, on which each black area corresponds in size and location to one of the set of windows required for the next diffusion stage. The mask is placed in near-contact with the photographic emulsion in a mask alignment machine, which is designed to allow optical alignment of the mask relative to previous diffusions on the slice, to ensure that successive layers of the circuit are in perfect registration with each other. When aligned, the mask is brought into contact and the slice is exposed to ultraviolet light (u.v.) via the mask. Areas of the resist that are located under transparent regions on the mask are polymerized and are unaffected by subsequent developments, but those under opaque parts are not affected by the u.v. and can be removed when the emulsion is developed, usually using a spray developer. Of course, all the operations described so far have to be carried out in a safe-light. The remaining resist is again hardened by baking and forms a series of windows through which the oxide can be etched selectively, using hydrofluoric acid. All that remains then is to remove the resist, using a mixture of sulphuric acid and hydrogen peroxide, and the slice complete with windows in the oxide is ready for the diffusion of dopants. The window-opening process, as described, uses *negative*-resist emulsion. *Positive* resists are available, which are used in much the same way, except that windows for etching are produced under transparent rather than opaque areas of the mask. This type of resist is most useful when relatively large areas are to be etched, for example, for defining interconnections.

It has so far been assumed that the various window-opening operations have been carried out using liquid etchants. These techniques have some disadvantages and are largely being superseded by *dry etching* methods. One difficulty with liquid etchants is that material is usually removed isotropically, the liquid attacking the material equally in all directions, which causes, for example, undercutting and subsequent changes in feature size, as shown in Fig. 10.8(a). The advantage of dry etching methods, such as *reactive plasma etching*, in which an ionized discharge is used to produce chemically active species that etch the material, is that material removal is preferentially in the direction of incident particles, which can be constrained to be normal to the surface. This ensures that patterns are etched anisotropically, straight into the surface, with no lateral removal of material under the lithographic mask, as shown in Fig. 10.8(b). Dry etching is also particularly useful for defining the metal

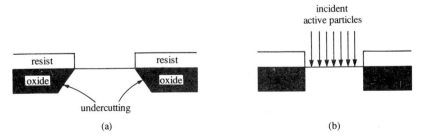

Fig. 10.8 **(a) Liquid etch of window in oxide showing undercutting and (b) reactive plasma dry etching of the same feature.**

interconnect layer in very large-scale integrated circuits, with little feature rounding and to the necessary high degree of precision.

The production of a set of photographic masks, each of which eventually determines the exact size and location of a particular part of every circuit component, all the emitters in all circuits on the slice say, is germane to the technology of integrated circuits. Each mask in the series not only is used to produce windows for one stage in the manufacture of an individual IC but simultaneously fulfils the same function for all circuits on the slice. It will also be apparent that many such masks are required, all of different configurations, one for each diffusion step. The production of one of a series of masks, to locate the buried-layer diffusion of a very simple hypothetical three-transistor circuit is illustrated in Fig. 10.9. A photographic master is prepared first, which is

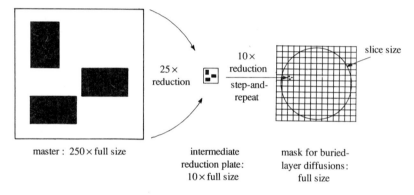

master : 250 × full size   intermediate   mask for buried-
                                 reduction plate: layer diffusions:
                                  10 × full size   full size

Fig. 10.9 **Photographic mask preparation, in this example to produce three buried layers per circuit (not to scale).**

250 × full size, say, and includes only the diffusions for one IC, as shown. The master, typically of order 1 m × 1 m, could in principle be drawn with ink but it is often more convenient to use *cut-and-peel* techniques, in which photographically opaque red plastic film on a transparent base is selectively removed by peeling after cuts are made through the red layer using a computer-controlled cutting head. This technique allows precise computer control of the master production. The master is then back-illuminated and photographed, usually with a 25 × reduction, to produce an intermediate reduction plate, which has exposed areas corresponding to those portions of the master that are transparent by virtue of the peeling off of opaque film. The final mask is produced in a *step-and-repeat* camera, which reduces the image of the reduction plate by a further 10 ×, to produce a 250 × reduced facsimile of the master on the diffusion mask. The camera then steps on and prints another replica alongside; this continues until the final plate is entirely covered with a matrix of identical diffusion patterns, one for each IC on the slice.

It will be recognized that the masks for each stage in the production of an IC are quite different. To illustrate this point, a possible set of masks for a very simple bipolar circuit are shown in Fig. 10.10(a)–(e). These correspond to isolation, base and emitter diffusion, contact opening and definition of

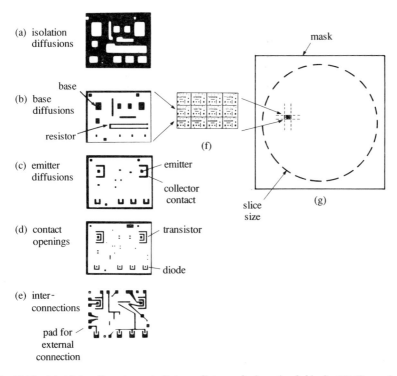

Fig. 10.10 (a)–(e) A rudimentary set of intermediate masks for a simple bipolar IC; (f) an enlarged section of an array of diffusion masks; (g) one of the complete set of final masks.

interconnection stages in the circuit fabrication. After step-and-repeat photography, each is reproduced in a matrix on a mask, as shown for the base diffusion mask, Fig. 10.10(g), an enlarged section of which is shown in Fig. 10.10(f). It will be obvious from this simple example that registration between successive masking stages is extremely important, to ensure that emitters, for example, are correctly located with respect to transistor bases. Optical indexing marks are therefore included on each mask in the set to facilitate precise registration. The rudimentary set of masks shown is only included as an illustration; a set of as many as 12 masks are required to produce a more typical bipolar IC.

It will now be clear why the technology is so powerful. Hundreds of identical integrated circuits, each containing possibly thousands of components, are all produced at the same time by a series of now established processing stages, on one semiconductor slice.

10.5 Bipolar integrated components

10.5.1 Integrated bipolar transistors

The principal component in a bipolar IC is, of course, the planar BJT. Bipolar planar transistors for inclusion in integrated circuits are basically similar to the discrete versions described in Chapter 8. An obvious difference is that the collector lead has to be accessible at the top surface of the slice. This is achieved as shown in Figs 10.11(a) and (b), which show a cross section and plan of an integrated planar transistor with diode isolation. The n^+-diffusion near the collector contact ensures that a low-resistance ohmic connection is made to the collector region. Since the collector contact is now made to the top surface, collector current is constrained to flow transversely to the active part of the transistor via a longer path than in the discrete version, with a corresponding increase in the series resistance of the collector, R_{CC}. This effect is often counteracted by a buried layer of n^+-material, which is formed during processing underneath the collector diffusion, as shown, and provides a low-resistance path for the collector current and effectively minimizes R_{CC}. Sometimes the n^+-collector contact diffusion is extended right through to the buried layer, as shown by the broken lines in Fig. 10.11(a), so as to reduce the collector access resistance still further. Such a diffusion is called a *collector wall* or *sink* diffusion. A possible equivalent circuit of the transistor plus parasitic components due to the isolation is shown in Fig. 10.11(c).

The active part of the npn transistor is formed by two successive diffusions into the n-type epilayer. Since the collectors are formed from the epi-material, its doping level during growth is carefully controlled to optimize transistor performance; fairly uniformly doped layers with around 10^{22} donors m^{-3} and a resistivity of 0.005 Ω m are typical. The first boron diffusion, which produces transistor bases (and resistors—see later), is a two-step process with deposition of p-type dopants followed by a drive-in stage, so as to produce a compensated p-type doped base with a concentration gradient as shown in Fig. 10.12(a). The base–collector junction occurs at the point where the p-type dopant just compensates the n-type in the epilayer, i.e. where the doping densities are equal (see Fig. 10.12(b)). The second diffusion into the compensated p-type base material to form emitters is designed to produce a heavily doped n^+-layer, for optimum emitter injection efficiency (see Chapter 8). Phosphorus is the usual dopant and this is kept in sufficient quantity at the surface throughout the diffusion to ensure that the doping level at the surface approaches the solid solubility limit of phosphorus in silicon, around 10^{28} m^{-3} (see Fig. 10.12(c)). Possible resulting doping density profiles for the entire transistor structure are shown in Fig. 10.12(d). Note that the concentration of acceptors in the base is

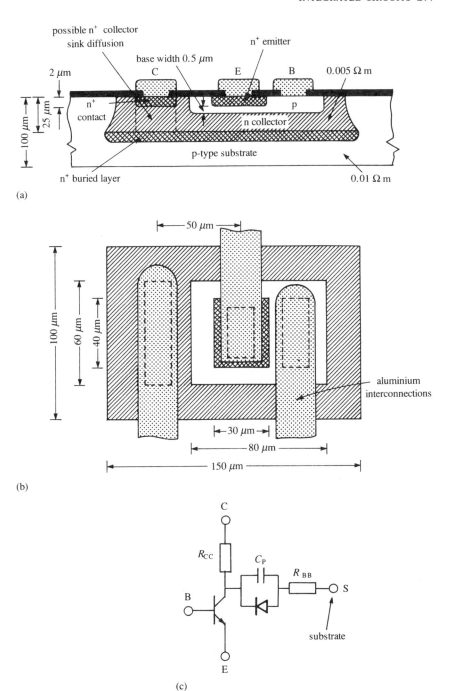

Fig. 10.11 General-purpose, npn planar bipolar junction transistor: (a) cross section; (b) plan; (c) a possible equivalent circuit.

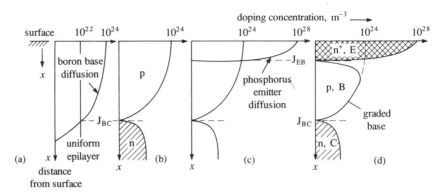

Fig. 10.12 Doping density profiles in a planar epitaxial BJT: (a) first, base diffusion into epilayer; (b) junction formation by compensation; (c) second, emitter diffusion; (d) final impurity profile.

not in practice precisely uniform, as has been assumed earlier, but falls off towards the collector. In such a *graded-base* transistor, a built-in electric field in the base is induced, which prevents diffusion of majority holes down the density gradient in equilibrium. When such a transistor is operating, the built-in field in the base is in such a direction as to accelerate injected minority electrons across it. This drift field on electrons crossing the base is additional to the usual diffusion forces, so the transit time from emitter to collector is reduced, which results in increased speed of operation and frequency response compared to the uniform-base BJT.

Since all the electrical contacts to the planar BJT are via the top surface, as shown in Fig. 10.11, it follows that currents flowing between a contact and the active transistor region have to flow through passive bulk semiconductor material, which behaves as a built-in parasitic resistance, called an *access resistance*. The *collector access resistance*, R_{CC}, would have a significantly high value, because of the relatively low doping levels and high resistance of the epilayer from which the collector is formed, but the highly conducting buried layer under most of the device, together with collector sink diffusion, as discussed, tends to bypass the collector current and hence reduce the effective value of R_{CC}. However, R_{CC} and to a lesser degree the *base access resistance*, R_{BB}, are often included in small-signal equivalent circuits (see Fig. 10.11(b)) to account for the inevitable resistive losses.

10.5.2 Integrated resistors

Monolithic integrated resistors are most conveniently formed in the bipolar technology by using the same diffusion that simultaneously creates transistor bases. This p-type semiconductor layer can provide a resistance that can be controlled by choice of geometry (see for example Fig. 10.13), as in all such bipolar circuits. An n-type island again serves to isolate the component

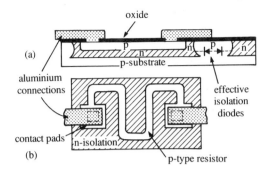

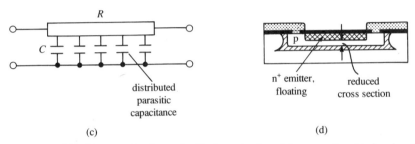

Fig. 10.13 **Diffused integrated resistor using bipolar technology: (a) cross section; (b) plan view; (c) a possible equivalent circuit; (d) a possible pinch-resistor.**

electrically from its neighbours, by virtue of the npn regions separating components, which behave as effective back-to-back diodes, as shown in Fig. 10.13(a). Under these conditions the only current possible between components under all relative bias conditions is very small and corresponds to the diode saturation current for one of the pn junctions. Since resistor diffusions and base diffusions are carried out at the same time, one obvious constraint of the technology is that the resistivity and thickness of the resistor layer are predetermined by the basic doping levels and junction depths chosen to optimize transistor performance and so cannot be altered at will. Since the resistivity and thickness of integrated resistors are fixed and their only design variable is their geometry in plan, a *sheet resistance*, ρ_s ohms per square ($\Omega/\square$), is usually quoted to characterize the layer. This is related to the bulk resistivity, ρ, by

$$\rho_s = \rho/t \ \Omega/\square \tag{10.1}$$

where t is the p-type layer thickness. Hence the resistance of an integrated resistor of length l and width w, formed in the layer, is

$$R = \rho l/wt = \rho_s l/w \tag{10.2}$$

It follows that the sheet resistance is the resistance of a unit square on the surface, which is independent of the size of the square. If ρ_s is quoted, resistance values can easily be determined using Eq. (10.2).

Actual sheet resistances for the base/resistor diffused layer are typically 100–250 $\Omega/\square$, so, since resistors with an aspect ratio, l/w, between 1:3 and 100:1 are practicable without fringing effects becoming too significant, resistors in the range 50 Ω–10 kΩ can be provided. The higher-value resistors are often folded, as in Fig. 10.13(b), to conserve chip area. For low-value resistors of order 1 Ω, the emitter n^+-layer, which has a sheet resistance typically in the range 1–10 $\Omega/\square$, can be used. For higher values of resistance than can be fabricated using the normal base layer, it is possible to use the epitaxial material, which usually has a sheet resistance of order 1 k$\Omega/\square$. Alternatively, a *pinch-resistor*, in which an additional floating emitter diffusion is incorporated as shown in Fig. 10.13(d), can provide higher resistance values. Current flow through the p-layer is restricted by the n^+ diffusion, which reduces the cross section of the p-layer and hence increases its effective sheet resistance.

Although integrated resistors of a variety of resistance ranges are possible using bipolar technology, as explained, they are all relatively extravagant in their use of chip area. For example, one 1 kΩ resistor might occupy the same area as 10 bipolar junction transistors. Since chip area is usually at a premium, for economic reasons, integrated resistors tend to be omitted altogether if possible in modern integrated circuits or replaced by the active resistance of a BJT.

10.5.3 Integrated capacitors

Integrated diffused capacitors are available in the bipolar technology by using the capacitance associated with one pn junction that is formed during transistor fabrication. A typical diffused junction capacitor, using the capacitance of the base–collector junction, is shown in Figs 10.14(a) and (b). Again the doping levels and junction locations are fixed for optimum transistor design, so cannot be changed for the capacitors, and the only design variable is the area of the capacitor. The chosen junction, in this example the base–collector junction, is usually reverse-biased, to ensure that leakage current across the junction is kept to a minimum. The capacitance of such a component is typically 100 pF mm^{-2} with reverse voltage of around 1 V. Obviously one disadvantage of such a capacitor is that its capacitance is a function of the applied reverse bias. Because the junction is not usually uniformly doped on either side, but is *graded*, as explained, the capacitance no longer varies as $V_j^{1/2}$ as in the ideal case (see Chapter 8). For example, if the slope of the impurity profile can be assumed to be approximately constant, the capacitance of such a *linearly graded junction* when reverse-biased can be shown to vary as $V_j^{-1/3}$, so the capacitance is not too sensitive to voltage variations. The integrated capacitor, C, shown in Fig. 10.14, has an access resistance, R_a, associated with built-in resistance of the epilayer between

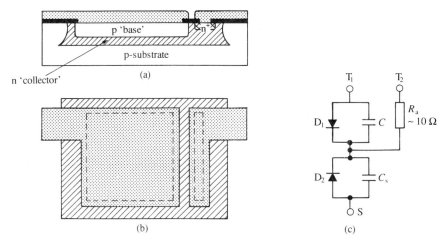

Fig. 10.14 Diffused integrated capacitor: (a) cross section; (b) plan view; (c) a possible equivalent circuit.

contact T_2 and the active region of the capacitance. The junction also behaves as a diode, D_1, and there is an additional spurious diode, D_2, and capacitance, C_s, due to the 'collector'–substrate junction, so a possible equivalent circuit for a relatively simple 'passive' component is as shown in Fig. 10.14(c).

It is also possible to use the emitter-base junction to form an integrated capacitor, and an increase of capacitance per unit area to around $10^3 \, \text{pF} \, \text{mm}^{-2}$ results, although this is at the expense of a reduction in breakdown voltage, which limits operating voltages to less than 5 V.

Again, whereas integrated capacitors are seen to be feasible using bipolar technology, they are relatively extravagant in their use of chip area, so are avoided wherever possible. This has led, for example, to the development of direct coupled circuits. All but the smallest values of capacitance are, wherever possible, connected externally to the circuit as discrete components.

10.5.4 Integrated inductors

A similar situation exists with regard to integrated inductors. Such components can be envisaged, for example, as a conducting spiral, but it would be difficult to produce a high Q and the device would be uneconomically wasteful of chip area. Consequently, inductors are either provided as an externally connected discrete component, for example an intermediate-frequency (IF) transformer, or small values are sometimes synthesized using a capacitor and integrated operational amplifier circuits.

10.5.5 Integrated diodes

Obviously, if a diode is required in a circuit, use can be made of any of the pn junctions present in the basic BJT structure, but the choice is limited by considerations of switching speed and breakdown voltage, as discussed in Chapter 7. The two most useful configurations are illustrated in Fig. 10.15. The base–collector pn diode shown in Fig. 10.15(a) is a relatively high-voltage device, the breakdown voltage of the junction being around 50 V; in the form shown, with emitter diffusion shorted to the base to form a common anode, A,

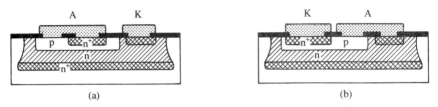

Fig. 10.15 Integrated pn junction diodes: (a) base–collector and (b) base–emitter configurations.

the switching time is a modest 100 ns. The base–emitter diode, Fig. 10.15(b), has a much lower breakdown voltage of around 5 V, due to the highly doped emitter, but, if this constraint can be tolerated, it has a much more useful switching time of around 5 ns; note that the collector is often shorted to the base, as shown, to eliminate the deterioration in switching time that would otherwise occur due to stored charge in the base–collector junction.

10.5.6 Integrated Schottky diodes and transistors

Schottky barrier diodes are also available using the standard bipolar technology described. They are extremely useful in switching circuits because of their fast switching time, typically of order 0.1 ns, which arises because of the almost negligible charge storage at the metal–semiconductor junction. A Schottky barrier is most conveniently formed between the epitaxial layer and the aluminium metallization used for interconnections, as shown schematically in Fig. 10.16. The aluminium anode, A, makes direct contact with the lightly doped n-type epilayer to form a Schottky barrier, but the cathode connection, K, is via the usual emitter-type n^+ collector contact diffusion, which ensures that a good ohmic contact is made.

It is also a relatively straightforward extension of the bipolar technology described to include a Schottky diode between base and collector of a BJT in

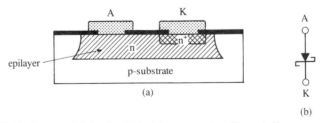

Fig. 10.16 Integrated Schottky diode: (a) cross section; (b) symbolic representation.

one integrated unit, called a *Schottky transistor*, as shown in Fig. 10.17. The aluminium base contact is allowed to overlap across onto the collector, as shown, to form the built-in diode. The role that the Schottky diode plays is to improve the transistor switching speed by preventing saturation.

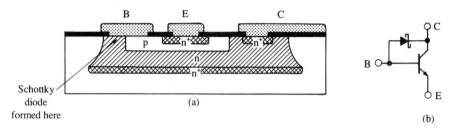

Fig. 10.17   The integrated Schottky transistor; (a) cross section; (b) the effective circuit diagram.

10.5.7 Integrated pnp bipolar transistors

Although npn transistors are standard in the bipolar integrated technology described, pnp devices are, of course, necessary for some circuit applications. Whereas it is feasible to produce complementary pnp devices either by diffusing an additional p-type layer, as in a triple-diffused transistor, or by using the p-type substrate as a collector, as in a *vertical pnp transistor*, both these solutions have severe technical and economic disadvantages that limit their usefulness. If pnp devices are required on the same IC chip as npn transistors, it is much more usual to incorporate *lateral pnp transistors*, in which the active currents flow parallel to the surface, or laterally. Such components can be fabricated using standard npn technology, as demonstrated in the cross section shown in Fig. 10.18. The p-type emitter and collector

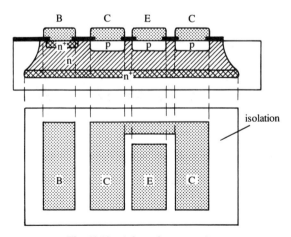

Fig. 10.18   A lateral pnp transistor.

diffusions are formed simultaneously by a standard npn 'base' diffusion and the epilayer constitutes the base region and also provides circuit isolation as usual. Minority holes are injected from the side walls of the emitter into the base and eventually reach the side walls of the surrounding collector, so providing the necessary transistor action. The lateral transistor, because it has a relatively long base region, often has an inferior gain compared with the npn BJT, of around 1–5 compared to, say, 100. However, since the emitter and collector are both produced by the same diffusion process, difficulties due to alignment of successive masks are avoided, and improved tolerance on the emitter–collector spacing becomes possible. Hence the base width can be controlled by the degree of lateral diffusion allowed, and reproducible, thin-based devices with gains approaching those of npn devices are now realizable.

10.6 MOS integrated circuits and components

The principal member of the insulated-gate family of integrated-circuit components is the induced-channel MOST shown in cross section in Fig. 10.19(i). Such MOS devices and associated integrated circuits are fabricated in silicon using very similar processing techniques to those used for bipolar circuits, but, because of the much simpler structure of the MOST, fewer processing stages are required. The basic processing procedure for producing an n-channel enhancement-mode MOST is outlined in Fig. 10.19.

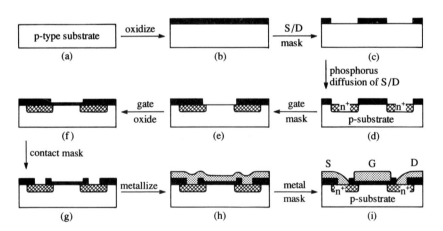

Fig. 10.19 Basic steps in the production of an n-channel enhancement-mode MOST.

The finished device, shown in Fig. 10.19(i), is self-isolating, since back-to-back pn diodes are automatically formed between adjacent components, so separate isolation diffusions are not required. Instead, windows are first opened in the oxidized p-type substrate slice, using a source/drain mask, Fig.

10.19(c), to locate the n^+-type, phosphorus source and drain diffusions, Fig. 10.19(d). A gate mask and the usual photolithographic etching routine is then used to open windows in the thicker, 1 μm, field oxide, to locate the thinner, 0.1 μm, gate oxide, Figs 10.19(e) and (f). A contact masking stage then opens windows for contacts to the source and drain, Fig. 10.19(g), followed by metallization of the entire slice, usually by vacuum evaporation of aluminium, Fig. 10.19(h). Finally, superfluous metal is etched away, using another masking operation, leaving contacts and interconnections between circuit components, as in Fig. 10.19(i).

It will be noted that the processing steps required for producing a MOST include only four masking operations and one diffusion, which can be compared with 12–14 masking stages and four diffusions typically necessary to fabricate a BJT. This relative reduction in processing complexity means that MOS integrated circuits can potentially be made with a better *yield, i.e.* fewer circuits that do not meet specification, than their bipolar equivalents, which has important economic advantages, as will be seen. The MOST also has the advantage of relatively small size; partly because of its lack of an isolation region, it only occupies a chip area of 5–10 per cent of that required for a BJT, which results in an enhanced device packing density in MOS ICs.

Diffused integrated resistors are not conveniently available in MOS technology, and if they were made would typically occupy more than 10 × the chip area required for an active device, so wherever resistance is required in a MOS IC, for example for a load resistor, use is made of the dynamic resistance that exists between the source and drain of a MOST. If the drain and gate of the MOS load are strapped together using the interconnecting metallizing layer, as shown in Fig. 10.20, the gate voltage equals the drain

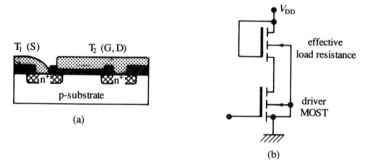

Fig. 10.20   An n-channel enhancement-mode MOST used as a dynamic load resistance: (a) a cross section; (b) the circuit configuration.

voltage and the device operates in the saturated mode. Under these conditions, the drain current is relatively insensitive to drain-voltage changes, as discussed in Chapter 9, and drain–source non-linear resistances typically in the kilohm range are produced. It is also possible to bias the gate of a MOST resistive load in such a manner that the device operates in the unsaturated region, to produce an ohmic resistance when required. Very high packing densities can be obtained if depletion MOSTs operated in the saturation region are used as load resistors for enhancement driver transistors, which makes such a circuit arrangement attractive for large-scale integration, for example in microprocessors. It will be noted that it is possible, and economically desirable, to produce large-scale integrated MOS circuits consisting entirely of active devices, and containing no passive components.

Passive capacitors can be most conveniently fabricated using MOS technology, although their relatively extravagant use of chip area precludes their extensive use. A typical arrangement is shown schematically in Fig. 10.21.

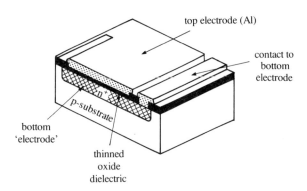

Fig. 10.21 Integrated MOS capacitor.

It is clear that such a capacitor can be made by the same processing steps as are used to produce a MOST, the bottom electrode being formed from a n^+-layer, diffused at the same time as sources and drains. Apart from a small access resistance in series, due to the finite resistance of the n^+-layer, the integrated MOS capacitor acts much more like a classical parallel-plate capacitor than does its bipolar counterpart, with capacitance per unit area given by the usual $\epsilon_r \epsilon_0 / t_o$ expression.

Although p-channel enhancement-mode MOSTs have disadvantages with regard to speed, compared to n-channel devices, complementary pairs of n- and p-channel enhancement transistors can be fabricated using a slight modification of standard MOS technology, to produce complementary MOS (CMOS) integrated circuits, which have very important applications. A cross section of a complementary pair of integrated MOS transistors is shown in Fig. 10.22. Since the components are formed on an n-type substrate, the induced p-channel device is self-isolating, but the n-channel MOST requires separate isolation, which takes the form of an additional p-type island or *well* as shown. Apart from the extra p-well masking stage and associated isolation diffusion, CMOS circuit processing essentially follows standard MOS practice, as outlined, for example in Fig. 10.19.

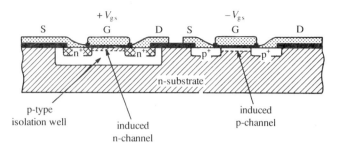

Fig. 10.22 Complementary integrated MOS transistors.

Integrated logic gates using CMOS circuitry draw only extremely small standing currents from the supply in either logic state. They are therefore most useful in portable, battery-powered systems and are also finding increasing application in large-scale integrated-circuit systems. Various examples of CMOS integrated circuits are illustrated in Figs 10.23 to 10.26.

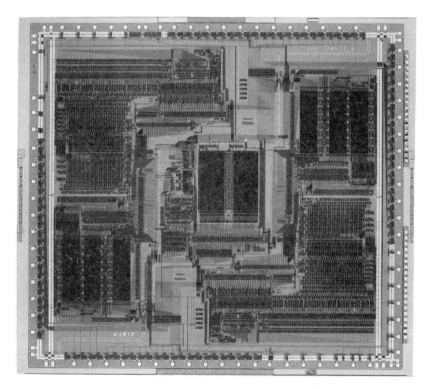

Fig. 10.23   A 24-bit block floating-point arithmetic unit. This 60 000-gate chip is a micro-programmable arithmetic unit for sonar digital signal processing. The 11.4 × 10.4 mm chip is fabricated by a 1.4 μm two-layer metallization CMOS process. (Courtesy of Plessey Research (Caswell) Ltd.)

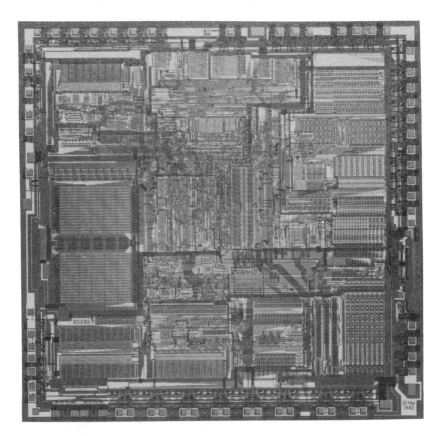

Fig. 10.24   The 80286 16-bit microprocessor slice. The device incorporates 130 000 transistors. (Courtesy of Intel Corp.)

10.7 Recent technological developments

In the standard MOS technology described, a certain amount of overlap of the gate electrode and source and drain diffusions is necessary so that registration tolerances and lateral diffusion can be allowed for. However, such overlap produces additional gate capacitance, which can reduce switching speed, so several new technologies have evolved to overcome this difficulty, which feature automatic gate-channel registration.

Ion-implanted MOS integrated transistors, for example, shown in Fig. 10.27, are fabricated up to the final stage by conventional MOS techniques, except that the p-diffusions are separated by slightly more than the final channel length required, so the gate electrode falls slightly short of them initially, as in Fig. 10.27(a). The entire slice is then flooded with p-type boron atoms, accelerated to sufficient velocity in an implantation chamber to penetrate

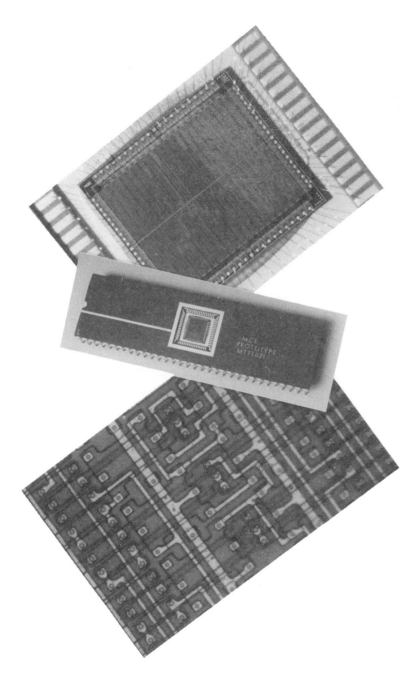

Fig. 10.25   Spatial filter for image processing. The 2850-gate array uses 3 μm CMOS technology. Gate array design by Final Year Project Student, University of Sheffield. (Courtesy of Dr M. Seed and Micro-Circuit Engineering.)

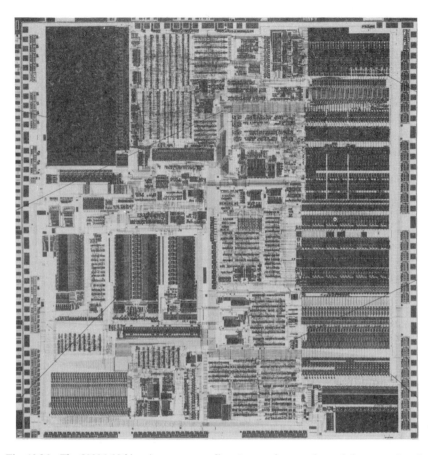

Fig. 10.26 **The 80386 32-bit microprocessor slice. A central processing unit incorporating the microprocessor is capable of processing 5 Mips (million instructions per second). (Courtesy of Intel Corp.)**

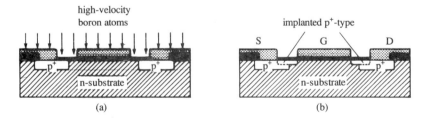

Fig. 10.27 **Ion-implanted MOS integrated p-channel transistor: (a) before implantation, with 'undersize' gate mask; (b) implanted transistor, with self-aligned gate.**

oxide layers and dope the silicon beneath. Ions striking metal regions are collected, so the gate metallization serves as an effective self-aligning mask, which precisely defines the implantations and hence the extremities of the source and drain, as shown in Fig. 10.27(b).

During ion implantation the crystalline region is converted into a disordered or amorphous layer due to energy exchanges between high-energy bombarding ions and the host atoms. Post-implantation annealing is necessary to re-establish the crystalline structure. Consequently, ion implantation is conventionally followed by a furnace anneal at 850°C to repair damage done to the lattice by the ion bombardment. Alternatively, especially for GaAs metal-semiconductor FETs (MESFETs) and LSI circuits (see Sec. 12.4) *rapid thermal annealing* (RTA) techniques are beneficial. RTA provides precise, relatively short, heating cycles for dopant activation (and incidentally for contact alloying). Furnace annealing, which is necessarily long, to avoid thermal stress, can affect the doping concentration profile and may also result in the thermal conversion of semi-insulating GaAs substrates. The peak carrier concentration is also limited to around 10^{24} m^{-3}. RTA techniques can improve the maximum carrier concentration without significant diffusion of dopants or thermal conversion of substrates. RTA annealing cycles can vary over the range 1 ns to 100 s. For the longer times, > 1 s, tungsten–halogen lamps or resistive heating strips heat the implanted slices. Flashlamps, lasers or electron beams are used to heat the slices for intermediate times up to 1 s. Very short annealing times are achieved by high-energy lasers, which melt the first micrometre or so of the semiconductor surface, which subsequently recrystallizes epitaxially.

Silicon-gate MOS technology also incorporates self-alignment of gates, but has an additional advantage of producing MOSTs with a reduced threshold voltage. In this process, the gate electrode is made from polycrystalline doped silicon, rather than aluminium. The basic processing steps are outlined in Fig. 10.28. The oxidized p-type substrate, Fig. 10.28(a), has windows opened in it

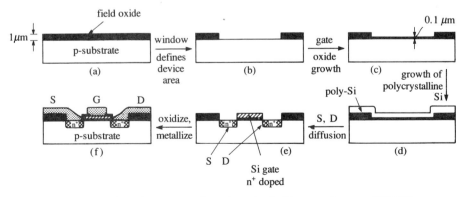

Fig. 10.28 **Fabrication sequence for n-channel polysilicon gate integrated MOST.**

photolithographically to locate each device, Fig. 10.28(b), and a thin gate oxide is then grown on the slice, Fig. 10.28(c). Next, a layer of polysilicon is deposited in an epitaxial furnace, using chemical vapour deposition techniques, Fig. 10.28(d). A further masking operation defines the gate area and the source and drain diffusions. The succeeding n^+ phosphorus diffusion, which produces source and drain, is self-aligned by the polycrystalline gate and simultaneously heavily dopes the polysilicon to produce a near-metallic silicon gate, Fig. 10.28(e). Finally, the slice is oxidized once again, contact windows are opened and aluminium contacts and interconnections are made using conventional technology, Fig. 10.28(f). Note that the use of the gate electrode as a self-aligning diffusion mask as described for the silicon gate would not be possible with an aluminium gate, which would melt during the subsequent diffusion stage.

Significant developments are also taking place in the technology for the fabrication of bipolar ICs, which will probably augment or supersede the conventional methods discussed so far. For example, trench isolation, discussed earlier, has sufficient advantages to merit its inclusion in most bipolar ICs. Another significant development is hyper-shallow *polysilicon emitter* fabrication, which is being incorporated in IC BJTs to improve their upper frequency and speed, particularly as devices are scaled to smaller dimensions. An outline of the basic fabrication technology is illustrated in Fig. 10.29. The first process is a relatively low-energy boron ion implant into

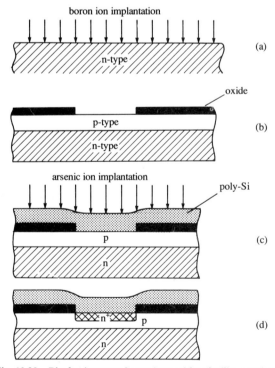

Fig. 10.29 Bipolar integrated transistor with polysilicon emitter.

the n-type layer, to form a shallow p-type base (Figs 10.29(a) and (b)). The slice is then oxidized and emitter windows opened, using conventional methods already discussed. There follows a polysilicon deposition and a subsequent arsenic ion beam implantation, Fig. 10.29(c). On annealing at around 950°C, the implanted arsenic diffuses through the polysilicon to form a shallow n^+ emitter region, shown in Fig. 10.29(d). The arsenic penetration, and hence the emitter thickness, is typically less than 0.1 μm. Hence the process is compatible with submicrometre emitter widths, while still retaining essentially one-dimensional emitter–collector current flow. The junctions formed can also be considered abrupt, which leads to unity gain frequencies as high as 25 GHz and a much improved speed performance.

Although modern photolithographic systems are now capable of producing 1 μm line resolution, because of diffraction effects such processes appear to be approaching an ultimate limit. Several new high-resolution techniques are being developed to overcome this difficulty, notably ion-beam, X-ray and electron-beam (e-beam) lithography, the latter being particularly successful, since they have a submicrometre capability, so providing at least an order-of-magnitude improvement in minimum linewidths. The e-beam technique used to create a mask by directly writing onto a resist-coated slice is shown in Fig. 10.30(a). A finely focused electron beam is deflected, and blanked off,

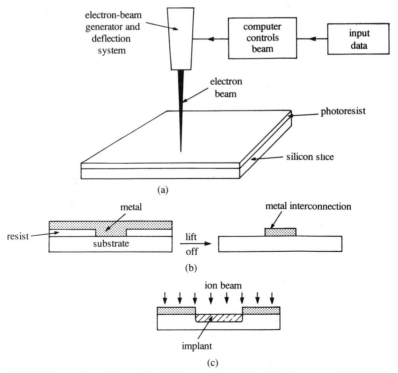

Fig. 10.30 Electron-beam lithography: (a) electron beam writing directly onto resist; (b) windows in resist used to create interconnections; (c) ion-beam inplantation.

under the control of a computer supplied with appropriate data, exposing the resist covering the slice wherever it impinges. Subsequent resist development selectively removes the exposed or unexposed resist, depending on whether the resist is of positive or negative type. Computer-controlled scanning of the electron beam can therefore be used to open windows directly in the resist, which are used in subsequent slice processing stages. For example, if a metal layer is deposited, say by evaporation, onto the resist layer with windows, the resist can then be dissolved and the surplus metal floated off, to leave a fine pattern of interconnections, as shown in Fig. 10.30(b); this lift-off process is now becoming quite common for defining the customized top-electrode configuration in uncommitted logic ICs. Alternatively, the windows in the resist can be used to locate regions for plasma etching of the substrate or under some conditions as a mask for selective ion implantation. The direct e-beam lithographic process described, although offering enhanced resolution, is relatively very slow, so throughput of slices is a distinct limitation on its universal application. However, e-beam lithography is now widely used to

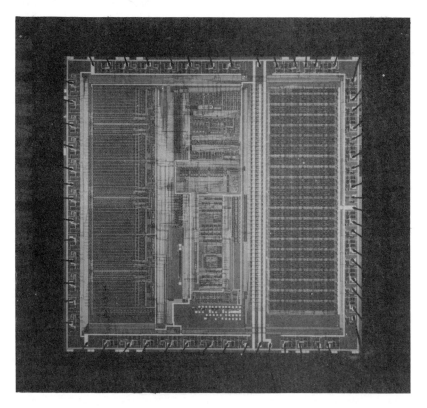

Fig. 10.31   A fully customized CMOS integrated circuit chip for encryption purposes. The chip contains 44 000 transistors and operates at encryption rates of up to 20 000 bits per second. (Courtesy of British Telecommunications Research Laboratories.)

make chrome master masks for subsequent replication and use in conventional photolithographic wafer processing.

10.8 Comments on integrated-circuit types

Although in general bipolar and MOS technologies have features in common, for example photolithographic masking, diffusion and so on, the two

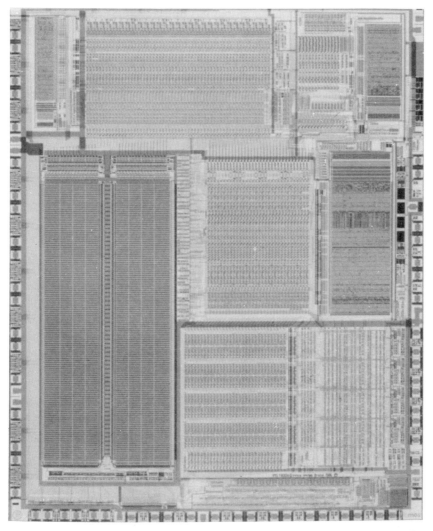

Fig. 10.32 **An Inmos transputer chip. The device is so named (short for transistor computer) since it is a computer on a single chip used as a circuit component. One transputer measures under 10 mm × 10 mm, contains over a quarter of a million transistors and can process 10 million instructions per second. (Courtesy of INMOS Limited)**

fabrication systems, as has been seen, are by no means compatible. Mixed technologies, in which MOS and bipolar active and passive components can be incorporated on the same circuit chip, have been devised, but these tend to be uneconomical for all but the most specialized application. Bipolar and MOS integrated circuits each have particular advantages; for example, bipolar circuits are usually faster and can handle more power, whereas MOS circuits can be more economical because of their potential for higher packing density and enhanced yield. Consequently, since compatible passive components are available in either technology, all-bipolar and all-MOS integrated circuits have been developed side by side. Further examples of a variety of integrated circuit chips are shown in Figs 10.31 to 10.33.

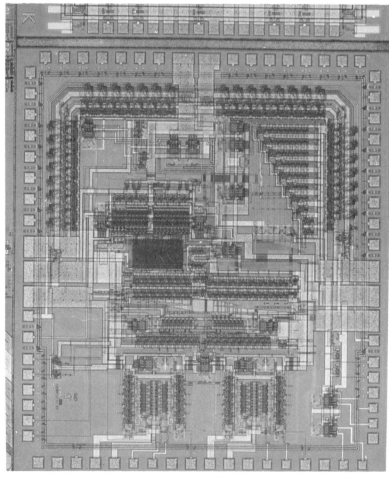

Fig. 10.33   Direct frequency synthesizer. This VLSI Si chip produces sine, triangular and square-wave outputs, both in phase and in quadrature. It is capable of fast (< 10 ms) frequency between 1 Hz and 500 MHz in 1 Hz steps. (Courtesy of Plessey Research (Caswell) Ltd.)

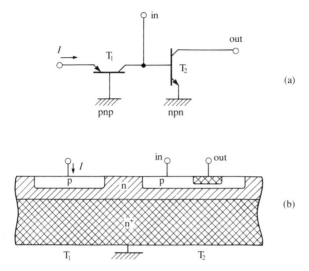

Fig. 10.34 Bipolar I²L logic cell: (a) circuit diagram; (b) realization on silicon slice.

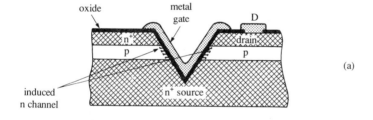

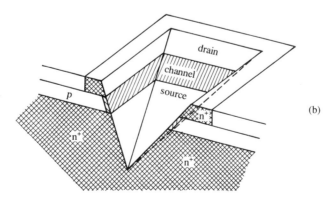

Fig. 10.35 Vertical induced n–channel MOS transistor: (a) a cross section showing short effective channel length; (b) a view showing the pyramidal crater and enhanced gate width.

Newer bipolar IC technologies, such as *integrated injection logic*, I^2L, which has no analogue in discrete form, have reduced the area of bipolar circuits, while retaining their speed advantage, so making them competitive with MOS with regard to packing density. A circuit diagram of an I^2L element and its integrated realization is shown in Fig. 10.34. T_1 is a pnp lateral device, and T_2 is an 'upside-down' npn transistor. The emitter of the npn is under its base and common with the base of the pnp. Also, the collector of the pnp is identical to the base of the npn. The two transistors are *merged* into an integrated logic cell.

Refinements of MOS techniques, for example in *vertical MOS* (VMOS) circuits, shown in Fig. 10.35, have produced reductions in size and an increased speed because of reduced gate length, which to some extent have offset advantages accrued by the newer bipolar circuits.

It seems that the two general classes of technology and circuits, bipolar and MOS, are likely to coexist for some time to come, each type being incorporated in systems that exploit its particular advantages.

Problems

1. Derive a general expression for a bend in a monolithic diffused resistor, constructed from circular arcs.

A particular process produces resistors with a sheet resistance of $200\,\Omega/\square$, and track widths of $10\,\mu m$ with spacing between tracks of $10\,\mu m$ are standard. Estimate the maximum value of resistance that can be produced in an available chip area of $200\,\mu m \times 200\,\mu m$, assuming semicircular turning points and end-contacts that have an effective resistance of 0.65 square. Sketch a plan view of the resistor.

Ans. $40\,k\Omega$

2. A particular folded diffused planar IC resistor employs a diffusion that has a sheet resistance of $100\,\Omega/\square$ and uses a resistor track width of $10\,\mu m$. Its geometry in plan consists of two parallel straight sections, each of length $100\,\mu m$ and with a space of $10\,\mu m$ separating them, joined by a semicircular section of the same $100\,\Omega/\square$ material, the track width being maintained throughout the turning point. End-contacts complete the component.

Estimate the resistance of the integrated resistor, assuming the end-contacts each effectively extend the length of the resistor by 0.6 square.

Ans. $2.4\,k\Omega$

3. A hypothetical integrated resistor, $40\,\mu m$ long by $2\,\mu m$ wide, is formed along with bipolar transistor bases in a p-type layer that is $3\,\mu m$ thick, and for the following estimations can be assumed to be uniformly doped with a sheet resistivity of $200\,\Omega/\square$. What is the resistance of the integrated resistor at this stage?

If an n^+-emitter-type diffusion is then diffused to a depth of 1 μm, to form a pinch-resistor, estimate the new value for the integrated resistance.

Finally, assuming that 10 V is applied between the n^+- and p-layers, with the former positive, estimate the value of the integrated resistance under these conditions and comment on the results. Assume that the relative permittivity of silicon is 12 and that the mobility of holes is $0.05\,\text{m}^2\,\text{V}^{-1}\,\text{s}^{-1}$.

Ans. 4 kΩ, 6 kΩ, 6.9 kΩ

4. A particular linearly graded pn junction in silicon has a capacitance of 300 pF when the applied reverse bias voltage is 10 V. Calculate the change in capacitance for a 50 per cent reduction in bias voltage, assuming that $C \propto V^{1/3}$ for the graded junction.

Ans. increase of 78 pF

5. A particular MOS integrated capacitor has a capacitance of 100 pF, an oxide thickness of 150 nm and is square in plan. The relative permittivity of the oxide is 3.9. The n^+-layer is 1 μm thick and, for the purposes of the following estimate, can be assumed uniformly doped with a donor density of $10^{25}\,\text{m}^{-3}$. Find the dimensions of the capacitor and estimate its access resistance, assuming an electron mobility of $0.13\,\text{m}^2\,\text{V}^{-1}\,\text{s}^{-1}$. Will the capacitance of such a device be dependent on the magnitude or polarity of the applied voltage?

Ans. 0.66 mm side, 2.4 Ω

6. A particular MOS capacitor is formed by growing 90 nm of oxide over an n^+-layer whose surface concentration of donors is $10^{25}\,\text{m}^{-3}$. Although the doping concentration decreases with depth in the silicon, assume it to remain constant for simplicity. Calculate the applied voltage at which the contribution of the depletion layer results in a decrease of the total capacitance to 90 per cent of the value of the oxide-layer capacitance, assuming relative permittivities for silicon and silicon dioxide of 11.7 and 3.9, respectively. The depletion-layer width may be assumed to be given by the usual expression

$$d = (2\epsilon V/eN)^{1/2}$$

If breakdown in the MOS capacitor occurs when the electric field in the oxide exceeds 600 MV m^{-1}, find the breakdown voltage of the capacitor, commenting on the result.

Ans. 69.6 V, 54 V

7. A certain MOS integrated capacitor, with nominal capacitance C_0, is of square plan. The artwork for the capacitor has dimensions of L per side, but these lengths are only reproducible to an accuracy of $\pm \Delta x$. Show that the capacitance of the integrated component including the contribution of such dimensional inaccuracies, C, is given approximately by

$$C = C_0(1 \pm 2\Delta x/L)$$

Hence design the artwork for a nominal 10 pF MOS capacitor with a tolerance due to dimensional inaccuracies of ± 5 per cent, estimating the accuracy to which the artwork must be cut. Assume that the dielectric has a relative permittivity of 4 and thickness 1 μm, and that the overall photo-reduction of the artwork is 250:1.

Ans. 133 mm square ± 3 mm

8. A monolithic transistor has the following specifications: collector resistivity, 2.5×10^{-3} Ωm; emitter depth, 1.5 μm; base depth, 2 μm; substrate depth, 15 μm; emitter area, 50×8 μm; collector contact area, 75×8 μm; number of collectors, 2; and distance between emitter and collector edge, 12 μm.

Estimate the collector parasitic resistance, assuming a buried layer having a sheet resistance of 15 $\Omega/\square$ and 5 $\Omega/\square$. Repeat the problem with an epitaxial layer thickness of 5 μm. Draw any conclusions from the result.

Ans. 40 Ω, 39 Ω, 14 Ω, 13 Ω

9. A JFET is included in a bipolar integrated circuit in which the 0.05 Ωm uniformly doped epilayer is 10 μm thick and transistor collector–base junctions are located 3 μm from the surface. Estimate the gate voltage on the JFET that will just cause pinch-off, assuming negligible drain voltages. Mention any assumptions made and comment on the result.

Take the relative permittivity of silicon to be 11.8 and the electron mobility to be 0.17 m^2 V^{-1} s^{-1}.

Ans. 27 V

11 Optoelectronic devices

11.1 Optical display devices

11.1.1 Introduction

Light can be emitted from a solid when it is stimulated in some way from an energetic source, a phenomenon that has the general title of *luminescence*. There are three principal distinguishing forms of luminescence, named according to their source of excitation energy. For *photoluminescence*, the incident energy is in the form of photons; whereas the application of an electric field to produce radiation is termed *electroluminescence*; and *cathode-luminescence* arises when a beam of electrons bombards the solid.

Most practical active display devices rely for their action on electroluminescent effects, particularly injection electroluminescence as manifest in certain pn junctions, principally light-emitting diodes (LEDs), which will be considered in detail later. However, a common feature of all types of luminescence is that radiation of a characteristic wavelength λ is emitted as a consequence of electronic transitions between two energy levels E_1 and E_2 (where $E_1 > E_2$ is assumed), related by

$$hf = E_1 - E_2 = hc/\lambda \qquad (11.1)$$

Usually E_1 and E_2 are components of allowed bands of energy, so the emitted radiation occurs over a spectrum of wavelengths.

11.1.2. Light output from pn junctions

Certain semiconductor pn junctions can be induced to emit light when biased in the forward direction. Under these conditions, as discussed in Sec. 7.3, the potential energy barrier is lowered and majority carriers from both sides of the junction are moved across to be injected as minority carriers, causing the local density of minority carriers to be increased there. Figure 11.1 illustrates this situation. The enhanced minority-carrier densities decay exponentially away from the junction because of recombination with majority carriers, as shown (again see Sec. 7.3). Such recombination of the injected minority carriers can be a radiative process, as illustrated, so the process is known as injection electroluminescence. Since, for the moment, we are assuming that the electrons

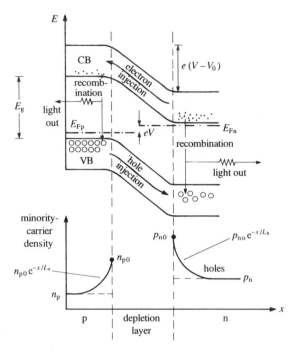

Fig. 11.1 **Light emission due to radiative recombination of injected carriers in a forward-biased pn junction.**

from the conduction band (CB) recombine directly with holes in the valence band (VB), so losing energy E_g, the characteristic wavelength of the output radiation, λ_g, can be found by rearranging Eq. (11.1), to give

$$\lambda_g = hc/E_g \qquad (11.2)$$

For example, GaAs has a gap energy of 1.44 eV, which would correspond to a radiated wavelength output of around 0.86 μm. There would of course be other frequencies emitted corresponding to larger energy losses by recombination of those injected electrons lying slightly above the edge of the CB, and for these the λ estimated by Eq. (11.2), would be smaller. Ideally, every injected minority carrier could recombine and cause a photon of radiation to be emitted, but, in practice, this situation does not prevail, which leads to the defining of a *quantum efficiency*, q.e., where

$$\text{q.e.} = \frac{\text{rate of photon emission}}{\text{rate of carrier injection}} \qquad (11.3)$$

The light output and quantum efficiency of a particular device is critically dependent on the junction material and, in particular, on the recombination processes that occur in it. Junctions in direct-band-gap materials, typified by GaAs, as discussed in Sec. 6.6.1, are usually most suitable for light-emitting diodes. This is because radiation recombination of a conduction-band electron directly with a hole in the valence band is possible with no change in momentum necessary, which makes it a statistically favourable process. Such a direct interband radiative transition is illustrated in Fig. 11.2(a). Interband

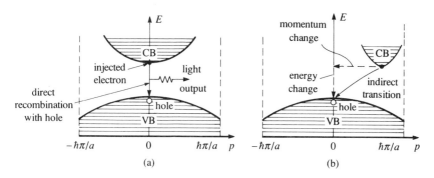

Fig. 11.2 (a) Interband recombination in a direct-band-gap semiconductor; (b) recombination in an indirect-gap semiconductor also involves a momentum change.

recombination processes are also possible in indirect-gap materials, as shown in Fig. 11.2(b), but these involve a momentum change in addition to the energy change, i.e. the transition is no longer vertical in the $E-p$ diagram. During such a process, both energy and momentum must be conserved simultaneously. The energy conservation is similar to the direct-gap situation and the emitted photon has energy $E_{ph} = E_g = hf$. However, for indirect-gap transitions, the recombining electron has to lose momentum of the order of $h\pi/a$ (see Fig. 11.2(b)), whereas the emitted photon has momentum, p_{ph}, given by

$$p_{ph} = mc = E_{ph}/c = hf/c = h/\lambda$$

Hence

$$\frac{\text{loss of momentum by electron}}{\text{momentum of photon}} \simeq \frac{h\pi/a}{h/\lambda} \simeq \frac{\lambda}{a} \qquad (11.4)$$

Including typical values of $\lambda \approx 10^{-6}$ m for visible radiation and a lattice spacing $a \approx 10^{-10}$ m, the ratio in Eq. (11.4) is of order 10^4. In other words, only a very small portion of the loss of electronic momentum can be accounted for by the momentum of the emitted photon. To preserve the momentum balance, a further quantum particle, a phonon, must also participate in each radiative recombination interaction to remove the excess electronic momentum. In other words, heat is generated by the recombination. Hence, the likelihood of a three-quantum-particle interaction of this type producing useful radiation is much less probable than for direct-gap transitions, which only involve two particles. Consequently, direct band-to-band radiative recombination of an electron and hole with photon emission is most efficient in direct-gap semiconductors such as GaAs.

It is also possible for radiative recombination to occur other than via direct interband transitions. The recombination may, for example, involve recombination centres located in the forbidden gap caused by impurities or dopants. Most recombination events in indirect-gap materials occur via such centres, as shown in Fig. 11.3(a). A possible sequence of events in such a recombination

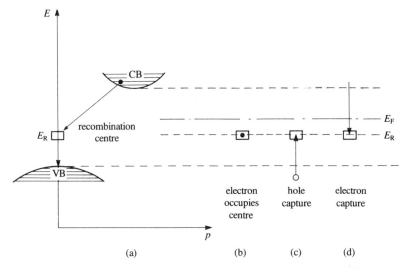

Fig. 11.3 (a) Recombination via a combination level in the gap; (b)–(d) a possible sequence of events in the recombination process.

process is also shown in Figs 11.3(b)–(d). If the recombination centre is assumed to lie below the Fermi energy $(E_F > E_R)$, it will initially be occupied by an electron, Fig. 11.3(b). The first stage in the recombination process is then that a hole from the VB is captured by the centre (Fig. 11.3(c)); this corresponds to an electron moving from the centre to the VB, thus heating up the lattice. Next, as in Fig. 11.3(d), an electron falls from the CB to the centre, giving up further energy as heat and so completing the recombination cycle by restoring the centre to its original filled state. Overall, an electron from the CB has moved via the recombination centre to recombine with a hole from the VB. Such a recombination process requires that the hole be captured at the localized centre for a sufficiently long time for the recombining electron to be captured with it. Otherwise the hole may be thermally excited back to the VB (which is the same as a VB electron being raised to the centre). The process is shown diagrammatically in Fig. 11.4. Whereas recombination results in the

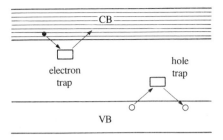

Fig. 11.4 Carrier trapping by localized centres in the gap.

permanent loss of electron–hole pairs, *trapping* describes the temporary removal of a carrier to a localized level, called a *trap* and again located in the forbidden gap, from whence, after a characteristic *trapping time*, it is re-excited to its orignal energy band.

Recombination and trapping centres thus occur at isolated localized energy levels in the forbidden gap. Such states are associated with various types of irregularity within the semiconductor or on its surface and are a consequence of crystal imperfections such as lattice defects, especially dislocations and vacancies, or, more usually, additional impurity or dopant atoms that have been deliberately introduced, for example, to enhance recombination. Whether a particular localized intergap level behaves as a recombination or a trapping centre depends on whether, after the capture of one carrier type, the next most probable event is the capture of the opposite carrier type or the re-excitation of the original carrier. Recombination or trapping times vary according to the times that centres have to capture a carrier before recombination or detrapping takes place. For example, *deep* electron traps, i.e. those far removed energy-wise from the CB, have slow release times because of the relatively large energy required to re-excite a trapped electron, whereas *shallow* traps near to the CB release electrons relatively quickly.

11.1.3 Light-emitting diodes

Equation (11.1) can be used to estimate that, for light emission in the visible spectrum resulting from interband recombination radiation events, the semiconductor must have a gap energy of around 2 eV. This rules out all elemental semiconductors (except diamond!) as suitable materials for LEDs. Although diodes can be readily made in Si, this is an indirect-gap material in which recombination is relatively difficult, which leads to very low efficiencies of radiation and its energy gap is such that the output would be in the infrared region. Consequently Si is reserved for those classes of electronic device for which a low recombination rate is a positive advantage, such as all types of transistors and, as we shall discover, solar cells. The III–V compound semiconductors (see Chapter 12) such as GaAs, GaP and $GaAs_{1-x}P_x$ are the most important active materials for efficient LEDs.

Although GaAs ($E_g = 1.44$ eV) devices only emit in the infrared region ($\lambda_g = 860$ nm), direct interband recombination leads to high luminescent efficiencies, which make them very useful for optical communications systems. Also, pn junctions can be readily fabricated by zinc diffusion into n-type material. One difficulty that arises with devices relying on interband recombination is that the radiation is of just the correct wavelength to be reabsorbed before being emitted, particularly in thick bulk regions, by promoting electrons across the gap from VB to CB. This problem can be considerably reduced in GaAs LEDs by doping with Si atoms, which can act as acceptor-like recombination centres. The principal luminescent transition then occurs between electrons in the CB and the Si acceptor level, as shown in Fig. 11.5. The resulting radiative output at wavelengths around 950 nm has

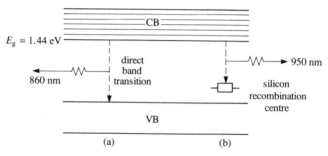

Fig. 11.5 Possible radiative recombination in GaAs LEDs: (a) interband and (b) via a Si impurity level.

a related energy that is too small (i.e. $< E_g$) to be reabsorbed by band-to-band electronic movement, which leads to much-improved efficiencies.

GaP ($E_g = 2.26$ eV, $\lambda_g = 549$ nm) is another useful material for LEDs that emit in the green or sometimes red part of the spectrum, depending on the doping. This is an indirect-gap material, so interband recombination is unfavoured, and dopants such as N or Bi are used to promote radiative recombination.

Gallium arsenide phosphide, a combination of GaAs and GaP, GaAs$_{1-x}$P$_x$, is particularly useful for LEDs in that the band gap can be adjusted from 1.44 eV for $x = 0$ to 2.26 eV for $x = 1$, corresponding to infrared to green emission. For compositions in which $0 < x < 0.44$, the band gap is direct, with high luminous efficiency. For example, diodes with GaAs$_{0.6}$P$_{0.4}$ ($E_g = 1.8$ eV) emit strongly in the red part of the spectrum and are used for a variety of displays. For $x > 0.44$ the band gap becomes indirect but useful emission in the yellow and green areas is still obtainable with N doping.

More complex ternary and quaternary semiconductor alloys such as (Al,Ga)As, (In,Ga)As, (In,Ga)(As,P) and multilayers of these with III–V semiconductors are used to tailor the band gap of LEDs for optical-fibre systems such that the wavelength of the infrared output corresponds to a minimum-attenuation window for the fibre type.

A schematic cross section of a discrete LED is shown in Fig. 11.6. The junction can be formed by liquid- or vapour-phase epitaxial techniques. This features a shallow junction so that the light generated mostly near to the

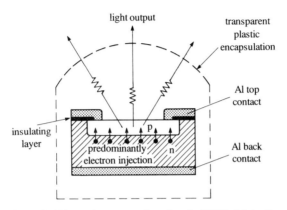

Fig. 11.6   Schematic cross section of typical LED fabrication.

junction only has to travel through a narrow bulk p-region before being emitted through an aperture in the top contact, so reducing the inefficient reabsorption processes. The useful light output originates from radiative recombination of injected electrons from the n-side when forward-biased. Under these conditions, non-useful light is also generated by holes injected into the n-region. This potentially inefficient situation can be minimized by ensuring that the junction current is predominantly electronic in nature. The ratio of electron to hole current at a junction for *any* bias condition, is, as we have seen in Sec. 7.6, approximately proportional to the conductivities of the n- and p-components, viz.

$$J_e/J_h \simeq \sigma_n/\sigma_p = eN_d\mu_e/eN_a\mu_h \tag{11.5}$$

The condition required, that $J_e \gg J_h$, is thus automatically achieved in most LED junctions since $\mu_e \gg \mu_h$ in many compound semiconductors, but the inequality can be further reinforced by differential doping such that $N_d \gg N_a$, to produce a pn$^+$ diode. Although the internal quantum efficiency can be high, not all the radiation generated is emitted, since a proportion of it strikes the semiconductor–air interface at angles greater than the critical angle, which is small because of the high refractive indices of many compound semiconductors, and is totally internally reflected. However, the external efficiency can be improved by providing dome-shaped transparent plastic encapsulation with a high refractive index.

11.1.4 Liquid-crystal displays

Active optical display devices require electrical energy to generate the light output. A common example of the type is the LED in which, at best, only about 10 per cent of the electrical energy is converted into usable light. For some applications, notably low-power portable systems, e.g. watches and calculators, a *passive* display device is to be preferred, which does not generate light but consumes negligible power in modifying an existing light source, which might, for example, be daylight. The most prevalent passive optical display device is the liquid-crystal cell, which is used in liquid-crystal displays (LCDs).

Liquid-crystal (LC) materials exhibit liquid-like properties at high temperatures, are crystalline solids at low temperatures, but, over an intermediate operating temperature range, typically between -10 and $60°C$, behave as semi-liquids while retaining some of the optical properties of their solid form. A characteristic of all LCs is their rod-like molecular constituents, which possess optical and dielectric anisotropies. Although there are various subgroups of LC, we shall only concern ourselves with currently the most useful, the *nematic*, in which the intermolecular forces are sufficiently strong to align the molecules parallel to each other. Furthermore, if the solid interface of, say, a glass-containing plate in contact with the LC is microscopically grooved in some manner, the rod-like molecules at the interface align parallel to the direction of the grooves, at the same time exerting aligning forces on molecules further from the interface, causing them to point in the same direction, as shown in Fig. 11.7. The important distinguishing electrical feature of LCs is

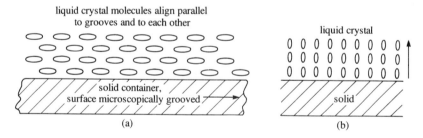

Fig. 11.7 **Liquid crystal/grooved interface: (a) with no field applied and (b) with an electric field.** $\mathscr{E} >$ **a critical value.**

that their relative permittivity depends on the orientation of the applied electric field: if $\mathscr{E}$ is parallel to the molecular axis, $\epsilon_r = \epsilon_{\parallel}$, and if it is perpendicular, $\epsilon_r = \epsilon_{\perp}$; and usually $\epsilon_{\parallel} \neq \epsilon_{\perp}$. The LC material is said to be *positive* if $\epsilon_{\parallel} > \epsilon_{\perp}$, and under the influence of an applied $\mathscr{E}$ field greater than a certain critical value, the molecules will be parallel to the direction of the field in order to minimize their potential energy, as shown, for example, in Fig. 11.7(b).

There are various possible forms of LCD device but only the most technologically useful one, the *twisted nematic* LC cell, as shown in diagrammatic form in Fig. 11.8, will be described here. The LC is contained between two textured glass plates, A and B, typically 10 μm apart, which are grooved in mutually perpendicular directions, creating a 90° twist in the molecular

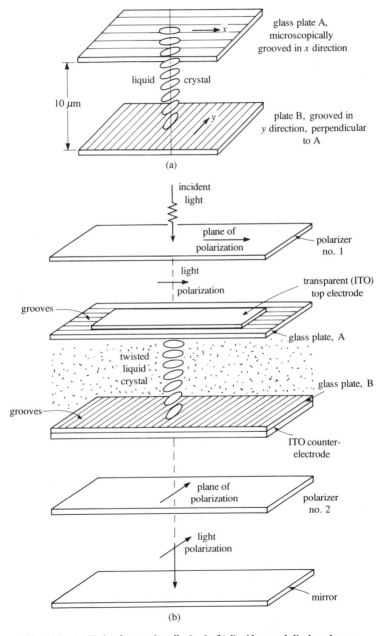

Fig. 11.8   (a) Twisted nematic cell. $\mathscr{E} = 0$; (b) liquid-crystal display element.

orientation in the LC, as shown in Fig. 11.8(a). A complete LC display element additionally incorporates two crossed light polarizing screens and a mirror, as shown schematically in Fig. 11.8(b). Light entering the top of the element is polarized by screen no. 1, whose plane of polarization is parallel to the grooves in the top plate, and enters the twisted LC cell via a transparent top electrode, usually indium-tin-oxide (ITO), and plate A. As the polarized light traverses the cell, the strong optical anisotropy of the twisted chains of molecules causes the plane of polarization of the light to rotate by 90°; the polarization of the light has been guided by the twisted molecules across the cell and emerges with polarization suitable for it to be transmitted by a second polarizer, which has its plane perpendicular to the first, and on to a mirror, as shown. On reflection, the light will pass through polarizer no. 2 and the LC cell again suffering a 90° polarization shift, finally re-emerging through polarizer no. 1. Hence, when viewed from the top, the element will appear as a bright reflecting surface.

Electric field can be applied across the cell by applying a voltage between a transparent, shaped, conducting top electrode, usually ITO evaporated onto glass plate A and defined by photo-etching, and a similar counter-electrode on plate B, as shown. Small applied AC voltages (5–10 V) create sufficiently high electric fields in the z direction to cause alignment of the LC molecules with the field and destroy their original light-guiding properties. The plane of polarization of incident light is no longer rotated through 90° on traversing the cell but is scattered and so is unable to penetrate polarizer no. 2. Very little light is reflected in this situation and the area covered by the shaped top electrode, over which the field exists, appears dark. Thus the application of a small a.c. voltage to the electrodes causes the top electrode to be displayed as a dark shape against a bright background. Alternating actuating voltages are used to enhance the LCD lifetime since d.c. fields tend to produce electro-chemical effects that degenerate the LC and shorten device lifetimes. Since the action of the device relies on the alignment of molecules by an $\mathscr{E}$ field and no corresponding conduction current flows, and also since no internal light source is required in the reflective type described, power consumption is minimal.

In the *transmission* version of LCDs, rear illumination replaces the rear reflector, with the obvious disadvantages of increased power dissipation, but otherwise their action is very similar to the reflective version.

11.1.5 Display arrays

The optical devices described so far have been shown to be capable of operation in an OFF/ON mode; indeed a principal application for LEDs is for use as low-consumption indicator lamps, but both LEDs and LCDs are potentially more useful as elements in alphanumeric display panels. Two possible arrangements of optical cells are illustrated in Fig. 11.9, the choice being based on display size, definition and allowed circuit complexity. The simpler seven-element array is normally used to show digits, for example in

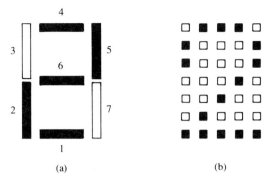

Fig. 11.9 Possible elements for LED and LCD optical display systems: (a) seven elements and (b) 35 elements.

LCD watches, whereas the 35 elements can show a complete alphanumeric set with more flexibility of design. The shape of each matrix element of such arrays is defined by the dimensions of the top electrode of the corresponding LCD cell or can be formed by grouping several LEDs in, for example, larger arrays.

Each element of the array has to be switched in response to a logic signal. For example, Fig. 11.9(a) shows the response to a logic signal corresponding to 2, in which elements 1, 2, 4, 5 and 6 have been switched ON and 3 and 7 remain OFF. Such *addressing* of display elements, i.e. providing them with appropriate logic signals to switch them ON and OFF, requires external circuitry and wiring to each element, which can become extremely complex in large arrays, so steps have to be taken to reduce the number of connections. In the sample system, such as that shown in Fig. 11.10(a), each element is provided with a separate lead to its top contact and there is a common bottom contact connection, giving a total of $n+1$ leads for an n-element array. This cumbersome arrangement can be considerably improved by, for example, using sets of parallel conductors to address the display matrix as shown in Fig. 11.10(b). All the top contacts are supplied by columns of addressing lines and all the bottom contacts are similarly interconnected in rows, as shown. Any particular element can then be addressed by applying suitable potentials to the common column and row contacts whose lines intersect at the element. For example, $+2$ V, say, on column 4, in Fig. 11.10(b), and -2 V on row d, would cause the LCD element shown to switch ON, assuming its threshold voltage is less than 4 V. It will be seen that the number of external contacts required in this addressing system is considerably reduced; for example, in an $n \times n$ array, to $2n$. However, this simplification is only achieved at the expense of an increased complexity of the addressing circuitry, since, to display a specific pattern of elements, the rows and columns must be continuously scanned by voltage pulses. A possible switching sequence might be as follows. Column 1 has $+$V applied and simultaneously $-$V is applied to any element of rows a–g in column 1 that is required to be ON. After an address time, t_a, the $+V$

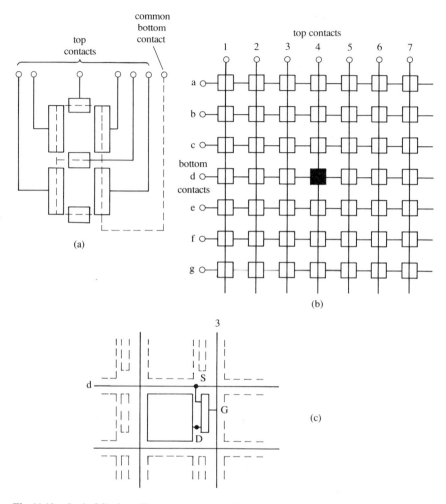

Fig. 11.10 Optical display addressing: (a) simple addressing: (b) matrix addressing; and (c) matrix addressing with MOSTs.

pulse is switched to column 2, a new train of $-V$ pulses is applied to the rows, and the process is repeated for column 3, etc. The complete scanning time before the display is replenished is thus (number of columns) $\times t_a$, which has to be kept below, say, 40 ms to avoid flicker.

One possible way of prolonging the ON time of a display element for longer than the addressing time, t_a, is to incorporate an individual MOST at each intersection of address lines, i.e. one MOST per element, as shown in Fig. 11.10(c). The element shown is addressed by applying a gate (G) turn-on voltage to column 3, at which point the voltage on the source, S, supplied by a pulse on row d, will be transferred to the drain D, all voltages being with respect to a common bottom connection to the cells, thus turning ON the LC element. The charge stored in the drain capacitance of the MOS takes some time to drain away through the LCD, because of its very low current consumption, which allows the device to remain switched on after the addressing pulses have been removed, which results in longer cell refreshing times. The additional MOSTs required by this method, of course, introduce more complexity, but the use of thin-film, amorphous silicon thin film transistors (TFTs) is being used in some arrays to alleviate this difficulty.

11.1.6 Semiconductor lasers

Semiconducting *junction lasers*, or *laser diodes*, which are small, highly efficient, light generators, in which the wavelength of the emitted radiation is ideally of a single frequency, are vital components of many optical communications systems. This is principally because the high frequency of the monochromatic emission can accommodate signals with relatively high bandwidths, around several gigahertz say, which can be pulse-modulated using conventional electronic techniques.

Most light sources emit radiation over a spectrum of frequencies; for example, a LED has a typical bandwidth of tens of nanometres. Moreover the phase of the emitted electromagnetic radiation corresponding to each photon is entirely random, which is called phase *incoherence*. In contrast, a feature of lasers is that the emitted radiation not only has a much smaller bandwidth, around hundredths of a nanometre for gas lasers and typically a fraction of a nanometre for solid-state lasers, but it is also *coherent*, i.e. the phase of the electromagnetic wave at some initial time is related to its phase at times immediately before or after, as in a simple sine wave.

In general, when a photon interacts with an electron, the electron can absorb energy to move to a higher excited state or alternatively, with equal probability, the electron can lose energy and in doing so emit a further photon, which leads to photon amplification. Which of these two possibilities, absorption or amplification, occurs, depends on the initial energy state of the interacting electron. If it is a low-energy state, E_1, as shown in Fig. 11.11(a), an impinging photon with energy $hf \geqslant E_g$ will be absorbed, elevating the electron to an available higher level, E_2. If, however, prior to the lasing occurrence, an unusually large number of electrons are present at the higher level, called a *population inversion*, which can be brought about by *pumping* the lower energy electrons with energy to raise them to the excited state, then, as illustrated in Fig. 11.11(b), an incident photon of the correct wavelength $\lambda_g = h_c/E_g$, can *stimulate* an electron to fall to the lower energy level and so emit another photon. Since the population inversion is a non-equilibrium condition, it is also possible for an electron to fall to a lower level in a random manner to produce incoherent *spontaneous emission* but such events occur relatively infrequently, and an upper energy electron can meanwhile be stimulated to emit coherent radiation by the presence of another photon of correct wavelength, λ_g. In such *stimulated*

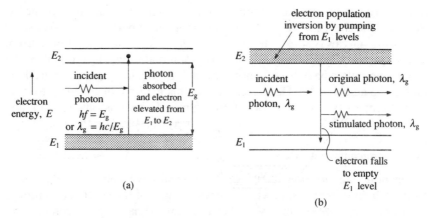

Fig. 11.11 **Interaction of photon and electron: (a) with no population invesion, photon is absorbed; and (b) with a population inversion, the original photon and a second stimulated one emerge.**

emission the emitted photon has the same wavelength, λ_g, corresponding to the gap energy, phase, polarization and direction of propagation as the incident one. Devices that rely on this mechanism to generate coherent radiation are called lasers, which is an acronym for light amplification by stimulated emission of radiation.

The pumping mechanism for population inversion and hence stimulated emission depends on the particular type of laser, but in the semiconductor laser diode, on which we will concentrate because of its relative importance in communications, inversion is achieved by forward biasing a pn junction that is degeneratively doped, for reasons which will become apparent. The energy band diagram for such a diode, with no bias voltage applied, is shown in Fig. 11.12(a). When a forward bias voltage, V_{FB}, is applied, such that $eV_{FB} > E_g$, all levels at the n-side are raised in energy, all those at the p-side are lowered, the barrier height is considerably reduced from its equilibrium value V_B, and the non-equilibrium band diagram becomes as shown in Fig. 11.12(b). The concept of Fermi energy is strictly no longer valid but it is constructive to consider the quasi-Fermi levels, shown as E_{Fp} and E_{Fn}. Large numbers of carriers are injected across the junction to create at active region of population inversion in the vicinity of the junction, as shown in Fig. 11.12(b), in which radiating transitions of electrons from the conduction band to empty states in the valence band can occur. For this active region to exist, with consequential stimulated emission, it will be seen that the forward bias voltage must be $eV_{FB} > E_g$. If the barrier voltage in the unbiased state, V_B, were not itself $> E_g$, then applying a forward bias voltage $> E_g$ would reduce the barrier height to zero and damagingly large currents would flow. The very heavy, degenerate, doping of each side of the junction ensures that eV_B is indeed $> E_g$, as can be seen from Fig. 11.12(a).

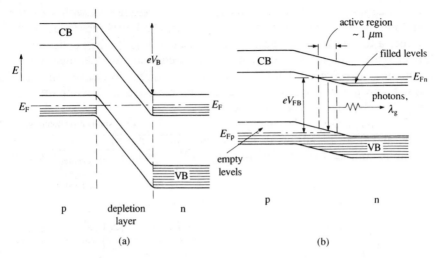

Fig. 11.12 **Energy band diagrams for a laser diode (a) in equilibrium and (b) with forward bias** $eV_{FB} > E_g$.

The electron–hole recombination transitions that occur in the active, inverted region cause photons to be emitted with energy almost equal to E_g and wavelength λ_g, but the process is only efficient in direct-gap semiconductors, so these are used exclusively for lasers. Photons are initially emitted spontaneously but these are then available to stimulate further emission and subsequent laser action. A diagrammatic view of the structure of a laser diode is shown in Fig. 11.13. It consists essentially of a p^+n^+ junction in GaAs or

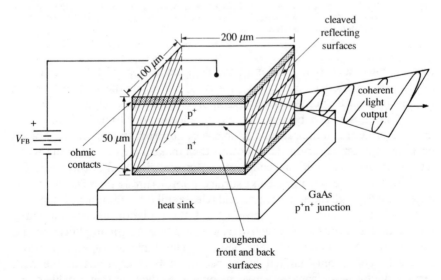

Fig. 11.13 **Diagrammatic representation of basic GaAs laser diode structure.**

other direct-gap semiconductor that is degeneratively doped, with acceptor and donor densities around 10^{24} m^{-3}, which are just below those required to produce a tunnel diode (see Sec. 7.13). Ohmic contacts for the application of forward bias and good heat sinking to accommodate the high power dissipation are provided. The GaAs crystal is very precisely cleaved on its end faces as shown to create smooth, parallel, semitransparent, reflecting surfaces, which are essential for the lasing action.

When small forward bias voltages, V_{FB}, are applied, no population inversion occurs and there is only weak incoherent light output due to spontaneous recombination; see Fig. 11.14(a). As soon as V_{FB} becomes $> E_g$ and a critical

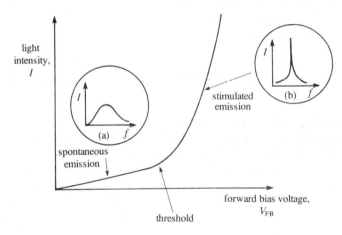

Fig. 11.14 **Radiation output from forward-biased laser diode. The inserts show the radiation spectra (a) for the spontaneous emission below the lasing threshold and (b) for stimulated emission above.**

current flows, such that the conditions depicted in Fig. 11.12(b) prevail, an inverted population of carriers is produced, which, in the first instance, is responsible for the spontaneous emission of photons. These photons reach the cleaved surfaces, which, because of internal reflection due to the high refractive index of the laser material, around 3.6, behave as mirrors and reflect the photons back into the active region. The photons then make a pass through the active region, stimulating additional coherent photon emission as they traverse it. On reaching the far cleaved mirror-like surface, photons are again reflected to pass once again through the active region and release still more photons, all coherent with each other. This process repeats itself, the radiation undergoing many reflections and traverses of the active region, resulting in light amplification and the generation of many coherent photons. This build-up of very powerful coherent light is only limited by reabsorption by transitions in the bulk semiconductor away from the active region and by the light that is deliberately allowed through the partially reflecting end mirrors, which typically reflect around one-third of incident radiation and transmit the remainder as useful light output. The light intensity stabilizes at a value where the light amplification is just balanced by these loss mechanisms, the population inversion being continuously replenished by the heavy forward bias. The front and back faces of the crystal are roughened to prevent build-up of radiation in this direction, which would provide a competing mode of oscillation.

The high threshold currents, around $10^7 \, \text{A m}^{-2}$ at 77 K rising to $10^9 \, \text{A m}^{-2}$ at room temperature, combined with forward bias voltages of order 1 V, indicate that, when lasing, large powers of 10^7–$10^9 \, \text{W m}^{-2}$ (10^{-1}–10 W for a $100 \, \mu\text{m} \times 100 \, \mu\text{m}$ device) would have to be dissipated by the heat sink for continuous-wave (CW) operation. This explains why laser diodes are usually operated in a pulsed regime, so as to reduce the mean power dissipation, which is especially suitable for digital optical communications, and often at reduced temperatures. Since the output light intensity can be changed readily by varying V_{FB} (see Fig. 11.14), transmission of optical information by pulse-height modulation is a further possibility.

A variety of direct-gap semiconductors have been used to fabricate lasers, such as GaAs, InGaP and AlGaAs. Most emit infrared radiation but $Ga(As_x P_{1-x})$ compound semiconductors can produce efficient output in the visible region, at 630 nm.

The simple lasers described so far are typified by large forward currents and high dissipation, with consequent requirements for cooling and pulsed operation. These disadvantages can be overcome by using *heterojunction* laser structures, which contain junctions formed between two or more *different* semiconductors (discussed more fully in Chapter 12), as opposed to *homojunctions*, which occur in the same semiconductor, e.g. a pn junction in Si or GaAs. One such single-heterojunction laser is shown in Fig. 11.15(a), in which a thin layer of p-type GaAs has been grown onto an n-GaAs substrate, followed by a p-type AlGaAs layer. The AlGaAs has a wider energy gap than the GaAs and a lower refractive index. The energy band diagram for the structure in equilibrium is shown in Fig. 11.15(b) and when heavily forward-biased to promote laser action in Fig. 11.15(c). Laser action results primarily from electron injection to the p-side due to the higher injection efficiency for electrons. The electrons injected into the narrow central p-region are confined to the region by the potential energy barrier at the p-GaAs/AlGaAs junction, as shown in Fig. 11.15(c). As a result of the electron confinement, the threshold current is substantially lower than for simple pn junctions. This is further enhanced by the optical containment of stimulated photons because of the change in refractive index at the heterojunction (see Fig. 11.15(d)), which causes them to be internally reflected back into the active region.

A further improving refinement is obtained with a *double-heterojunction* structure in which an active GaAs layer, either n- or p-type, is sandwiched between two layers of higher-band-gap AlGaAs, as shown in Fig. 11.16(a).

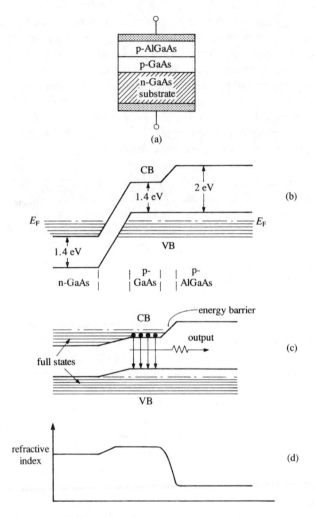

Fig. 11.15 (a) Basic construction of single-heterojunction laser and its band structure, (b) in equilibrium and (c) when forward-biased above threshold; (d) the variation of refractive index across the device.

Both heterojunctions constrain the generated light to the active region by virtue of the refractive index differences at the interfaces, see Fig. 11.16(c). The band-gap differences simultaneously create a potential barrier in the conduction band to prevent injected electrons diffusing out of the active region, together with one in the valence band, to confine injected holes, as in Fig. 11.15(b). By such means, the threshold current density can be reduced to around $10^5 \, \text{A m}^{-2}$, which is low enough for CW operation at room temperature.

The threshold current is often reduced still further by restricting the current

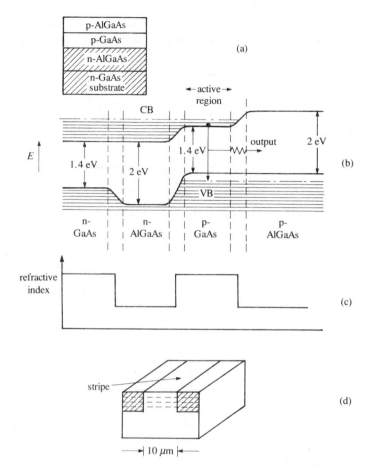

Fig. 11.16 **(a) Double-heterojunction laser; (b) its band structure when forward-biased above threshold: (c) the variation of refractive index; (d) stripe-geometry laser.**

to flow in a narrow *stripe* of the structure, typically a few micrometres wide, as shown in Fig. 11.16(d). This can be achieved, for example, by etching away the hatched regions shown, to create a *mesa* structure, or by proton bombardment of the hatched regions, which converts both semiconductors into a semi-insulating form.

11.2 Optical detectors

11.2.1 Photoconductivity and photoconductive detectors

As a consequence of the relatively small number of carriers created thermally by the electron–hole pair generation process, an intrinsic semiconductor has a low conductivity at temperatures such that $kT < E_g$. The pair generation rate at a particular temperature can be enhanced, however, by illuminating the semiconductor with radiation of a suitable wavelength. When radiation with a wavelength less than some critical value, λ_{max}, is incident on the surface of the semiconductor, it is absorbed and excess carriers are generated within the solid, which leads to a change in the conductivity. Such optically induced *photoconductivity* changes can form the basis of an optical detector.

If the energy radiant on a semiconductor is of such a wavelength that the photon energy, hf, is greater than the gap energy, E_g, then each absorbed photon releases sufficient energy for an electron–hole pair to be generated. Thus there is a lower limit of f below which no absorption occurs given by

$$hf_{min} \geq E_g \tag{11.6}$$

If hf is less than E_g, an electron in the valence band can only gain sufficient energy to take it into the forbidden gap, which is an unallowed transition, so no absorption occurs and the material is transparent to radiation at such frequencies. The lower frequency limit for absorption, f_{min}, corresponds to an upper wavelength boundary, λ_{max}, given by

$$E_g = hf_{min} = hc/\lambda_{max}$$

or

$$\lambda \leq \lambda_{max} = \frac{hc}{E_g} = \frac{1.24}{E_g(eV)} \mu m \tag{11.7}$$

A table listing the value of λ_{max} for several common semiconductors is included as Table 11.1. Since the visible light spectrum occurs at wavelengths between 400 and 700 nm it will be noted that these materials can form the basis for detectors in the visible to near-infrared range. Provided the wavelength criterion in Eq. (11.7) is satisfied, radiation incident on a semiconductor surface can create additional carriers and the conductivity can be substantially

Table 11.1 Maximum wavelength of incident radiation for carrier photogeneration in various materials.

Semiconductor	λ_{max} (nm)
Ge	1800
Si	1200
GaAs	880
GaP	550
CdS	520

increased over that of the sample in the dark. If the photon frequency f is greater than f_{min}, then optical absorption still occurs, each photon generating an electron–hole pair and the excess energy is dissipated as heat, which is a less efficient process.

Carrier pairs are simultaneously being generated thermally, which is manifest as an unwanted noise signal in a photoconductive optical detector. Thermal generation has therefore to be kept to a minimum relative to carrier generation by incident light. We have seen that the number of thermally produced electron–hole pairs in an intrinsic semiconductor, n_i, is an exponential function of gap energy, E_g, and temperature T

$$n_i \propto \exp(-E_g/2kT)$$

Hence, if a semiconductor is chosen with a low E_g, so as to be able to detect long-wavelength radiation, the thermally induced carriers would create an unacceptably high noise background, unless operated at low temperatures so as to reduce n_i. For example, InSb has $E_g \simeq 0.2\,eV$, so that it can detect radiation around 6000 nm, but such a small energy gap would cause the thermally generated carriers to swamp those produced optically, at room temperature, i.e. the dark conductivity would be high, becoming little changed under illumination. Hence the InSb detector is operated at 77 K, liquid-nitrogen temperature, when n_i is reduced by a factor of over 10^6 from its room-temperature value and the signal-to-noise ratio becomes manageable.

Suppose the power of the incident radiation is $W\,\mathrm{W\,m^{-2}}$. Since each photon has energy hf, the photon flux (i.e. the number per second per unit area, n_{ph}) is given by

$$n_{ph} = W/hf = W\lambda/hc\ \mathrm{s^{-1}\,m^{-2}} \tag{11.8}$$

so for constant power, W, the photon flux increases linearly with wavelength.

Let us now consider the penetration depth of the absorbed radiation. Consider an elemental slice of semiconductor, thickness δx, distance x from the illuminated surface, as shown in Fig. 11.17. If $g(x)$ is the carrier generation rate in the elemental volume, then the number of carriers generated in it per

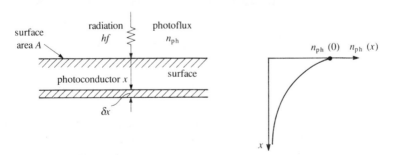

Fig. 11.17 (a) Absorption of radiation by photoconductor and (b) exponential decay of photon flux away from the surface.

second, $g(x)A\,\delta x$, can be equated to the number of photons absorbed per second, since each photon generates one carrier pair. So

$$g(x)A\delta x = [n_{ph}(x) - n_{ph}(x + \delta x)]A$$

which in the limit reduces to

$$g(x) = -dn_{ph}/dx \qquad (11.9)$$

On the other hand, in equilibrium the generation rate in the element must be equal to the recombination rate, which is proportional to the number of carriers present, which in turn is proportional to the number of photons. So

$$g(x) \propto n_{ph}(x) = \alpha n_{ph}(x) \qquad (11.10)$$

The coefficient of proportionality, α, is called the absorption coefficient, for reasons that will become apparent if we solve the two equations for $g(x)$ to give

$$n_{ph}(x) = n_{ph}(0)e^{-\alpha x} \qquad (11.11)$$

i.e. the photon flux, $n_{ph}(x)$, falls off exponentially from its value at the surface $n_{ph}(0)$, as shown in Fig. 11.17(b), becoming $1/e$ of its surface value at depth α m. The absorption coefficients, which have units of reciprocal length, m^{-1}, for two characteristic materials are sketched in Fig. 11.18. For wavelengths

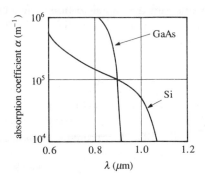

Fig. 11.18 Absorption coefficients of two semiconductors as functions of wavelengths.

greater than λ_{max} the absorption coefficients are negligible; the material is transparent and there is no absorption. When λ is just less than λ_{max}, very strong absorption occurs in GaAs, mostly very close to the surface, with α values rising sharply to around 10^6 m^{-1}, but the increase in absorption with decreasing λ is slower in Si. This arises because GaAs is a direct-gap semiconductor and electron–hole pairs can be generated relatively easily by the direct elevation of electrons from conduction to valence band. Si, on the other hand, is an example of an indirect-gap material, which, as we have seen, requires the participation of additional phonons for pair generation, which consequently is much less likely to occur.

The simplest form of photoconductive detector consists essentially of a slab of semiconductor, with a large collecting area for the incident radiation, which has $\lambda \leqslant \lambda_{max}$, where λ_{max} corresponds to the particular E_g, as given by Eq. (11.7) and is provided with ohmic contacts at the ends; see Fig. 11.19. The operating

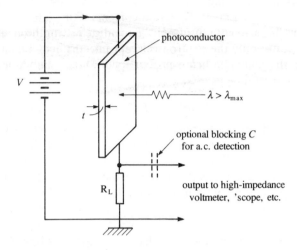

Fig. 11.19   Schematic of photoconductive optical detector.

temperature is chosen such that thermal carrier generation is low, so that the semiconductor has a low dark conductivity, and very little current flows in response to the applied voltage. When illuminated, the photogenerated carriers cause the conductivity and hence the current flowing to increase appreciably, so that the voltage across the load resistance, R_L, rises to indicate the presence and strength of the radiation. The thickness of the photoconductor is usually much greater than α, to maximize the amount of absorbed radiation.

In equilibrium in the semiconductor, the generation rate will equal the recombination rate, which will be proportional to the number of carriers of each type. So, in intrinsic material

$$g = rnp = rn_i^2$$

Since the conductivity of the intrinsic material is

$$\sigma = n_i(\mu_e + \mu_h)e$$

then the current density in the illuminated semiconductor when an electric field $\mathscr{E}$ is applied to it is

$$J = \sigma\mathscr{E} = (g/r)^{1/2}(\mu_e + \mu_h)e\mathscr{E} \tag{11.12}$$

Incorporating Eq. (11.10) shows that

$$I \propto n_{ph}^{1/2} \tag{11.13}$$

We see from this equation that the photocurrent, in this instance, is proportional to the square root of the intensity of the incident light.

So far, we have assumed that each absorbed photon generates an electron–hole pair, causing a single electron to flow in the external circuit. This direct process is relatively rare and occurs only in pure crystals with large applied fields and low light intensities. Other processes are much more probable and can yield greater external currents. For example, consider the following possible sequence of events. An incident photon produces a carrier pair as before, but only the electron is swept into the external circuit by the applied field, the generated hole being trapped. The trapped-hole density, p_t, will then vary with time as

$$dp_t/dt = -p_t/\tau_{Lth} + g \qquad (11.14)$$

where τ_{Lth} is the lifetime of a trapped hole. The trapped holes create a positive space charge, $p_t e$, in the semiconductor and an equal density of mobile electrons flows in from the external circuit to maintain charge neutrality. The injected electrons cause the material to be conducting while the holes remain trapped, and the conductivity in this situation is given by

$$\sigma = en\mu_e = ep_t\mu_e \qquad (11.15)$$

The ability of a semiconductor to detect radiation, its photosensitivity, is measured in terms of its photoconductive gain, G, which is defined as the ratio of the photocurrent to the rate of generation of electron–hole charges by the incident light; if we consider a crystal of cross section A, length l and carrying a photocurrent I, then

$$G = I/eg(Al) = J/egl$$

We then express the current density, J, in terms of the applied electric field, $\mathscr{E}$, and the conductivity given in Eq. (11.15) to obtain

$$G = \frac{\sigma\mathscr{E}}{egl} = \frac{(ep_t\mu_e)\mathscr{E}}{egl} = \frac{p_t\mu_e\mathscr{E}}{gl}$$

Now, the steady-state density of trapped holes is given by Eq. (11.15). Hence

$$G = \tau_{Lth}(\mu_e\mathscr{E}/l)$$

Further, the bracketed term is the reciprocal of the transit time of electrons along the length of the crystal, T, say. Therefore

$$G = \tau_{Lth}/T \qquad (11.16)$$

We see from this expression that the photosensitivity increases with increasing lifetime, but since the speed at which the photoconductor can respond to a light stimulus is also a function of τ_{Lth}, if τ_{Lth} is made large to increase the photoconductive gain, this can lead to a sluggish response.

While photoconduction occurs in single-crystal semiconductors, some of the most successful light-sensing devices employ polycrystalline materials, for example cadmium sulphide, selenides and oxides, whose high defect density is a source of trapping states, which we have seen can often be most beneficial.

We have seen that for intrinsic materials the wavelengths of the incident radiation must be below some maximum value, λ_{max}, for photoconduction to occur. In many common semiconductor materials, detection of far-infrared radiation, with wavelengths of tens of micrometres, is precluded since the gap energies are such that λ_{max} is less than the infrared wavelength; the materials are transparent. However, it is possible to dope the intrinsic material with deep-lying impurities, for example, copper in germanium, which can have $\lambda_{max} = 30\ \mu$m, to produce extrinsic photoconductors, which have been used in successful infrared detectors. If such a material is held at low temperatures, all the impurity levels remain occupied but free carriers can be excited if infrared phonons whose energies correspond to the energy depth of the impurity level are absorbed by the crystal.

11.2.2 Photodiode

The photodiode detector consists essentially of a pn junction with a window through which incident radiation can illuminate it, as shown diagrammatically in Fig. 11.20. If the wavelength of the radiation is less than λ_g, then

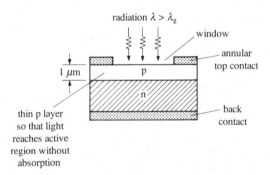

Fig. 11.20   Elements of a junction photodiode.

electron–hole pairs are generated, and those near to the junction can cause a *photocurrent* to flow in the reverse-biased device, which is additional to the usual leakage current due to thermally generated minority carriers, and so serves to detect the radiation.

The radiation incident on the diode will, of course, generate electron–hole pairs all over the device, but in the bulk p- and n-regions these light-generated minority carriers quickly recombine with majority carriers. However, for those photogenerated electron–hole pairs created in the depletion layer, the $\mathscr{E}$ field there, due to contact potential and bias voltage, tends to separate the components, the holes drifting in the direction of the field and the electrons in the opposite direction, as shown in Fig. 11.21, so as to reduce their recombination and to produce a usable photocurrent. If the diode is reverse-biased the light-induced minority carriers in the depletion region cause an increase in the current flowing round the circuit, which is additional

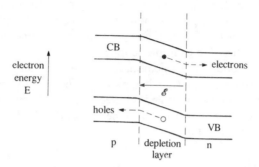

Fig. 11.21   Separation of photogenerated electron–hole pairs by electric field in depletion layer.

to the leakage current, I_0, and can be used to detect the presence of the radiation. The detector is usually operated in the reverse-biased mode so that the volume of the depletion region in which the effective carrier generation and separation occurs is maximized and also the $\mathscr{E}$ field is large under these conditions, so as to enhance the separation. Photogenerated minority pairs outside the depletion layer but near to the junction can also contribute to the photocurrent, provided that such carriers can survive recombination and diffuse to the edge of the depletion layer, to be swept across the junction by the $\mathscr{E}$ field there. As has been seen, the average distance that a minority carrier can travel in a 'hostile' environment of majority carriers without recombining is the diffusion length, L, so it follows that only those carriers generated within distances L_e and L_h of the depletion-layer edge can contribute effectively to the photocurrent. The situation is shown schematically in Fig. 11.22. Carriers photogenerated outside the active volume, length $(w + L_e + L_h)$, cross-sectional area A, do not contribute to photoconduction, but those created inside produce a photocurrent, I_L, in addition to the reverse-biased diode dark current.

Suppose that the volume generation rate of minority carriers when light is shone on them is G_L electron–hole pairs per unit volume, per second. Then for example:

photocurrent flowing due to holes = useful number of holes generated per second × charge = $G_L(Al_h)e$

where Al_h is the useful volume of generated holes. There is a corresponding component of the photocurrent due to photo-induced minority electrons, so

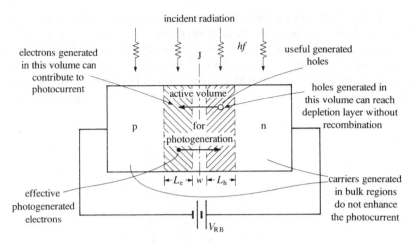

Fig. 11.22 **Diagram showing active region for photogeneration in a photodiode.**

the net photocurrent is

$$I_L = eAL_hG_L + eAL_eG_L = eAG_L(L_h + L_e) \qquad (11.17)$$

It has been assumed that the width of the depletion layer $w \ll L_{e,h}$.

The photocurrent, I_L, is in addition to the usual diode current I_D

$$I_D = I_o[\exp(eV/kT) - 1]$$

so the total current flowing in the external circuit of the biased diode is

$$I = I_D - I_L = I_o[\exp(eV/kT) - 1] - I_L \qquad (11.18)$$

The minus sign arises because, conventionally, the diode current I_D is considered to be positive for the forward-biased case with V positive and the photocurrent I_L is in the same direction as that for a reverse-biased diode. It follows that a possible equivalent circuit for a photodiode might be a current generator, I_L, in parallel with a diode, as shown in Fig. 11.23. In the reverse-biased operating condition, as shown, the diode voltage V is negative, so only a very small diode *dark current*, I_o, flows, typically of order 1–10 nA, so the external current is $I \simeq I_L$, which is proportional to the generation rate G_L,

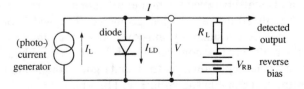

Fig. 11.23 **Equivalent circuit of a photodiode and the usual bias arrangements.**

from Eq. (11.17). Hence the detected output voltage across the load resistor R_L is proportional to I_L and the incident radiation.

The I–V curves corresponding to this situation are shown in Fig. 11.24. The photocurrent progressively shifts the unilluminated diode I–V characteristic downwards by an amount I_L, which is proportional to the light intensity. The operating point for photodetection is determined by the intersection of the

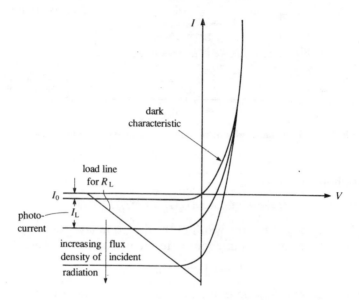

Fig. 11.24 I–V **characteristics of a photodiode.**

load line corresponding to the load R_L and the characteristics in the third quadrant, as shown. As the light intensity is varied, the operating point moves along the load line and, over a wide range, the current is largely independent of the bias voltage so the output voltage varies almost linearly with the flux density of the incident radiation.

11.2.3 Pin, avalanche and Schottky diodes and phototransistor photodetectors

The effective light-generating region of a photodiode and hence its sensitivity can be substantially improved by the introduction of an intrinsic semiconducting layer, i, sandwiched between the n- and p-regions, to form a *pin* structure. Figure 11.25(a) illustrates a possible geometry. The width of the active generating region is considerably improved over the simple diode, since, because of the relatively low conductivity of the i-region, only a few volts of reverse bias cause the depletion layer to extend right through it until it joins up with that in the n-region, as shown in Fig. 11.25(b).

If a pin photodiode is heavily reverse-biased into an avalanche condition, the photocurrent can be enhanced by the carrier multiplication process. Such

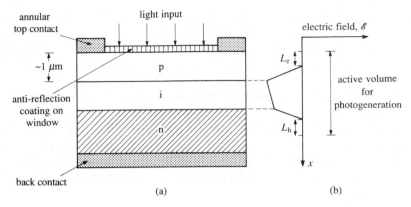

Fig. 11.25 (a) Structures of pin photodetector and (b) the $\mathscr{E}$ with reverse bias.

avalanche photodiodes (APDs), in addition to an increased sensitivity, have a faster switching time, by at least an order of magnitude over pin diodes operated in the non-avalanche mode, but this is at the expense of a higher operating voltage, usually greater than 100 V. The reverse voltage is limited to a value below the self-sustaining avalanche regime that would cause large currents to flow in the absence of illumination.

Another way to separate photogenerated minority carriers in a semiconductor and so prevent recombination and produce a useful photocurrent is to use the electric field in the depletion layer of a metal–semiconductor Schottky junction. A schematic diagram of such a *Schottky photodiode* is shown in Fig. 11.26(a) and its energy band structure is shown in Fig. 11.26(b).

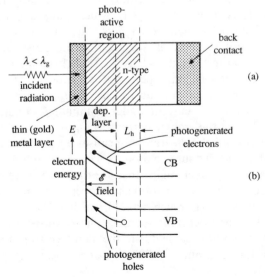

Fig. 11.26 (a) Schottky photodiode and (b) carrier separation by the $\mathscr{E}$ field.

It will be seen that the carrier separation mechanism is much the same as that for the pn junction already discussed. The principal extra advantage of the Schottky device is that the sensitivity in the near-ultraviolet region is enhanced since the metal can be made thin enough to transmit radiation of such wavelengths.

Another possible mechanism for providing extra internal amplification of photocurrent, as described for an APD, is to use a *phototransistor* detector. This consists essentially of a npn transistor structure with an integral window to allow illumination of the collector–base junction region by the incident illumination. The transistor base has no external connections, i.e. it is operated open-circuited. The operation of a phototransistor is illustrated in Fig. 11.27.

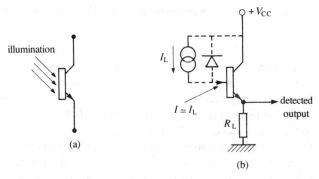

Fig. 11.27   **The phototransistor: (a) symbolic representation of open-circuit base type and (b) photo-induced base current and external circuit.**

The illumination causes a photocurrent, I_L, to be generated and the collector leakage current, I_{co} is increased to $I_{co} + I_L$, which is approximately equal to I_L since $I_L \gg I_{co}$. The photogenerated carriers thus effectively produce a base current, I_L, flowing into the nominally open-circuit transistor base, which is directly proportional to the light intensity. Apart from the manner in which the base current is derived, the operation of the device is the same as for a conventional npn transistor. The transistor amplifies the induced base current and the collector current under illuminated conditions therefore increases to $h_{fe}I_L$ and the emitter current to $(h_{fe} + 1)I_L$. Since h_{fe} is typically > 100, the internal gain of the transistor causes the photocurrent in the base to be magnified by such a figure, to produce a detected output current flowing through R_L that is very much bigger than for the corresponding photodiode. A disadvantage is that, at small base currents, corresponding to low illumination levels, h_{fe} can be much reduced, which tends to reduce the internal amplification of the photocurrent. Also, the phototransistor is typically three orders slower than a photodiode due to the increased transit time of carriers diffusing across the base, but this difficulty, of course, is compensated by the increased sensitivity.

11.3 Solar cells

The pn junction photodiode described in the previous section operated with an applied reverse bias voltage in the *photoconductive mode*. If, however, the illuminated diode were to be used without external bias, in an open-circuit condition, a measurable forward voltage would be apparent between p- and n-regions. This is called the *photovoltaic effect*.

When the open-circuited pn junction is illuminated, electron–hole pairs are created near to the junction, as described before, causing an extra minority-carrier photocurrent to flow in the junction. Consequently, the balance of drift and diffusion currents usually maintained by the electric field corresponding to the contact potential, V_0, is upset. A voltage, V_{oc}, appearing across the depletion layer reduces the effective contact potential to $(V_0 - V_{oc})$ and the consequent reduction in junction field and barrier height, as shown in the band diagram of Fig. 11.28, ensures that equilibrium is restored.

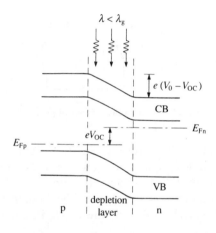

Fig. 11.28 Band diagram of illuminated, open-circuited pn junction.

The pn diode can be used as a photodetector in this *photovoltaic mode* or for the direct conversion of solar energy into electrical energy in a *solar cell*. If an illuminated pn junction is shunted by a load resistor, a forward voltage is induced across the diode and a reverse current flows, with magnitude depending on the load resistance, R_L. The diode I–V characteristics in this situation are as developed before but this time the operation is confined to the fourth quadrant, as shown in Fig. 11.29. The equivalent circuit for cell and load is shown in Fig. 11.30. It will be noted that the direction of the load current and the polarity of the photovoltaic voltage are such that power is delivered by the junction to the load.

As R_L is changed, the load current can be varied between a maximum value

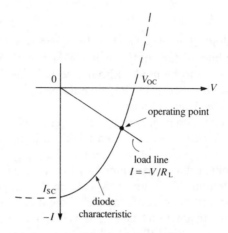

Fig. 11.29 *V–I* **characteristics of solar cell.**

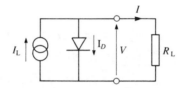

Fig. 11.30 **Equivalent circuit of photocell.**

I_{sc}, when $R_L = 0$ and $V = 0$, to zero, when $R_L = \infty$ and $V = V_{oc}$, as shown in Fig. 11.29. At both these extremes, no power is transferred to the load since either V or I is zero, but the usual operating point for the cell occurs at the intersection of the load line $I = -V/R_L$ with the diode $I–V$ characteristics, as shown in Fig. 11.29, when both V and I are also finite and power is transferred to the load.

The maximum voltage obtainable from a solar cell, V_{oc}, can be found using the equivalent circuit, Fig. 11.30, assuming that $V = V_{oc}$ and $I = 0$. Then

$$I_L = I_D = I_0[\exp(eV_{oc}/kT) - 1]$$

which gives

$$V_{oc} = (kT/e)\log_e[(I_L + I_0)/I_0] \qquad (11.19)$$

V_{oc} usually has values around half the gap energy, typically 0.5 V.

The maximum current from the illuminated cell, I_{sc}, occurs when it is short-circuited and $V = 0$. Again using the equivalent circuit, it is evident that under these conditions

$$I = I_{sc} = I_L - I_D = I_L - I_0(e^0 - 1) = I_L \qquad (11.20)$$

$I_{sc} = I_L$ is usually tens of milliamps.

However, as we have seen, the solar cell is usually operated with a load across it, which will lead to a voltage $V < V_{oc}$ across the load and a current $I < I_{sc}$ flowing through it. The power delivered to the load, P, is then

$$P = IV$$

which is maximum when

$$dP/dI = V + I \, dV/dI = 0$$

or

$$dV/dI = -V/I$$

In other words, the output power is maximum when the slope of the load line is equal to minus the slope of the I–V characteristic at the operating point, as shown in Fig. 11.31. The maximum output power is then

$$P_{max} = V_m I_m$$

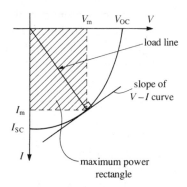

Fig. 11.31 Conditions for maximum power output from solar cell and the maximum power rectangle.

which is less than $V_{oc} I_{sc}$ and is shown as the maximum power rectangle in the diagram. A *fill-factor*, f.f., which shows how closely $V_m I_m$ approaches $V_{oc} I_{sc}$ and acts as a useful figure of merit for solar-cell design, is often defined thus

$$\text{f.f.} = V_m I_m / V_{oc} I_{sc} \tag{11.21}$$

Although the photodetector diode and the solar cell have fundamentally similar structures, the latter is designed specifically to exploit the direct conversion of solar energy. Solar cells are designed in such a way as to maximize the conversion efficiency, which is given by

$$\eta = \frac{I_m V_m}{P_s} = \frac{V_{oc} I_{sc} \times \text{f.f.}}{P_s} \tag{11.22}$$

where P_s is the incident solar radiation, which is of order $1\,\text{kW m}^{-2}$ (at the equator at noon — called air-mass 1, AM1).

Considering first the I_{sc} term, we have seen, Eq. (11.20), that this equals the photogenerated current, I_L, which in turn has components

$$I_L = eGA(L_e + L_h) = eGA[(\tau_h D_h)^{1/2} + (\tau_e D_e)^{1/2}]$$

In the usual np^+ cell, the lifetime of holes in the n-type, τ_h, is much greater than that of the electrons in the p-region, τ_e, because of the relatively high recombination in the p^+ material. In these circumstances, $L_h \gg L_e$ and

$$I_{oc} = I_L \simeq eGA(\tau_h D_h)^{1/2}$$

For high conversion efficiency, this term must be maximized by (a) making the collection area, A, as large as possible and (b) choosing a high-life time semiconductor in which recombination is relatively small; this condition can be satisfied, as we have seen, by using an indirect-gap semiconductor, such as silicon, in which recombination is a relatively slow mechanism. Such design considerations lead to solar-cell structures typically as shown in Fig. 11.32. The cell area is limited only by the maximum size of available single-crystal silicon slices, which is currently around 150 mm diameter. For higher powers, single cells are connected together in series/parallel *arrays*, to increase the effective radiation catchment area. The front contact, often gold or aluminium,

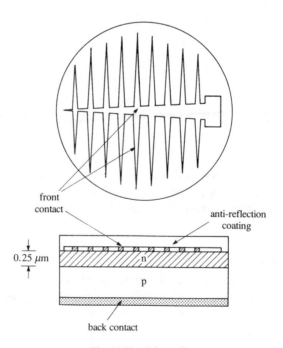

Fig. 11.32   Solar cell.

has the characteristic 'finger' geometry so as to combine maximum current collection with minimum light inteception. The top n-layer is made extremely thin so as to reduce the absorption of light before it reaches the active junction region. An added refinement, which is quite common, is the provision of a transparent anti-reflection coating, as shown.

As indicated by Eq. (11.22), another possible way to increase the efficiency of a cell would be to choose a semiconductor with a high V_{oc}. Since V_{oc} has a value approximately equal to half the gap energy in volts, it would appear that materials with high E_g values would be favoured. That this is so, if only partially, can be seen from the graph of Fig. 11.33, which sketches the

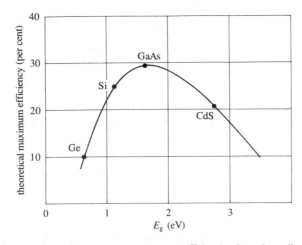

Fig. 11.33   **Theoretical maximum room-temperature efficiencies for solar cells constructed in various materials.**

theoretical maximum efficiency versus E_g for a number of materials. It will be noted that the maximum efficiency for GaAs, around 28 per cent, is marginally greater than for Si, at 25 per cent, which is much greater than that for Ge. However, GaAs is much more expensive and is not available in slices with such large cross-sectional areas, so Si is the preferred material for most applications, GaAs cells being reserved for applications where cost considerations are secondary and for *concentrated cells*, which use lens systems to concentrate the incident light onto a small-area cell.

In practice, Si solar cells have efficiencies in the range 10–20 per cent. Typically, for example, a small 200 mm^2 cell might have an efficiency of 15 per cent, delivering 10 mW at 0.6 V in AM1 light. However, a series/parallel array of 50 mm diameter cells, for space-vehicle power say, might for example provide 28 V at 1 kW.

Problems

1. A certain semiconductor has a gap energy of 2.3 eV. Calculate the range of wavelengths that are (a) absorbed and (b) transmitted by the semiconductor. To which regions of the optical spectrum do these ranges correspond?

Ans. (a) $\lambda < 540$ nm, (b) $\lambda > 540$ nm

2. Show that for a solar cell the short-circuit current, I_{sc}, is related to the open-circuit voltage, V_{oc}, by a relationship of the form

$$I_{sc} = A[\exp(BV_{oc}) + C]$$

where A, B and C are to be determined.

When operating at a temperature of 300 K a certain silicon cell of 100 cm^2 area gives a short-circuit current output of 3.3 A under 1 kW m^{-2} illumination. Assuming a saturation current under these conditions of 0.3 nA, determine: (a) the open-circuit voltage; (b) the energy conversion efficiency, assuming the cell behaves ideally; and (c) how the efficiency is affected if a more realistic fill factor is assumed.

Ans. $A = I_0$, $B = e/kT$, $C = -1$, (a) 0.6 V, (b) 19.7 per cent, (c) 11.8 per cent for f.f. $= 0.6$

3. The leakage current, I_0, of a 200 mm^2 silicon p$^+$n junction solar cell at 300 K is 1.0 nA. The short-circuit current of this device exposed to the noonday sun is 20 mA and the electron–hole pair generation rate in the silicon is 3×10^{24} m^{-3} s^{-1}. Assuming the diffusion coefficient for holes is 1.25×10^{-3} m^2 s^{-1}, estimate: (a) the diffusion length and hence the lifetime of minority holes in the n-region of the device, assuming that the electron lifetime in the p-region is very small because of the high impurity level there; (b) the resistance to be connected across the cell in order that 10 mA of load current is delivered to this load; and (c) the power delivered to the load under the conditions of part (b).

Ans. (a) 2.1×10^{-4} m, 32 μs, (b) 42 Ω, (c) 4.2 mW

4. A circular silicon wafer, 75 mm diameter, is used to fabricate a solar cell. In bright sunlight the short-circuit current of the cell is 500 mA when operated at 30°C. Calculate the open-circuit voltage provided by the cell for the same illumination and temperature, if, under dark conditions, the reverse saturation current is 0.3 μA.

The responsivity of a cell is defined as the ratio of the photocurrent to the power incident on the cell. If the irradiance at the cell surface is 100 W m^{-2}, what is the responsivity of the cell?

Ans. 374 mV, 1.13 A W^{-1}

5. Show that the condition for maximum power output from a solar cell is given by

$$(1+eV_m/kT)\exp(eV_m/kT)=1+I_{sc}/I_o$$

where V_m is the voltage for maximum power and the other symbols have their usual meaning.

Assuming that $I_{sc}\gg I_o$ and $V_m\gg kT/e$, write the above equation in the form

$$\log_e x=c-x \qquad \text{where } c \text{ is a constant}$$

and hence find the voltage at maximum delivered power V_m, for a solar cell with a reverse bias saturation current of 1 nA which is illuminated such that the short-circuit current is 100 mA at 20°C.

What is the maximum power output of the cell at this illumination?

Ans. 396 mV, 37.1 mW

6. A particular solar cell, operated as a normal Si diode in the forward direction, in the dark, at 30°C, has a saturation current I_0, of 3 nA. When the cell is illuminated by sunlight, the measured short-circuit current is 40 mA. (a) Find the open-circuit voltage of the cell. (b) Derive expressions for the optimum voltage V_m, and current I_m, at which the maximum output power, P_m, is obtained. (c) Estimate the values of V_m, I_m and P_m. (d) Hence estimate the cell efficiency, assuming an effective cell area of 2 cm^2 and an illumination of 1 kW m^{-2}.

Ans. (a) 0.43 V, (c) 0.358 V, 37.3 mA, 13.4 mW, (d) 6.7 per cent

7. Two solar cells differ only in that their saturation current densities, J_o, are 10^{-9} and 10^{-8} A m^{-2}. The photocurrent density of each, J_L, when operating at 290 K in AM1 light (equivalent to 1 kW m^{-2}), is 250 A m^{-2}. Each has an area of 10^{-3} m^{-2}.

When the cells are connected in parallel under AM1 illumination and connected to a load resistance, a current of 400 mA is supplied to the load. Find the voltage, load resistance, power produced and efficiency under these conditions.

When the two cells are connected in series, again under AM1 illumination, and a load current of 200 mA is drawn, find the voltage drop across each cell and the power out from the combination.

Ans. 0.574 V, 1.43 Ω, 229 mW, 11.5 per cent, 0.616 V, 0.558 V, 234 mW

8. In a particular rudimentary array, two cells are connected together in parallel and operated at 300 K. One cell has an open-circuit voltage of 0.55 V and a short-circuit current of 1.3 A. The second cell has corresponding values of 0.60 V and 1.0 A. Assuming that both obey the ideal diode law, calculate the open-circuit voltage and short-circuit current of the combination. Comment on the results.

Ans. 0.562 V, 2.3 A

12 Compound semiconductors and devices

12.1 Introduction

So far, our principal attention has been directed towards the semiconductor, silicon. This is because silicon components, devices and integrated circuits have, until relatively recently, acquired an almost total, well deserved market dominance. However, in some applications, albeit specialized for the moment, silicon is not the optimum material to use and other semiconductors, notably the *compound semiconductors* GaAs and to a lesser extent InP, are finding a progressively increasing usefulness. For example, many of the optoelectronic components discussed in Chapter 11, including notably a family of solid-state lasers, rely on GaAs for their active medium. There is also emerging an increasingly important requirement for other GaAs active electronic devices and related integrated circuits. As will become apparent later, these can have superior performance to their silicon counterparts, particularly important properties being an enhanced speed, increased thermal capacity and better radiation resistance. There are also classes of device that rely on the particular properties of GaAs for their operation, which are not replicated in Si. An example is the Gunn diode microwave oscillator, which, as we shall see, depends for its operation on a negative slope in its electric field–electron velocity characteristic, which is not present in Si.

In spite of the inherent advantages of compound semiconductors for certain currently specialized applications, there seems to be no likelihood of GaAs ousting Si as the principal material for the manufacture of the majority of integrated circuits. The reasons for this are economic. GaAs is relatively rare and consequently expensive; slice diameters are limited by the difficulties in growing single crystals. The inability to find a native oxide and the presence of large inherent impurities are further factors that limit its usefulness. Nevertheless, in spite of these shortcomings, GaAs and related materials have sufficient positive advantages in certain areas to ensure that they are being used for an increasingly significant number of ICs and devices.

In this chapter we discuss the structure and basic electronic properties of compound semiconductors and the various devices and circuits incorporating them that are currently available.

12.2 Compound semiconductors

Silicon is the prime example of an *elemental semiconductor*. As we have seen, it is located in column IV of the periodic table, has four electrons and four vacancies in its outer shell, and forms covalent bonds with its neighbours to form a tetrahedral single-crystal solid. By contrast, the principal compound semiconductors have components drawn from columns III and V of the periodic table, a section of which is reproduced in Fig. 12.1. Elements in

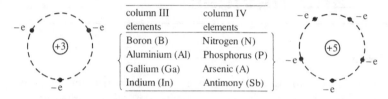

column III elements	column IV elements
Boron (B)	Nitrogen (N)
Aluminium (Al)	Phosphorus (P)
Gallium (Ga)	Arsenic (A)
Indium (In)	Antimony (Sb)

Fig. 12.1 The more important component elements of III–V compound semiconductors and their outer-shell configurations.

column III have three electrons in their outershell and five vacancies and those in V have five electrons and three vacancies. In principle, all combinations of pairs of such elements are possible, but the compounds with the most important electronic applications are GaAs and, to a lesser degree, InP. Whereas, as for Si, the crystalline structure is still tetrahedral, their crystal bonding is entirely different, being *ionic* in character, due to an electron lost by the group V element being acquired by the group III component. A two-dimensional diagrammatic representation of the tetrahedal structure of GaAs is shown in Fig. 12.2. The gallium atoms, which have three valence electrons, combine with arsenic atoms, with five valence electrons, by forming an ionic bond, and each occupies alternate sites in the manner illustrated.

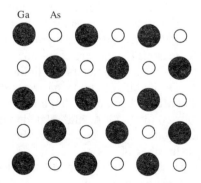

Fig. 12.2 Diagrammatic representation of the tetrahedral structure of GaAs.

Some of the more important electronic properties of the main compound semiconductors are contrasted with those for Si in Table 12.1 The significant values to note are that the electron mobility measured at low fields is typically an order of magnitude higher in GaAs than in Si, although the hole mobilities are similar, and that, above a field of around $3 \times 10^5 \, V \, m^{-1}$ in GaAs, the

Table 12.1 A comparison of typical characteristics of the more important compound and elemental semiconductors.

	GaAs	InP	Si
Gap energy, E_g(eV)	1.39	1.29	1.09
Low-field electron mobility, μ_e ($m^2 \, V^{-1}s^{-1}$)	0.4–0.9	0.46	0.05–0.15
Low-field hole mobility, μ_h ($m^2 \, V^{-1} \, s^{-1}$)	0.040	0.015	0.045
Electron saturated drift velocity, v_s($m \, s^{-1}$)	2×10^5	–	10^5

saturated electron velocity is double that in Si, which requires a higher critical field of around $10^6 \, V \, m^{-1}$ for saturation to occur. Several important general conclusions can be drawn from these comparisons. First, it is evident that, under all conditions, electrons travel faster in GaAs than in Si, assuming the same drift field. This implies that, all else being the same, the transit time of electrons in GaAs devices is much shorter than in Si, with a corresponding decrease in switching speed or a higher frequency cut-off. The speed benefit may not be as great as suggested by the difference in low-field mobilities, a factor around 10, since carriers often reach saturated velocities in operation, but under these conditions there is still a 2:1 advantage in velocity with a corresponding speed increase. These considerations of speed also apply to logic gates, since their propagation delay is approximately proportional to carrier velocity. Hence a logic gate using GaAs devices has the potential to operate typically twice as fast as the corresponding gate in Si.

A further point to note from Table 12.1 is that the hole mobilities are all much the same, so there is no advantage to be gained by using p-type GaAs devices, for example. Consequently n-channel GaAs field-effect devices are the norm.

Also worth noting is the wide discrepancy in gap energies between GaAs

(1.4 eV) and Si (1.1 eV). These, together with the direct-gap properties of GaAs compared to the indirect gap for Si, ensure the prominence of GaAs as the optimum material for certain optoelectronic devices, as discussed in Chapter 11. Another advantage is that, when GaAs is used as the active material for devices or ICs, then, because of its relatively enhanced energy gap width, components can operate over a relatively wide temperature range, typically $-200°C$ to $+200°C$. Also related to the high gap energy is the reported comparatively high immunity to radiation damage of GaAs ICs, which has obvious implications for space and other applications. The high E_g also means that GaAs devices when operated at room temperature have relatively few thermally generated carriers and low leakage currents, which leads to a low noise figure.

Another advantage of using GaAs for ICs is that it can have a relatively high bulk intrinsic resistance. Indeed, it can be produced in a semi-insulating form with resistivity around $10^6 \, \Omega m$, which is ideal substrate material, since neighbouring circuit components can be conveniently isolated electrically simply by physical separation while still maintaining crystallinity between substrate and active GaAs epitaxial layer. The high substrate resistance also minimizes the effect of parasitic capacitances.

Finally on the plus side, since many optoelectronic components are fabricated in GaAs, the possibility of their complete and efficient integration with GaAs integrated circuits is particularly attractive. Two advanced GaAs monolithic microwave ICs (MMICs) are illustrated in Figs 12.3 and 12.4.

In spite of these very real advantages of GaAs components and circuits, the overwhelming consideration as to their deployment, at this stage in their development, is their relatively low yield in production. This, with other factors such as slice size, leads to a much higher cost than for a Si alternative of the same complexity. The low yield also tends to inhibit the degree of integration of GaAs circuits. The reliability of GaAs circuits, at the present stage of their development, is also not as good as for Si. Other factors to be considered are that complementary circuits are undesirable because the low hole mobility in GaAs excludes p-type devices, and that MOS technology is unobtainable due to the absence of native oxide of GaAs and the otherwise very high interface charge levels.

12.3 Doping compound semiconductors

Extrinsic doping of a compound semiconductor such as GaAs is not quite as straightforward as for an elemental semiconductor. In GaAs the conductivity type, whether n- or p-type, is dependent not only on the dopant but additionally on the position it eventually occupies in the GaAs lattice. For example, a possible dopant for GaAs is Si, which has, as we know, four electrons in the outer shell. If the Si dopant atoms were to occupy a Ga (three-electron) site, the extra electron from the Si could be thermally ionized to

Fig. 12.3 **Miniature radar: a miniature broad-band (3–6 GHz) phased-array radar transmit/ receive module using GaAs monolithic microwave ICs (MMICs). (Courtesy of Plessey Research (Caswell) Ltd.)**

create an electron in the conduction band, so the Si behaves as an n-type impurity. If, on the other hand, the Si dopant atom were to occupy an As site, the Si atoms deficiency of an electron would create a vacancy, which could be thermally ionized to produce a hole, the Si in this instance behaving as a p-type dopant.

Often dopants are introduced into GaAs by ion-implantation methods, as discussed in Chapter 10. The highly energetic beam of, say, Si ions tends to disrupt the lattice and deposit the Si interstitially, as shown diagrammatically in Fig. 12.5(a). In such positions, the Si atoms are not electrically active. A subsequent anneal cycle, typically at 800°C, not only restores the regular crystalline structure of the parent GaAs, but causes the silicon atoms to move substitutionally to a lattice site, as shown in Fig. 12.5(b), and become available as dopants. Since the Si dopant atoms are bigger than the As atoms, yet smaller

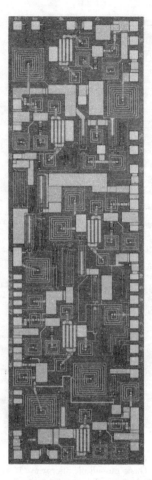

Fig. 12.4 A GaAs monolithic microwave integrated circuit (MMIC). This 10 mm^2 chip, probably the world's most complex analogue GaAs MMIC to date, provides transmit and receive functions for advanced phased-array radar systems. Several thousand such chips will be used in each radar. (Courtesy of Plessey Research (Caswell) Ltd.)

than Ga, there is a tendency for them to occupy Ga sites, as shown, so behaving as n-type dopants.

In the earlier stages of development, pure intrinsic GaAs was unattainable, and its conductivity type was governed by its inherent impurities. Semi-insulating GaAs could then be made by, say, doping the initial p-type material with chromium, which creates deep-level donor sites that effectively compensate the p-type and form near-intrinsic material (see Chapter 6). With improved growth technology it is now possible to produce high-resistivity, semi-insulating, intrinsic material in an original undoped state, which eliminates difficulties present in the compensated form on subsequent doping. It is possible to increase the resistivity of intrinsic GaAs even further by

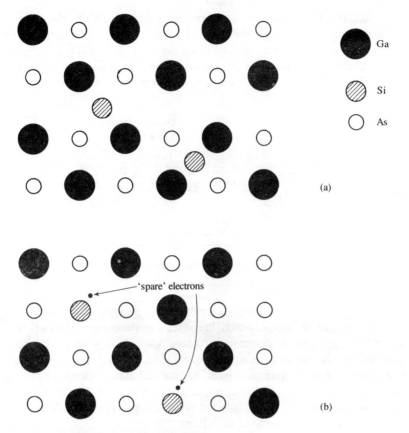

Fig. 12.5 **(a) GaAs implanted with Si ions; (b) after annealing the Si substitutes Ga atoms in the lattice and behaves as an n-type dopant.**

bombarding it with high-energy protons, which disrupt the lattice, to form an amorphous structure, which has inherent high resistance.

12.4 GaAs Schottky diodes and MESFETs

The application of compound semiconductors to active devices is constrained, as we have seen, principally by their low hole mobilities, which has inhibited the development of bipolar transistors and complementary logic circuits. A further influence is the lack of a native oxide to GaAs and the difficulty of producing satisfactory charge-free interfaces with other insulators, which effectively precludes the development of MOS devices. Effective Schottky barriers can be produced on GaAs, however, using a wide range of metals; of particular interest are Al, W, Ti and Pt. This is the key to the, at least initial, dominance of the GaAs MESFET (metal–semiconductor FET), which has become the workhorse of GaAs ICs.

The MESFET behaves in many respects like a JFET, the principal difference being that the control gate is a metal, which forms a Schottky barrier to the GaAs, rather than a diffused junction as in a Si JFET. A schematic diagram of the cross section of the essential elements of a GaAs MESFET is shown in Fig. 12.6. Source and drain electrodes, which consist of heavily doped

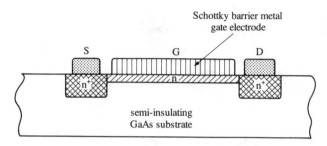

Fig. 12.6   A cross section of the essential features of a GaAs MESFET.

n$^+$ stripes in the semi-insulating GaAs slice, are connected by a more lightly doped n-type channel region, which, in contrast to the induced channel JFET, is an in-built, pre-grown, permanent feature. Current flow in the channel is controlled by the metal gate electrode, which forms a Schottky barrier in conjunction with the n-channel. The n$^+$ source and drain regions are provided with ohmic contacts, which, as usual, can also serve as interconnections to other IC components. To give some feel for size, typical gate width and hence channel lengths are between 0.2 and 1 μm.

A MESFET can be fabricated that operates in either depletion or enhancement mode, depending on the channel thickness and degree of doping. The depletion type, called a D-MESFET, was the first to be developed. It has a comparatively thicker and more highly doped channel. It is a normally ON device, electron drain current flowing freely along the channel when a positive drain voltage, V_d, is applied and the gate voltage, V_g, is zero. Negative voltages applied to the gate control the drain current, I_d, by causing the depletion layer in the channel to widen, thus reducing the effective channel thickness, increasing the channel resistance and so reducing I_d. This behaviour, which is illustrated in Fig. 12.7, is analogous to that described for the JFET (Chapter 9). The depletion layer, as shown, has two components, one corresponding to the potential in the channel due to V_d, the other due to V_g. If V_d is increased, I_d increases, in a manner much as predicted by Eq. (9.9), but eventually the depletion layer widens sufficiently to fill the channel almost completely, which becomes pinched-off (as for a JFET—see Sec. 9.3), and thereafter the drain current saturates. Increasing the negative gate bias causes the depletion layer to widen, pinch-off occurs at lower V_{ds} and the saturated drain current is lower. This behaviour results in a family of drain characteristics as depicted in

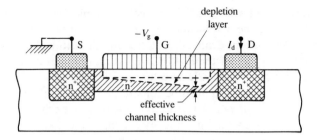

Fig. 12.7   Formation of depletion layer in a D-MESFET.

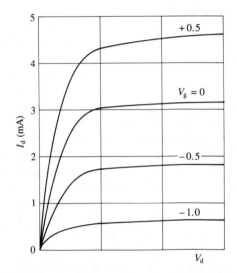

Fig. 12.8   Drain characteristics for a D-MESFET.

Fig. 12.8. When $V_g = 0$, the normally ON device has a conducting channel, as explained, but a narrow depletion layer is still present in it, corresponding to the in-built contact potential of the Schottky barrier. It is possible to narrow the depletion layer further, and so enhance the drain current for a given V_d, by the application of positive bias to the Schottky gate. A drain characteristic similar to the top curve in Fig. 12.8 would result. The positive excursions of V_g must be restricted to a maximum value of around 0.7 V, above which the gate currents, corresponding to a forward-biased Schottky diode, become excessive.

The enhancement metal–semiconductor FET (or E-MESFET) is very similar in construction to the D-MESFET. Its distinguishing feature is its shallower channel, which is more lightly doped. This ensures that the depletion layer at $V_g = 0$ occupies the entire channel region. Under these conditions the channel is essentially pinched-off, so the E-MESFET behaves as a normally OFF device. A positive V_g, above a lower threshold value corresponding to the turn-on voltage V_T in a MOSFET, reduces the depletion-layer width to less than the n-layer thickness, thus allowing drain current to flow. Progressively increasing V_g *enhances* the drain current flowing. The drain characteristics are much the same shape as for the D-MESFET (Fig. 12.8) but of course the gate voltage excursions are only effective for positive V_g above threshold. The maximum gate voltage again must be constrained to around 0.8 V, in order to avoid excessive gate current. A further point is that, in both classes of device, channel geometries are so small, typically of order 1 μm, that, even for modest drain voltages of a few volts, the longitudinal fields are above the minimum for the electron velocities to saturate. Under these conditions, the expressions developed to describe the performance of the JFET (Chapter 9) no longer apply, strictly, to the MESFET.

Choice between D- and E-type MESFETs has largely been determined by differences in fabrication procedures and integrated-circuit application. D-type devices require fewer processing steps, typically seven, so the potential yield is higher. On the other hand, since V_g is of the same polarity as V_d in the E-type version, only one power supply line is required; their power consumption is also lower. E-types also have a circuit advantage in being more suitable for direct-coupled logic since D-types need voltage shifting between stages to comply with turn-off requirements of the subsequent stage. The limited signal swing between V_T and around 0.8 V gives E-MESFETs a lower noise immunity than D-MESFETs in ICs. A computer-generated mask set for a prototype GaAs IC incorporating a MESFET is shown in Fig. 12.9.

Both types of MESFET rely heavily on ion-implantation doping technologies in their manufacture. Some of the stages in fabricating a planar D-MESFET are illustrated in Fig. 12.10. A semi-insulating GaAs slice, which has a silicon nitride passivation/encapsulation layer pre-deposited on it, is illuminated with high-energy Si^+ ions, which are implanted through the Si_3N_4 to form the high-resistance channel region, Fig. 12.10(a); photoresist normally serves as the masking medium. An anneal cycle, typically at 800°C, activates the dopant, the doping level, N_d, being typically $10^{23} \, m^{-3}$. A low-resistance n^+ implantation follows, Fig. 12.10(b), often using S ions, to form drain and source stripes, $N_d \simeq 10^{24} \, m^{-3}$. Ohmic contacts to S and D regions are next provided by an Au–Ge–Ni alloy evaporation and delineation, Fig. 12.10(c). After a further Si_3N_4 deposition and gate window-opening stages, a Ti–Pt–Au layer is then deposited over the slice to form the Schottky gates, Fig. 12.10(d). Another insulating layer is followed by window opening over S, D and G metallizations and the final interconnection metal layer, which could again be Ti–Pt-Au, Fig. 12.10(e). Additional electrical isolation between components can be provided by high-resistance boron implantations, as shown in Fig. 12.10(e).

The E-MESFET is sometimes fabricated by a variant in the technology known as the self-aligned gate process. Some of the alternative steps are shown in Fig. 12.11. After the usual n-channel implantation, the Schottky gates, often

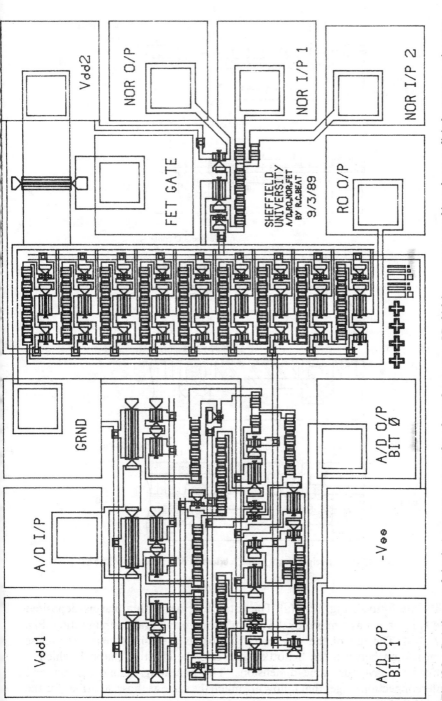

Fig. 12.9 GaAs integrated circuit. A computer-generated mask set of a prototype IC which incorporates a two-bit analogue-to-digital converter, a nine-stage ring oscillator, a two-input NOR gate and a separately connected MESFET. This computer-aided design formed part of a Final Year Undergraduate Project at Sheffield University. (Courtesy of Dr J. Woodhead and Mr R. C. Beat.)

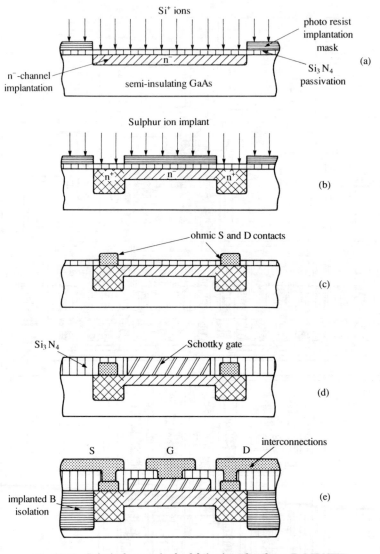

Fig. 12·10   **Principal stages in the fabrication of a planar D-MESFET.**

TiN, are formed, Fig. 12.11(a). The gates, together with a plasma-deposited SiO_2 layer, act as a mask for the next n^+ source and drain implantation, Fig. 12.11(b). The use of the gate metallization as a self-mask ensures precise registration between the Schottky gate and the effective channel, which is forced to have the same dimensions. The annealing, ohmic contact and interconnection stages that follow are much the same as those described for the D-MESFET.

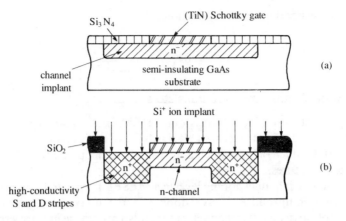

Fig. 12.11 Important fabrication steps for E-MESFET with self-aligned gate.

For a very thin conducting channel relative to the channel diffusion depth a, the saturation current for an E-MESFET can be obtained much as for the enhancement-mode MOST, Eq. (9.31), which gives

$$I_{ds} = \frac{\mu_e \epsilon_{rs} \epsilon_0 w}{la} \frac{(V_g - V_T)^2}{2} \qquad (12.1)$$

where a has replaced the oxide thickness in the MOST and V_T is the turn-on gate voltage, which is just sufficient for a conducting channel to form and allow conduction between S and D. The transconductance is defined in the saturated current region, as usual, and is given by

$$g_m = \left(\frac{dI_{ds}}{dV_g}\right)\bigg|_{V_d} = \frac{\mu_e \epsilon_{rs} \epsilon_0 w}{la}(V_g - V_T) \qquad (12.2)$$

In the special case of very short gate lengths or when a D-MESFET is operated in its saturated current region and most of the channel is pinched-off, the effective channel length can become very short with a high longitudinal $\mathscr{E}$ field imposed, such that the carrier velocities become saturated, $v = v_s$. The carrier transit time from S to D, τ, is then approximately l/v_s and the device cut-off frequency, f_T, is given by

$$f_T \simeq 1/(2\pi\tau) = v_s/(2\pi l) \qquad (12.3)$$

If, for example, $l = 1 \, \mu m$, $v_s = 2 \times 10^5 \, m \, s^{-1}$ (for GaAs), the cut-off frequency is in the microwave spectrum, around 30 GHz. For these large electric field conditions, when carrier velocities become saturated, the saturation drain current becomes

$$I_{ds} = \frac{\epsilon_{rs} \epsilon_0 v_s w_s}{a}(V_g - V_T) \qquad (12.4)$$

with a corresponding transconductance

$$g_{\mathrm{m}} = \frac{\partial I_{\mathrm{ds}}}{\partial V_{\mathrm{g}}} = \frac{\epsilon_{\mathrm{rs}} \epsilon_0 v_{\mathrm{s}} w}{a} \qquad (12.5)$$

12.5 High-electron-mobility transistors

The high-electron-mobility transistor (or HEMT), also known as the modulation-doped field-effect transistor (MODFET), is a more highly developed form of MESFET, which offers a much improved speed of operation, particularly at lower temperatures.

The choice of channel doping level in a MESFET tends to be, as is usual, an engineering compromise. Heavy doping in the n-layer can provide large numbers of carriers in the channel, but this also causes the impurity scattering centres, which control the carrier mobility and electron velocity, to be enhanced. Hence, although the number of carriers in the channel may be increased, their mobility is reduced, which could lead to an overall reduction in the channel conductivity. The HEMT is a multilayer device incorporating several semiconductors in which this difficulty is avoided by carriers that are copiously generated in one material being transferred into a neighbouring material that has few scattering centres, which can operate as an effective channel region.

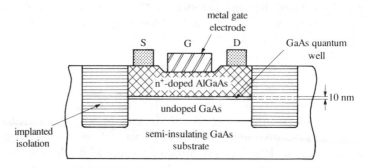

Fig. 12.12 Essential elements of a HEMT.

A much-simplified cross section of a HEMT is shown in Fig. 12.12. The large energy-gap difference between the AlGaAs and the GaAs determines that an electron's potential energy is lower in the GaAs. Hence electrons in the heavily doped AlGaAs layer tend to spill over into a thin interfacial layer of the undoped GaAs. This region, which behaves as a quantum well for electrons, is very lightly doped, so has few scattering impurities and the electrons in it behave essentially like a two-dimensional high-mobility electron gas, thus providing a most efficient highly conducting channel between the source and drain, which can be controlled by the gate voltage.

Improved f_T values in the range 40–100 GHz have been reported for HEMTs. However, the technology used for their fabrication is complex, involving metal-organic chemical vapour deposition (MOCVD) and molecular beam epitaxy (MBE), so they are only currently being developed for applications in which speed is an overriding consideration.

12.6 Heterostructure bipolar transistors

Bipolar transistors are preferred to FETs for some applications because of their large current capability and high transconductance. The difficulty with realizing a conventional npn transistor in a compound semiconductor arises from the very low hole mobilities in such materials, which would cause unacceptably high base resistance. A *heterojunction* between emitter and base, that is a junction between dissimilar semiconductors (which contrasts with the more usual *homojunction* formed in the same material, as in a silicon pn junction, for example), can be used to enable high base doping and improved base conductivity, so reducing the access resistance.

A high emitter efficiency, which implies high current injection across the emitter–base junction and suppression of hole injection from base to emitter and hence high gain, is achieved in a conventional npn bipolar transistor by high differential doping of the emitter with respect to the base, which can cause the device to be less effective at high speed. In a heterostructure bipolar transistor (or HBT) differences in energy band gap, E_g, between emitter and base materials, which form a heterojunction, are used to cause electrons to experience a smaller barrier to injection across the junction than that seen by holes, which results in an emitter efficiency, η_E, that approaches unity regardless of the doping ratio between emitter and base.

Let us consider the heterojunction between n-type AlGaAs, which eventually might form the emitter of an HBT, and p-type GaAs, which is the usual base material. E_g for the AlGaAs is around 1.8 eV and for GaAs is 1.4 eV, so the band structure for the two materials before a junction is formed will be of the form shown in Fig. 12.13(a). As the two materials are joined together to form a heterojunction, electrons will initially spill over from high energy levels in the CB of the AlGaAs into lower available states in the CB of the GaAs and holes will move in the opposite direction, until electronic equilibrium is achieved and the Fermi level is continuous across the junction, as shown in Fig. 12.13(b). Depletion layers now occur at either side of the junction because of the initial movement of carriers, within which the bands bend to form energy barriers that prevent further movement of charge in equilibrium. The bands bend upwards in the AlGaAs to provide a barrier to electrons and downwards (i.e. in the increasing *hole* energy direction) to prevent hole movement from the GaAs, as shown in Fig. 12.13(b). These differences lead to discontinuities in the band diagram at the junction, as indicated. It will also be noted that, because of the differing gap energies, the energy barrier for electrons is smaller than that for holes. Hence, when the junction is forward-biased to reduce the barrier height

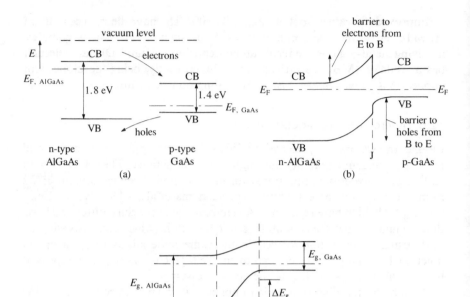

Fig. 12.13 **Band diagrams for an n-type AlGaAs/p-type GaAs heterojunction: (a) before junction formation: (b) an abrupt junction; and (c) a compositionally graded junction.**

and allow junction current to flow, the junction behaves as a Schottky barrier and electrons can be injected into the GaAs by thermionic emission over the barrier but holes in the GaAs still face a considerable energy barrier so relatively few are transferred to the AlGaAs. Most of the junction current is transported by electrons, therefore, which leads to high emitter efficiencies, when the combination forms the emitter–base junction of an HBT, which are independent of doping levels at either side of the junction. Since carrier injection is exponentially dependent on barrier energy, only a small difference in band gap between the components of the heterojunction is necessary to create a significant change in barrier heights at either side and to ensure that carrier flow is predominantly from the larger band-gap material.

The 'spike and trough' discontinuity in the CB of the heterojunction shown in Fig. 12.13(b), which is typical for an abrupt junction, can create a trapping well for electrons injected over the barrier, which increases recombination and in some cases causes a deterioration in performance. Such discontinuities in the band edges can be avoided by compositional graded doping of the junction, to create a band structure as shown in Fig. 12.13(c). The energy bands now blend smoothly but the differences in gap energies still ensure that a difference in barrier heights, ΔE_g, occurs, so maintaining the inherent advantages of the heterojunction.

Physical realization of HBTs depends to some extent on the technology employed, but the general features of a planar device will be identifiable in the schematic diagram of Fig. 12.14. The n-type AlGaAs emitter is lightly doped,

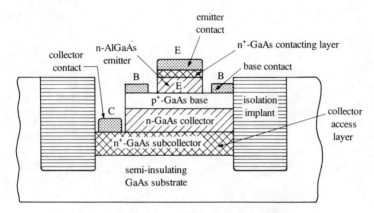

Fig. 12.14 Essential features of an HBT.

$n \sim 10^{23}\,\mathrm{m}^{-3}$, but the GaAs base can be more heavily doped, $p \sim 10^{25}\,\mathrm{m}^{-3}$, while still maintaining a high η_E. The additional n$^+$-GaAs layers provide low-resistance contacts to the metal electrodes. A photomicrograph of a prototype heterojunction bipolar transistor is shown in Fig. 12.15.

HBTs are emerging as strong contenders for high-speed microwave or digital applications. Reported f_T values of over 100 GHz have stimulated their further development. An alternative InP/InGaAs technology for producing HBTs, although less mature, offers potential speed advantages. Such devices are also compatible with long-wavelength optical communication systems, which is a further incentive to their development.

12.7 Gunn diode oscillator

In spite of the implications in its name, a Gunn diode contains no junctions but is a two-terminal device consisting essentially of an n-type GaAs region sandwiched between n$^+$-GaAs contacting layers and metal electrodes, as shown in Fig. 12.16. The device is used principally as a low-voltage local oscillator in microwave circuits. As such, its applications are somewhat specialized, but some details of its mode of operation are well worth attention here.

The Gunn diode, named after its discoverer, relies for its operation on a unique, essential feature of GaAs that it exhibits a negative resistance at high

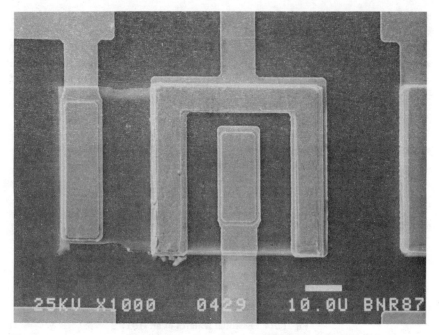

Fig. 12.15 A prototype GaAs/AlGaAs heterojunction bipolar transistor. (Courtesy of Dr P. A. Houston. University of Sheffield and Bell Northern Research.)

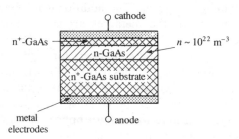

Fig. 12.16 Basic structure of a Gunn diode.

applied electric fields, $\mathscr{E}$. At low fields, its behaviour is perfectly conventional, as for Si; the electron drift velocity with n-type GaAs material, v_D, is proportional to $\mathscr{E}$, the constant of proportionality being the mobility, μ_e. Since the current, I, is proportional to v_D, and $V \propto E$, then, at low fields, $I \propto V$ and the material exhibits a resistance, R. However, as $\mathscr{E}$ is increased above a threshold value (around 0.3 MV m^{-1}, which is easily achieved in very short devices), the drift velocity is found to *decrease* with increasing field, as shown in Fig. 12.17. The current, in this region, decreases with increasing voltage, which is characteristic of a *negative R*. Now, energy is dissipated in a positive R, so it

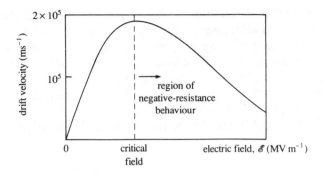

Fig. 12.17 Drift velocity in GaAs as a function of the electric field.

follows that the negative-resistance region can be associated with power generation, which can be utilized to generate or sustain oscillations.

But why does this unusual negative-R behaviour occur? It is principally due to the presence of two separate valleys in the conduction band of GaAs (see Fig. 12.18), which do not occur in Si. At low $\mathscr{E}$ fields, electrons reside in the

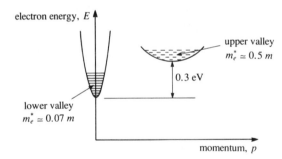

Fig. 12.18 Conduction-band energies in GaAs.

lower central valley of the conduction band, possessing an effective mass, m_e^*, that is a small fraction of the free-electron mass, so the electron mobility is high and the drift velocity is proportional to the field and is correspondingly high. As the field is increased, electrons eventually gain sufficient momentum to transfer to the upper, satellite valley where, since the radius of curvature of the band and hence m_e^* is increased to be of the same order as m, the electron mobility is considerably reduced. Electrons in the upper valley therefore have a much reduced drift velocity. As $\mathscr{E}$ is increased further, a larger proportion of electrons are transferred to the upper valley so their average drift velocity is reduced. This progressive reduction in v_D with increasing $\mathscr{E}$ field, as illustrated in Fig. 12.17, leads to the differential negative-resistance effects mentioned earlier.

Although detailed description of device operation is beyond the scope of this book, it is worth while pointing out that the general operating principle of a Gunn diode, in common with all negative-resistance devices, is to act in conjunction with some form of oscillatory circuit, a high-Q microwave cavity in the GaAs case, in such a manner that the effective negative device resistance negates the residual resistance of the oscillatory circuit. Thus, sustained oscillations can be maintained in the 'tank' circuit instead of the damped oscillations that would prevail were the resistance in the circuit to remain positive.

13 New technologies and devices

13.1 Multicomponent semiconductors—energy band engineering

Significant improvements, such as much higher electron mobility and saturation velocity, achievable by the choice of compounds such as GaAs rather than Si have already been discussed. Ternary and quaternary multi-component semiconductors can offer such further advantages that their development and application are being vigorously pursued. Of particular interest at the moment are the $Al_xGa_{1-x}As$ and the $In_{1-x}Ga_xAs_yP_{1-y}$ families.

For example, $Al_xGa_{1-x}As$, while remaining a direct-band-gap semi-conductor over much of its compositional range, which is distinctly advantageous for its application to optoelectronic devices, as was discussed in Chapter 11, also has a lattice constant that is well matched to GaAs, which is useful for multilayer devices such as the HBT. Moreover, by choice of x during formation, the gap energy, E_g, can be deliberately adjusted from around 1.4 eV corresponding to $x=0$, pure GaAs, to 3 eV at the AlAs extreme, $x=1$.

Similarly, $In_{1-x}Ga_xAs_yP_{1-y}$ has a band gap that can be *tailored* between 0.75 eV and 1.35 eV, with corresponding photon emission wavelength range, again for direct band recombination, of 1.66 to 0.92 μm. The implication is that an injection laser incorporating such a semiconductor can have its emission varied over a most useful wide range by initial choice of composition. For example, it is possible to tailor the output wavelength such that it corresponds to that at which signal attenuation in optical fibres is minimum, around 1.5 μm.

The term *energy band engineering* has been coined to describe the activity of adjusting E_g to provide operational advantages by changing the compositional parameters of multicomponent semiconductors. It is envisaged that such techniques will become increasingly important.

13.2 Multiple quantum well structures—wavefunction engineering

In conventionally grown AlGaAs, the Al and Ga atoms occupy group III sites in the lattice in a random configuration. However, if the de Broglie electronic wavelength $\lambda_e = h/p$ (see Sec. 1.5) is much longer than the lattice constant, a, this is of little consequence since then an electron in the solid is unaffected by the randomness of position of the Al and Ga atoms and experiences a potential that is averaged over many atomic sites. A typical λ_e at room temperature might be 20 μm and the lattice constant, a is around 0.5 μm, so the condition that $\lambda_e \gg a$ is easily satisfied.

However, using modern growth technology, it is perfectly possible to eliminate the randomness of atomic position and produce regular predetermined structures of the Al, Ga and As atoms. For example, it is now possible to fabricate alternate layers of AlAs and GaAs in a multi-sandwich heterostructure, as shown in Fig. 13.1(a), in which the characteristic layer thickness,

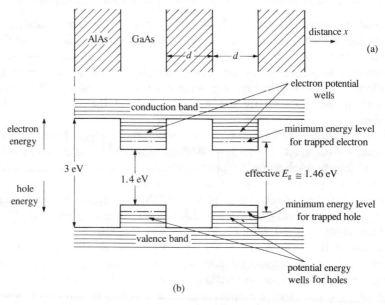

Fig. 13.1   Multiple-layer AlAs/GaAs semiconductor and a simplified energy band structure.

d, can be comparable to λ_e, which gives the material unique and useful electronic properties. Since the AlAs layers have a band gap of 3 eV and the corresponding figure for GaAs is 1.4 eV, a simplified band structure for the sandwich might be as shown in Fig. 13.1(b). The precise degree of sharing of the E_g differences, 3–1.4 eV, between conduction- and valence- band edges is still being debated but, in any event, the band structure will always exhibit periodic wells in the conduction-band edge, as shown, corresponding to the GaAs sections. If the characteristic length, d, is long compared to the electronic wavelength, λ_e, then an electron in a GaAs layer behaves as it would in bulk material. However, if the layers are made thin, such that $d \sim \lambda_e$, electrons in the conduction band of the GaAs become trapped in the potential well, since they cannot surmount the potential energy barriers that exist at the GaAs/AlAs interfaces. Holes in the valence band are similarly constrained by the potential energy wells in the GaAs. The trapped carriers are only allowed to occupy certain energy levels, which are determined by the dimensions of the well, as discussed in Sec. 2.2. For example, Eq. (2.6) shows that the lowest possible energy that an electron can have when contained in a one-dimensional potential energy well of width d is

$$E_{n=1} = h^2/8m_e^* d^2 \qquad (13.1)$$

which is about 50 meV for 10 μm layers in GaAs. This minimum energy level is measured from the bottom of the conduction band in our band model of Fig. 13.1(b). Trapped holes have a similar minimum energy, which is reduced since $m_h^* > m_e^*$, measured from the top of the valence band in the GaAs, as shown. The overall effect of the potential wells in the GaAs region is that the effective energy gap for direct recombination is increased by more than 50 meV, as shown, to say, 1.46 eV. In effect, the band gap in the GaAs has been changed by a few per cent, not by compositional modifications, but by choice of layer thickness, d. The output wavelength of a laser or LED formed in this material could, for example, be controlled by choice of layer thickness, d. For instance, if d is decreased, E in Eq. (13.1) increases, so the effective gap energy is greater and photon output from direct recombination occurs at shorter wavelengths.

Tailoring the effective band gap by dimensional rather than compositional changes has been called *wavefunction engineering*, and the *low-dimensional* (i.e. $d = 1 - 10 \mu$m) multilayers are known as *multiple quantum well* (MQW) structures. Such small geometries are only realizable by very modern epitaxial growth techniques such as molecular beam epitaxy and metal-organic chemical vapour deposition.

MQW structures are already being incorporated in a variety of components, such as in the emitter–base junction of an HBT to produce an oscillator, in lasers and other optoelectronic devices and in optical switching devices and modulators. Since many other novel applications for this relatively new semiconductor material are being devised, the future of the technology is most promising.

13.3 Amorphous semiconductor devices

In recent years, a growing awareness of the potential of amorphous semiconductors, for use as the active element in a range of electronic devices, has developed. Of particular interest is amorphous silicon, a-Si, which has been incorporated in thin-film solar cells, now widely used to power calculators, and also in thin-film transistors (TFTs), integrated arrays of which are used to address a liquid-crystal display, to provide a flat-screen video display; details of these and other applications will be discussed later. Meanwhile it will be instructive to consider the properties of amorphous semiconductors in order to discover their unique electronic characteristics, which can be exploited in a variety of novel devices.

Amorphous semiconductors, of which elemental silicon is by far the most important practically, differ in an essential aspect from all the semiconductors discussed so far, which have essentially to be single crystals to be able to function successfully in devices. On the contrary, as the name applies, an amorphous semiconductor has *no* crystalline form; its constituent atoms have no apparent structures and, ostensibly, possess random spacing and orientation. Although such long-range atomic disorder is a feature of a–Si, atoms are not completely randomly positioned. The covalent bonding between adjacent atoms in crystalline and amorphous silicon is similar, so a *short-range* or *nearest-neighbour order* exists, even in a–Si. In other words, neighbouring atoms are spaced equally in either form of Si, by an amount characteristic of the length of a covalent bond. However, in a–Si, the ordered periodicity of atomic spacing that is present in the crystalline form is no longer apparent over distances greater than a few atomic spacings and atoms are randomly positioned at such *long ranges*.

In crystalline Si, covalent bonding between an atom and its four neighbours in a tetrahedral array results, as discussed in Sec. 6.2, in a crystalline solid with a diamond structure. The two-dimensional representation of such a semiconductor, Fig. 6.6, shows a well ordered distribution of atoms and a regular lattice constant, *a*. In a–Si, since the binding is still the same, the unit cell remains tetrahedral, but cells are now packed with arbitrary rotations around a common bond, as shown diagrammatically in Fig. 13.2(a). The corresponding two-dimensional representation of the atomic orientation in a–Si might be as shown in Fig. 13.2(b). All bonds are satisfied but the atoms are randomly oriented; there is no longer a fixed lattice constant in any direction and disorder is evident after a few atomic spacings. However, this assumes an ideal situation in which each Si atom is bonded covalently to four neighbouring atoms. In a more realistic situation, not all the bonds in the a–Si can be satisfied; a proportion of the Si atoms do not have the full complement of neighbouring atoms with which to bond, so their four bonds are not all filled and *dangling bonds* are introduced as a consequence, as illustrated in Fig. 13.2(c).

As might be expected, because of the short-range order that is common to crystalline and amorphous Si, behaviour similar to that in an intrinsic semiconductor might be observed in a–Si, and this is so. For example, a light absorption edge occurs in both materials at (almost) the same wavelength, corresponding to the photon energy, hf, being equal to the gap energy in the crystalline material, E_g, which confirms the short-range ordering of atoms in the a–Si. We would therefore expect to find some sort of energy gap associated with the band structure of a–Si, which will be discussed later.

Further evidence for intrinsic-like behaviour of a–Si is obtained from conductivity, σ, versus T measurements. As we have seen, the conductivity of

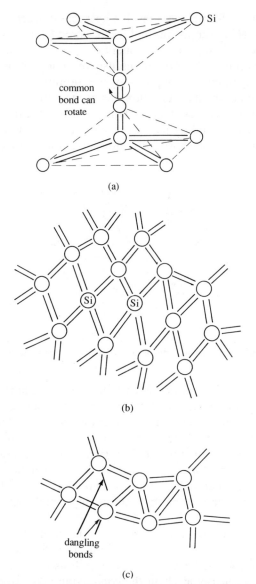

Fig. 13.2 (a) Atomic structure of a–Si; (b) a possible two-dimensional representation of the idealized material; (c) a more realistic model, including unsatisfied, *dangling*, bonds.

intrinsic silicon varies exponentially with T, thus

$$\sigma = \sigma_0 \exp(-E_g/2kT)$$

so a graph of $\log_e \sigma$ versus T^{-1} would yield a straight line of slope $-E_g/2k$. Similar linear $\log_e \sigma$ versus T^{-1} plots are obtained from a–Si, so confirming that this form of Si behaves, in this respect at least, like an intrinsic semiconductor. However, as we shall see, it would be quite wrong to assume that the energy band structures are identical.

Amorphous semiconductors are insensitive to dopant impurities, so that they cannot easily be made n- or p-type by the addition of dopants. (We will see later how this fundamental limitation has been overcome, so as to produce successful electronic devices.) One explanation is that the random structure of a–Si can easily accommodate additional bonds due to, say, an n-type dopant, as shown in Fig. 13.3. The phosphorus atom substitutes for a Si atom, as usual,

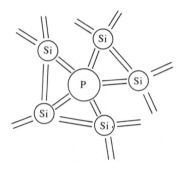

Fig. 13.3   A phosphorus 'donor' atom accommodated in the a-Si lattice.

but all its bonds can be accommodated by five neighbouring Si atoms and thus there is no extra loosely bound electron, as in the crystalline n-type material, suitable for promotion into the conduction band. Another, more likely, explanation concerns the dangling bonds that are present in large numbers in the amorphous semiconductor. Any free electrons or holes that are introduced by donor or acceptor dopants are trapped by the dangling bonds. In other words, amorphous materials contain so many electron and hole traps that they can accommodate dopants without any significant change in their conductivity.

The electron and hole traps in a–Si are numerous, localized and, in the energy band description, form additional discrete states at levels that correspond to the forbidden gap of crystalline Si, as shown in Fig. 13.4(a). The quasi-continuous energy levels in the conduction and valence bands of the crystalline material, their density, $S(E)$, varying as $E^{1/2}$ (Eq. (4.18)), are augmented by localized levels, which are represented by $S(E)$ having *tails* that extend into the energy gap region, as shown in Fig. 13.4(a). In some amorphous materials the band tails can overlap: the density of states, $S(E)$, is then continuous across the gap region of the crystalline counterpart, as shown in Fig. 13.4(b), corresponding to localized levels spread throughout the 'gap'.

If a–Si, say has a continuum of energy levels, such that no forbidden energy gap exists, as described, then how can this be resolved against its

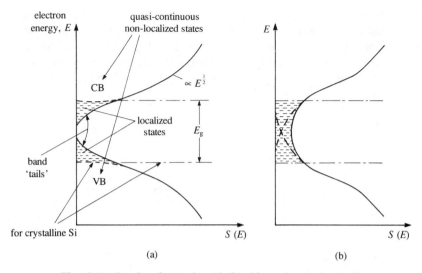

Fig. 13.4(a) Density of states in a-Si; (b) with overlapping band tails.

experimentally observed intrinsic behaviour, manifest for example in its $\sigma(T)$ characteristic? The answer lies in the mobilities of the carriers at the various energy levels in such materials. Carriers in the conduction- and valence-band regions are situated, as we have seen, in quasi-continuous, closely spaced levels in which they can move relatively easily under the influence of an electric field; such carriers have a high mobility corresponding to that in the crystalline material. However, levels in the 'gap' are localized in space, and the further away from the band edges the more likely they are to be localized; carrier transport can they only occur by a *hopping* process, which describes thermally assisted tunnelling through the potential barrier that separates each trapping centre, as illustrated in Figs. 13.5(a) and (b). The probability of such processes

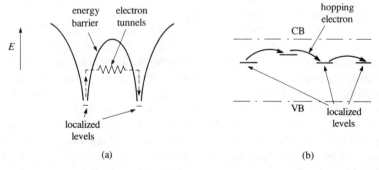

Fig. 13.5 (a) Carrier transport between localized levels by thermally assisted tunnelling (hopping); (b) electronic hopping between levels in the gap.

occurring is low, which leads to mobilities of order 10^2–10^3 times lower than in the *extended states* of the conduction and valence bands. These considerations lead to the concept of a *mobility gap*, shown in Fig. 13.6, where the carrier mobilities drop to relatively very small values around the band edges. Another factor is that, due to the disorder in an amorphous solid, the carriers, even in

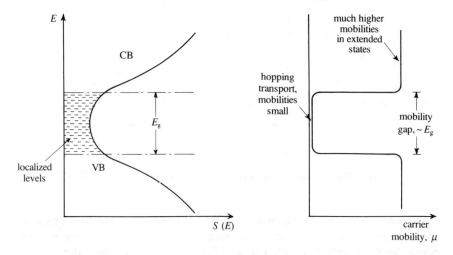

Fig. 13.6 Illustration of the concept of a *mobility gap* in amorphous semiconductors.

the extended bands, suffer more collisions than in the crystalline state; their mean free path between scattering events is therefore much reduced, and mobilities are typically only a few per cent of those in the crystal.

Since the conductivity is proportional to the product of the number of carriers, and hence $S(E)$, and their mobility, $\sigma = ne\mu_e$, the presence of a mobility gap in a–Si demonstrates that the contribution to the overall conductivity due to carriers in the gap is negligibly small. This accounts for the intrinsic behaviour, which occurs in spite of the carriers in localized levels that exist throughout the normally forbidden energy gap.

The electronic states in the gap, with density typically of order 10^{26} m^{-3} eV^{-1} for conventionally produced material, initially dominated the electronic and optical behaviour such as to make early amorphous semiconductors of little usefulness for practical devices. A major development was the discovery that the a–Si films produced by the flow-discharge decompositon of silane (SiH$_4$) possess a relatively very low density of gap states and exhibit a large photoconductive effect, which was not formerly present. The presence of hydrogen in the reaction chamber, as a by-product of the SiH$_4$, is essential to the reduction of gap states and the enhanced electronic properties of the silicon. That this is so has been confirmed subsequently by producing a–Si films by conventional r.f. sputtering or electron-beam evaporation in the presence of hydrogen, which causes a dramatic decrease in the gap states to around $10^{20}\,\mathrm{m}^{-3}\,\mathrm{eV}^{-1}$. This change in the density of states is accompanied by a corresponding decrease in the conductivity by many orders of magnitude and other electronic and optical variations, dependent solely on the presence or otherwise of hydrogen in the material. The precise role of hydrogen is complex and not entirely resolved, but one possibility is that hydrogen combines with dangling bonds in the a–Si, thus passivating or *compensating* such bonds. The situation is complicated by the fact that much more hydrogen than is necessary to compensate the dangling bonds is present in the films, typically between 3 and 15 atomic per cent, which leads some researchers to believe that the observed properties belong to those of a different alloy semiconductor, referred to as hydrogenated amorphous silicon, or a–Si:H. In spite of these theoretical difficulties, such semiconductors have considerable potential in a variety of electronic applications.

An important stage in the development of a–Si:H devices was the discovery that such materials could be doped p- or n-type, so making possible a variety of junction devices. The doping can be achieved by introducing diborane gas (B_2H_6) for p-type or phosphine (PH_3) for n-type to the reaction chamber, for example as shown diagrammatically in Fig. 13.7. An alternative

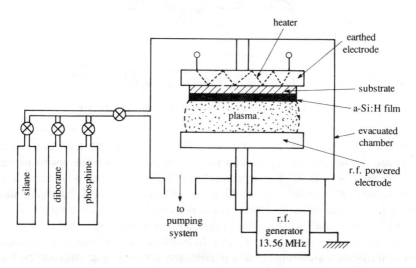

Fig. 13.7 Glow-discharge plasma deposition of a-Si:H and its gas-phase doping.

method is to use ion-implantation doping techniques. Doping either n- or p-type increases the conductivity of the intrinsic a–Si:H from, typically, $10^{-14}\,\mathrm{Sm}^{-1}$ to around $10^{-4}\,\mathrm{Sm}^{-1}$. Such behaviour can be attributed to substitutional doping by B atoms for p-type or P for n-type, made possible by the compensation of the dangling-bond gap states by hydrogen, but against this conclusion may be a little oversimplified. Nevertheless, a–Si:H can provide extremely useful electronic and optical properties intrinsically, and the ability to produce doped regions and junctions in the semiconductor has led to a number of potential applications.

One of the first applications of a–Si:H was in an amorphous-silicon solar cell. A significant portion, approaching one-half of currently produced cells, use a–Si:H. The absorption coefficient of a–Si:H is much higher than for crystalline Si over the visible region, so most of the sunlight incident on a cell is absorbed within a micrometre or so of the surface. Hence only a thin layer of a–Si:H is necessary and thin-film devices, which use little Si in their manufacture, are standard. A further potential advantage that a–Si:H possesses is that large-area cells are reasonably straightforward, particularly using established sputtering techniques, whereas the size of crystalline cells is constrained by the maximum single-crystal slice available. A typical a–Si:H n^+ip^+ solar cell is shown in outline in Fig. 13.8. Briefly, photogenerated

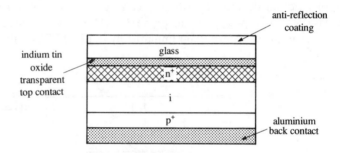

Fig. 13.8   A hydrogenated amorphous silicon, pin solar cell.

carriers in the i region are separated by the high electric field that exists between the heavily doped n^+ and p^+ regions to produce the photocurrent output (see Sec. 11.3). The open-circuit voltage of a cell is around 1 V. Conversion efficiencies approaching 10 per cent are still relatively low compared to the crystalline version (20 per cent), but the advantages of the thin-film cell ensure its continued production and development.

Another important application of a–Si:H is its use as the active semiconductor in thin-film transistors (TFTs), particularly for large-area electronics. The most usual configuration is as an inverted, insulated-gate, enhancement-mode, field-effect (IGFET or MOST) device, as shown in Fig. 13.9. Arrays of such devices can be produced by now conventional thin-film and photolithographic techniques, the silicon nitride gate insulator, for example, being deposited by the introduction of ammonia into the same glow-discharge reactor used to create the a–Si:H films. A graph of I_d versus V_g for a typical early device is shown in Fig. 13.10. It shows that such TFTs have a large dynamic range, with I_D changing by over six orders for gate voltage swings of

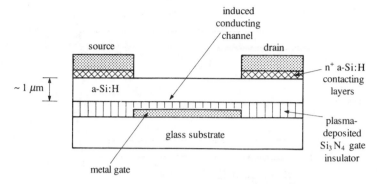

Fig. 13.9   Cross section of an a-Si:H thin-film MOST.

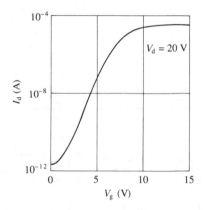

Fig. 13.10   Transfer characteristic of an early a-Si:H TFT.

a few volts. Arrays of TFTs with such characteristics are suitable for use as switching elements in large-area liquid-crystal colour video display panels. A cross section of a possible geometry of a single pixel is shown in Fig. 13.11(a). The liquid-crystal display (LCD) consists essentially of two glass plates coated with indium tin oxide (ITO), which act as transparent contacts, with a twisted nematic liquid crystal between, as in Figs. 13.11(a)–(c). The bottom ITO layer is photolithographically formed into square pads to create picture elements, *pixels*, which can be selectively and individually switched by incorporating an a–Si:H TFT switching device at one corner of each element in the array, Fig. 13.11(b). The bottom square ITO drain pad is connected electrically to the drain of its associated switching transistor and the upper common ITO electrode is earthed. Each element acts as a parallel-plate capacitor having transparent electrodes and filled with liquid crystal, which is in series with the drain of an amorphous switching transistor. The gates and sources of the TFTs are provided with buses G_n and S_m as shown, via which an individual switching

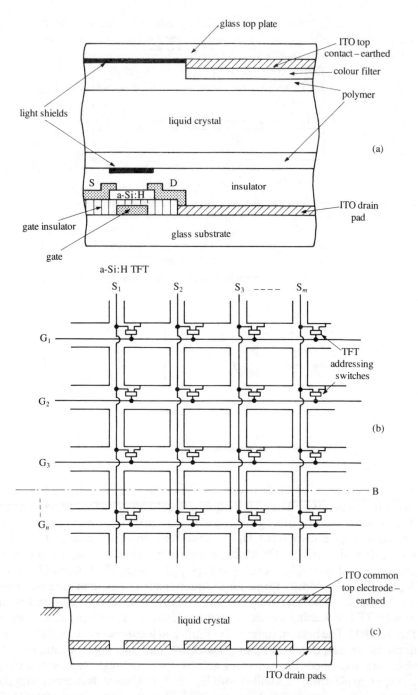

Fig. 13.11   A thin-film liquid-crystal display: (a) 'cross section of one pixel; (b) arrangement for addressing individual elements using a-Si:H TFTs; (c) a simplified cross section at A–B on (b).

element can be addressed. A variety of addressing schemes are possible but the simplest provides one row of common gate connections, G_2 say, with a voltage pulse above threshold, V_T, while at the same time the picture information is supplied by voltages to the source buses, S_1–S_m, a positive V_s to switch a pixel ON and 0 V for OFF. The array is then scanned by transferring a pulse $> V_T$ to the succeeding common gate bus, G_3 in this case, while simultaneously repeating the application of information voltages to the source buses. As a particular transistor is switched ON, its associated liquid-crystal capacitive element is charged and becomes opaque. The display is normally viewed by back-lit illumination, in a transmission mode, to provide adequate contrast in colour displays, the colour being provided by colour filters associated with each picture element (see Fig. 13.11(a)). The attractiveness of the technology derives from the fact that all components of the display, liquid crystal and switching transistors, can be fabricated using existing thin-film micro-electronics technology and even the addressing circuits can be made compatible by incorporating a–Si:H TFT drive circuits. Large-area colour video displays are now becoming available, to provide the first truly flat screen televisions.

Index